# Survival & Rescue Equipment of World War II:

## Army Air Forces and U.S. Navy

## VOLUME 1

Multi-Person Life Rafts
One-Man Life Rafts
Life Preservers

Dustin Clingenpeel

ECP

ISBN 978-1-943492-60-2

Book and cover design © Mick J. Prodger

Elm Grove Publishing
San Antonio, Texas, USA
www.elmgrovepublishing.com

# Contents

# Introduction

Historically, air-sea rescue had its inception and greatest advances in emergency equipment, procedures and techniques amid war. What was accomplished in less than five years of conflict probably would not have been accomplished in decades of ordinary development, and retains its relevance in the new millennium. Each article developed had its own specific background or purpose, and was often included as a compliment to other items. Specific requests occurred as a direct result of service activities, an individual's idea or resulted from continuous development.

The US Navy Bureau of Aeronautics' development was carried out at a rapid, high priority rate, when a new item was requested from the fleet. It would start with an unproven idea or a modification of existing equipment and go through stages of development and distribution. Once a piece of equipment was standardized, every effort was made to expedite its production and availability. The process began at the Flight Gear and Safety Equipment Section, Equipment and Materials Branch, Engineering Division, or the BuAer directly associated with the Naval Aircraft Factory. Here the item was taken through the technical stages including the design, preparation of specifications, initiation of test projects, preparation of technical orders, and notes in new or changed equipment. Once the item was ready for manufacture, procurement action was initiated by the Airborne Equipment Section of Maintenance Division NAF, or by the Aviation Supply Office. Quantities were established by the Military Requirements Division of the Bureau of Aeronautics. In some instances, a new type or construction might require manufacturing experience, delaying delivery. No two items followed the same pattern. To serve a practical purpose, limited quantity contracts were awarded to service test, and in some cases procurement was expedited.

Once an item was placed on procurement, delivery schedules were established and quantities were outlined for prescribed supply points. The central supply point was the Naval Aviation Supply Depot, Philadelphia, which housed the Aviation Supply Office. Two primary supply points were maintained, one on each coast. These were Aviation Supply Annex in Oakland, California, and Aviation Supply Annex in Norfolk, Virginia. The supply annexes had the responsibility of maintaining supplies for Naval Air Stations under their jurisdiction. Supply points maintained at an air station could requisition supplies organic to aviation supply from the annexes, or directly from the ASO. Ordnance was typically requisitioned from Naval Supply Depots, and medical items from the Bureau of Medicine & Surgery, however some of these items were maintained at aviation supply points. Each air station had purchasing power through purchasing agents who could make direct procurement locally, or from approved contractors. In some cases, due to urgency, an item would be placed on expedited delivery directly from the contractor to air stations (specifically to NAS Pearl Harbor) to assure it became available as soon as possible.

The USAAF had a more complex system, and when the Wright Field Personal Equipment Laboratory, in Dayton, Ohio, received a request or suggestion for modification, it would undergo a preliminary review before being presented to AAF Headquarters. When the idea was submitted to Operations, Commitments and Requirements Assistant Chief/Air Staff at AAF Headquarters, it was first checked for its potential value. Upon approval from OC&R, Wright Field was authorized funds for development to the point of service testing. The item was then sent to the AAF Proving Ground Command at Eglin Field, Florida, for a series of rigid service tests. When completed, a report and the item would be forwarded to the AAF Board in Orlando, where a decision would be made to recommend the item for standardization, rejection or require further service testing. Their conclusions and recommendations would be submitted to AAF Headquarters for the final decision to approve or disapprove the item, and if approved, standardization and procurement followed. This authority allowed Wright Field to prepare drawings and specifications and to proceed with the negotiation of contracts. The basis of issue and distribution to the various supply points and theaters was determined by AC/AS, Material and Services, under the authority of the Air Technical Service Command. The Emergency Rescue Branch, Personal Equipment Laboratory of the Air Technical Service Command at Wright Field was charged with the responsibility for development of emergency rescue equipment used by the USAAF. The Emergency Rescue Branch was established in August 1943.

The USAAF Air Service Command controlled, handled, and distributed all air forces property. In the continental United States, distribution was coordinated through eleven control area commands and control depots, then through the lowest echelon sub-depot located regionally. ASC depots received a constant stream of material from the factories and channeled it to operational forces. From storage depots or direct from the factory, supplies were moved to control area depots which were responsible for maintaining sub-depots. Sub-depots were the principle suppliers of materials to flying units located at air fields or bases.

Once an item reached a sub-depot or air station responsible for the supply of a squadron or group, they were divided into classification groups being stowed, maintained and distributed accordingly. Individuals were assigned responsibility for the management of those departments. For personal equipment, a Supply Officer was accountable for flying clothing, a Parachute Officer for parachutes and a Material or Safety Officer for life rafts and oxygen equipment.

From experience in early operations a need arose for more efficient maintenance and intelligent use of oxygen equipment. The BuAer were the first to designate the position of an Oxygen Officer authorized under BuAer Circular Letter No. 5-41 dated 19 March 1941. This letter detailed that one officer in each squadron and at each base be designated as Oxygen Officer. This officer would be responsible to the commanding officer for the condition of the oxygen equipment and assure that all flight personnel fully understood its operation prior to use in flight. The USAAF inaugurated the position of Oxygen Officer under AAF Regulation No. 55-7 dated 30 May 1942. It detailed that the first assistant operations officer in each group and squadron be designated as a Unit Oxygen Officer. Both services established the position with existing personnel, but soon found it inadequate for this responsibility. However, it temporarily served the initial demands to prepare pilots and aircrew for high altitude operations. A formal course for the training of oxygen officers for the AAF was established at the School for Aviation Medicine in July 1942 as a five-day course. To meet wartime demands the BuAer established a six-week training course in April 1943 at NAS Jacksonville, and upon graduation, personnel were designated as Oxygen Officers.

The overall training curriculum on safety equipment in 1942 could be best described as deplorable, and survival knowledge severely lacking. It was assumed that most men had some level of outdoor skills, but this proved not to be the case, and by the close of that year it became apparent that there was an urgent need for more training and maintenance of safety equipment. One principle attempt to remedy the situation was by the 8th AF in October 1942, when it was recommended to extend the responsibility of the Oxygen Officer to a new designation of Equipment Officer. This recommendation was based on evidence of numerous reported deficiencies and was to include the proper use and care of flying clothing, adequate knowledge of crash, ditching or oxygen procedures, and life rafts/safety harness procedures. Additionally noted was the improperly equipped crews as a result of theft and hoarding of equipment, ignorance of bail-out procedures, lack of proper crew indoctrination, and lack of cooperation between the officers responsible for supply. Coupled with similar reports from other combat zones regulation 55-7 was revised 4 May 1943 to make the Unit Oxygen Officer an Equipment Officer.

The first Equipment Officers appointed were from group headquarters and each squadron of the 8th Bomber Command. As part of this selection most of those chosen for the positions of both the oxygen and equipment officer were from flying personnel, as it was felt that pilots and aircrews were more likely to respect advice from an individual with actual flying experience. It soon became apparent that officers on flying status were often too preoccupied to properly execute the duties they were delegated to perform. Regulation 55-7 was again revised 28 October 1943, restricting the position to ground officers and assigning a new designation of Personal Equipment Officer.

In November 1943, the course for Unit Oxygen Officers at the School of Aviation Medicine, Randolph Field, Texas, was replaced by a two-week instruction course for the newly created Personal Equipment Officer. This course was eventually transferred to the AAF Tactical Center in Orlando, Florida, increasing the training period to three weeks in all aspects of flying safety equipment and air-sea rescue. Regulation and supervision of the PEO was under the jurisdiction of the Air Surgeon, but due the major expansion of duties it was felt that the program was now out of the scope of Air Surgeon activity. In the new AAF Regulation 55-7A dated 4 March 1944, OC&R AC/AS would be charged with supervising the activities of PEO and made responsible for disseminating information to them. The structure of assigned PEOs falls into four tiers. The top tier was the Command Personal Equipment Officer, who supervised and coordinated all PEOs within the command and principally maintained an administrative roll. On the second tier was the Wing PEO, who recommended the personal and emergency equipment to be used on combat missions and was involved in more hands-on inspection, supervision and coordination of the Group and Squadron PEO, as well as assisting the Command PEO in maintaining records. The Group PEO supervised and coordinated directly with Squadron PEOs and was more directly involved with inspections and continued training. On the bottom tier was the squadron PEO who worked directly with flying personal and assured they were properly trained and equipped for any combat mission.

By 1943 the Pacific fleet was expressing a concern that more training was required in the use and maintenance of emergency rescue equipment. In response, the BuAer went directly to the source of the issue by inaugurating survival training to begin in preflight schools, continuing through to the intermediate and primary courses. Soon after the opening of the Oxygen Officer School at NAS Jacksonville, the syllabus was expanded to include training in emergency equipment such as life rafts, life preservers and other related survival gear. A new designation, Aviation Equipment Officer, was assigned, superseding the position of Oxygen Officer. By January 1944 a new training course was established for the Aviation Equipment Officer at NAS Pensacola. To help compensate for the lack of, and need for, such an officer for fleet activities, the Chief of Naval Operations drafted and distributed Aviation Circular Letter No. 17-44 dated 4 March 1944 which directed squadron commanders to appoint an officer to assume the duties of the Aviation Equipment Officer in units where a trained officer had not yet been assigned. As operations increased in the Pacific, more specialized training was required, in large part due to the increase in specialized emergency equipment, basic survival skills and indoctrination for the health and safety of individuals. As part of the responsibility of the AEO, they were required to disseminate information and keep up with current events conducting survival training in the operational area. It was felt that the survival training aspect was not reflected in their title, and in October 1944 their designation was changed to Aviation Equipment & Survival Officer (AESO). One AEO or AESO would be assigned to a Carrier Air Service Unit (CASU) acting as a maintenance officer in the Assembly & Repair departments. When assigned to a Naval Air Station, they would take on the role, primarily, of instructor. One AESO would be assigned to each squadron who would take on the roles of both maintenance officer and instructor. Prior to the institution of the AEO, the squadron parachute rigger had principally taken on that role. Within a CASU there would be separate facilities that inspected and maintained parachutes, life rafts, flight clothing and other emergency equipment. Aboard a carrier these facilities would be consolidated into one department maintained by the squadron parachute rigger. His responsibility, in large part, was to cater to all the squadron members and their needs regarding personal equipment, which would include training in its use. The AEO would take on this role, but still work in close coordination with the parachute rigger.

For both the AAF and BuAer, the position of Equipment Officer could be divided into custodial, training and inspection duties. Under the custodial duties, he would organize equipment storage and make sure everything needed for a mission was prepared. Training duties consisted of instruction in the care and use of parachutes, oxygen equipment, clothing, flak suits, life rafts and emergency kits. He would also give instruction on emergency procedures relating to bailing out, ditching, fires in flight and survival on land and sea. Before and after each mission, all equipment would be inspected for damage, staining or wear, to assure it was in perfect working order. This included items in storage during mandatory inspection periods. The EO's job was to coordinate with the supply, engineer, parachute and intelligence officers and not assume the responsibility of any one of their jobs, and to coordinate with the flight surgeon sharing the responsibilities in medical and survival training. The Equipment Officer would be responsible for the dissemination of information by any means available to him, the record keeping and reporting of the performance of all this equipment. Most importantly, his duties were not to conflict with other established positions.

*The terms MARK-, MK-, Mark- and Mk- (followed by a number) are used interchangeably throughout the text, because they were used interchangeably by the various departments and divisions of the military at different times and in different lists, catalogs and publications, even when referencing the same item. Similarly, Mark- numbers are used both as Roman numerals (I, II, III, etc) and the more familiar Arabic numbers (1, 2, 3, etc). The same item might appear in one reference as Mk-III and as MARK-3 in another. Terms are used according to their source.*

"Our aviators and seaman, with confidence that they will fly and sail again tomorrow, dare to face greater odds in the war today. Vision our Services, ingenuity of industry has provided survival and rescue equipment, which has lessened the hazards, improved the safety, and given our men greater courage. May we never be content with present equipment, but constantly improve it with experience, continued study and cooperative effort."

*Admiral Russell R. Waesche,*
*Commandant of the United States Coast Guard*

# LIFE RAFTS

## Part I.
### Multi-Person Types

Arguably, the most vital piece of air-sea rescue equipment was the life raft.

Statistically speaking, a survivor's likelihood of rescue increased dramatically, when using a life raft, by roughly 40-50%. In one monthly report, 82% of those using life rafts were rescued and only 48% without. Overall statistics during WWII cannot be studied in depth, as records of incidents were not always kept, but enough data is available to lend a perspective. 70% of recorded survival cases occurred over the world's oceans and 75% of those were in the Pacific, followed by the North Sea and English Channel. These can be distinguished as warm and cold water regions directly affecting rescue. 60% of survivors were rescued in warm water and only 30% in cold. These averages would fluctuate monthly during the war. An estimated 5,000 ditchings involving multiple people were recorded from 1942-1945, including the ETO, which had approximately 4,250 ditchings and bailouts, most of which included multiple survivors. Bailouts comprised the bulk of the recorded incidents from Europe.

Each respective theater presented its own issues in stark contrast. In the ETO there was only a 1 in 3 chance of rescue if forced into the water, versus a greater than 50% chance in the Pacific. One example of a specific record was submitted by the XXI Bomber Command, stating that from November 1944 to August 1945 there was a loss of 361 B-29s with 3,104 crew members, of which 1,424 were known to go down at sea. Only 687 were rescued.

With the outbreak of war, air travel exponentially increased to an estimated rate of one flight every 13 minutes across the Atlantic and every 1.5 hours in the Pacific. With this expansion came an increase in-flight emergencies. The most common in-flight emergency was combat damage, followed by engine and other mechanical failures, then fuel exhaustion coupled with weather and navigational errors. When there was a choice, it was considered preferable to ditch at sea, rather than bailout over land, to avoid capture (a stronger factor in the Pacific where the Japanese were known to kill). The probability of rescue was higher when at sea than on land, and steadily increased as the war progressed. In the Pacific, air-sea rescue organizations would be on standby waiting to coordinate between aircraft, surface vessels and submarines that formed rescue or picket lines to and from targets. The closer a pilot could get to these known operational areas, the more his chance of rescue increased. After engagements over Europe, attempts were made to get to the English Channel or North Sea, same for the Mediterranean, but because of the extremely cold waters between Britain and France many perished after they bailed out. After the invasion of France, crews only needed to make it to allied territory, which made bailing out a realistic alternative. The life preserver would be the first item implemented, but a life raft would afford long term survival.

In the post-WWI era, float type airplanes did the bulk of long duration flying over water. They were used in patrol and reconnaissance, proving vital in patrolling for German U-Boats during the great war. Long flights using land-based planes often ended in disaster, resulting in the delegation of seaplanes and flying boats to the task.

On 17 October 1922, the first US Navy aircraft carrier was commissioned as the USS Langley, followed by two more commissions of the USS Lexington and USS Saratoga in 1927. Land based planes were now able to be launched from surface vessels lending support in bombing other navies establishing air power as a viable asset in fleet activities. Float type aircraft no longer dominated the skies over the ocean.

New types of aircraft were developed for carrier activities, designed without buoyant characteristics, promoting performance and offensive capabilities. In time, carrier task forces would consist of fighter, bomber and torpedo planes, which remained the standard upon entering WWII. Prior to, and in conjunction with, the introduction of aircraft carriers, the US Navy operated from land-based stations, though some vessels were equipped with launches for advanced patrol purposes. Air-sea rescue activities during this era were negligible and would remain so up t the outbreak of WWII.

On the rare occasion an emergency situation arose, the aircraft was usually relatively close to land, and shore-based vessels, airships and/or float planes could be dispatched to assist in the rescue in a relatively short amount of time. For the USAAC most over water accidents happened on lakes, rivers, bays and off-coast lines. As operational areas expanded to Alaska, Philippines, Hawaii and Panama so did the need for water survival concepts.

The development of life rafts progressed from a peacetime environment to wartime, establishing a whole new set of unforeseen demands.

*(Left) The bulk of ditchings for the US Navy consisted of carrier type aircraft holding 3 persons or less. This TBF/ TBM Avenger was forced to ditch after a failed catapult launch off the USS Anzio, while operating in the Marianas, December 1944. The crew safely evacuated the aircraft then boarded their raft until they were picked up by a destroyer. (National Archives)*

*(Right) Some planes were better suited structurally for ditching than others. A ditching is unpredictable with larger aircraft, which are prone to flood quickly, allowing anywhere from 15 seconds to 15 minutes for the crew to evacuate and retrieve any emergency supplies. This B-29 near Iwo Jima has ditched and has successfully remained afloat. (National Archives)*

*(Left) The USAAF reported the majority of ditching incidents, due to its operational size and amount of aircraft with large crew capacities. Sealed, air-tight compartments in the fuselage and wings assisted in keeping the plane afloat, allowing time for the crew to safely exit and obtain emergency provisions. This B-17 has successfully ditched and deployed its complement of life rafts. The crew could board the raft and safely float away to await rescue. England February 1944. (National Archives)*

*(Above) Even once aboard the raft, safety, survival and rescue were not completely assured. Provisions were continually added to the complement for providing comfort to survivors. Occupants had to battle immersion sores, sunburn, dehydration, nausea, vertigo and boredom, often affecting moral. These images give some perspective of the helpless feeling a survivor may encounter being adrift in an endless ocean, especially if alone. The sun was also a consistent issue, creating glare off the water affecting eyesight and mental stability. (National Archives)*

*(Left) Two naval aviators await rescue in a two-man life raft with a one-man parachute type tethered to it, April 1944. (National Archives)*

*(Right) USAAF aircrew members afloat, somewhere in the Pacific Ocean, in E-2 life rafts. As the para-raft became available, each crew position would be issued one, having multiple means of individual survival and group emergency provisions. (National Archives)*

Standardization of life rafts for the aeronautical services commenced in the 1920s. Inflatable devices of any kind during that era could be best described as novelties with dependability issues. In the post-World War I era, companies and entrepreneurs experimented with varying design principles. Companies with experience in fabricating airships and balloons presented the best results. One such example was presented to the US Navy by the Connecticut Aircraft Co. in 1920, based on an Italian design that had been removed from one of their airships for experimental purposes. Other companies such as the New England Airship Co., Airships Inc. and Air Cruisers Inc. submitted designs to both the US Navy and USAAC through the 1920s. These rafts then entered the commercial market, orientated towards sportsmen. During this era, neither service conducted extensive flying over water, as aircraft capabilities were minimal and presented no real service requirement. Kapok life jackets were initially provided.

The 1920s were a progressive era in aircraft design, allowing planes to fly farther. With expanding service areas via floating airfields they were now operating in vast open areas over the world's oceans. By the end of the decade, life saving devices became more relevant, prompting standardization for aeronautical use. On 20 April 1929, the BuAer released a standard life raft specification titled Rafts, Life, Inflatable as NavAer Specification M-3. Three sizes of gas inflated rafts were detailed for capacity ratings of one-man, three-man and five-man. They were described as oblong in shape, colored orange-yellow, with bladders fabricated utilizing rubber coated airplane cloth. The heavy-duty cloth was to be coated on both sides of the fabric.

Each raft was to be provided with a basic compliment of two two-sectional spruce oars and a brass hand pump. The 3 and five-man rafts were equipped with two seats, only differing in size. The dimensions of the three-man was 7' x 4' with a 12" diameter bladder and the five-man was 9' x 5' with a 15" diameter bladder.

The rafts would be folded and rolled and contained in a khaki duck carrying case, 29" x 15" for the three-man and 32" x 15" for the five-man.

This original draft outlined basic detail requirements. Specification M-3 would ultimately undergo 19 revisions by VJ-Day. Upon acceptance, it became a requirement to equip all aircraft with pneumatic life rafts when engaged in over water flying.

*(Above left) Commander Richard Byrd poses with the life raft he used when he was forced to ditch on his trans-Atlantic flight off the coast of France in July 1927. This raft meets the general description of the standardized model of the 3 and five-man type. One feature as noted by BuAer specifications was the use of metal rings for the oar locks. The raft is likely a commercial model but they did influence initial designs by the BuAer and USAAC. (Above right) Four sailors are utilizing an early generation M-3 life raft, the notable feature is the metal ringed oar locks, c.1930.*

The subsequent amendments became more specific in the construction details, implementing Federal and Navy Department specifications for materials. By June 1931, five amendments had been applied, increasing the original page count from 5 to 11, as of Amendment-E. Refinements continued at a steady pace, implementing the first significant change to specification under Amendment-F 15 October 1931. Under this amendment, the capacity ratings of the rafts were changed to one-man, two-man, four-man and seven-man. The 1, 2 and four-man rafts would continue to carry a compliment of two, two-sectional spruce oars, but the seven-man would be upgraded to carry 4 two-sectional oars (one per oar lock) due to its size increase. At the time of the amendment, a repair kit had already been added, secured to the raft bladder in a sealed pouch. Two additional pockets were added under this amendment, secured to the raft bladder, for the oars and general supplies. The new series of rafts retained similar dimensions only changing the capacity, the two-man was 7.5' x 4' with 13" diameter bladder, the three-man was 9' x 5' with 15" diameter bladder and the seven-man was 12' x 6' with 16" diameter bladder. Packaged in their carrying cases, the rafts were now 34" x 18" for the two-man, 35" x 18" for the four-man and 36" x 24" for the seven-man. Following the new capacity ratings, engineering of the rafts' designs was integrated as part of Amendment-G of 27 January 1932. The rafts were no longer oblong in shape, changed to a pointed bow that better equalized pressure and flotation. Inflatable seats were added on the 4 and seven-man rafts. The independent, manually-inflated seats prevented excessive draw or collapsing of the sides.

The rafts provided were based on contractors' interpretation of specifications, leading to a variety of types delivered. To maintain consistency and quality control, as well as to assure formality continued improvement on designs, the BuAer drafted additional specifications for life raft oars and hand pumps. These specifications, detailed in M-162 and M-197, were included in the next significant change under Amendment-J dated 21 November 1933, under the new title of Rafts, Life, Pneumatic. Specification M-3 now included a list of NavAer, Federal, and Navy department specifications for overall construction of the rafts. Amendment-J additionally details that all exposed surfaces were to be orange-yellow, which now included the carrying case. The bladders were to be constructed of a new lightweight rubber-coated fabric, and the metal ring oar locks were deleted and replaced with layered fabric tabs. As detailed in Amendment-G, the seats in the 4 and seven-man rafts were inflatable, a detail which now changed to an inflatable cross tube.

Clerical revisions continued, leading to the next significant change with Amendment-M, dated 2 December 1935, which assigned Mark designations to each raft and a rated pound capacity. M-3 rafts were now referenced as the MK-I, 700 pounds; MK-II, 1000 pounds; MK-IV, 1800 pounds; and MK-VII 2500 pounds. The mark number signified the rated seating capacity.

This designation system remains the standard into the present day.

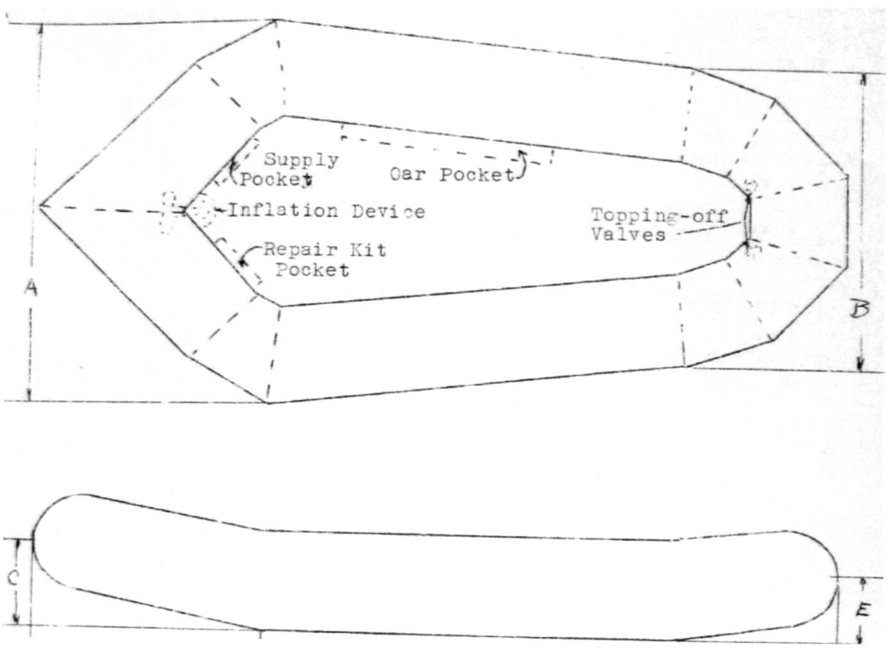


*(Left) A drawing that accompanied Amendment-J, from November 1933. showing the new pointed bow. Unfortunately, no drawings were included in specifications prior to Amendment-G illustrating the original "oblong" design. Seat schematics are not included, so as to be universal to all capacity rafts. At the bow is the CO2 inflation system and the manual valves are at the stern. Note the three pockets for the repair kit, oars and general supply, which also doubled as a bailing bucket. The appearance of the M-3 as detailed in this drawing would remain unchanged until the complete redesign under Amendment-R in 1944.*

*(Opposite page, bottom left) Photographs of early BuAer life rafts have proven difficult to obtain, but some have surfaced to lend some perspective on their designs. Pictured is a seven-man raft that best conforms to Amendment-G as it includes inflatable seats and is displaying its basic compliment of sectional spruce oars and hand pump. A standard feature was the provision for a life line that encircled the bladder. By this date two pockets were to be provided: one for the repair kit and one for the pump, and a pocket can be seen attached to the bladder, below the nozzle of the pump. The bow of the raft includes the CO2 bottle and at the stern the two manual valves can be seen. The raft is divided into two sections, separated by internal vertical bulkheads at the bow and stern, which can be inflated separately. This raft is not utilizing the metal oarlock rings. The items in front of the raft are articles carried aboard the aircraft, specifically a Martin PM-1 flying boat, December 1932. (National Archives)*

*(Left) A PM-1 being loaded at NAS Pearl Harbor in January 1936. Supervising the transfer of equipment is an officer in an early four-man raft.*

*(Left) A two-man raft used in fleet aircraft being demonstrated in Los Angeles, CA, October 1933. (National Archives)*

*(Below) A worker at the Goodyear Tire & Rubber Co. holds the seven-man raft as detailed in Amendment-J c. 1934. The two forward seats comprised an inflatable cross tube and a separate sheet of waterproof fabric as a bench seat. This detail would be the same with the four-man raft, but only two bench seats. The rear seat would remain a suspended section of fabric. The basic engineering seen here would remain consistent until 1944. (National Archives)*

*(Left) Goodyear workers conduct final assembly of two-man rafts c.1934. The 2 and 4 man are virtually identical, differing slightly in dimensions. The four-man included a cross tube bladder under the bow seat.*

*(Below) A stack of completed two-man rafts can be seen below both inflated and bundled in their carrying case. (National Archives)*

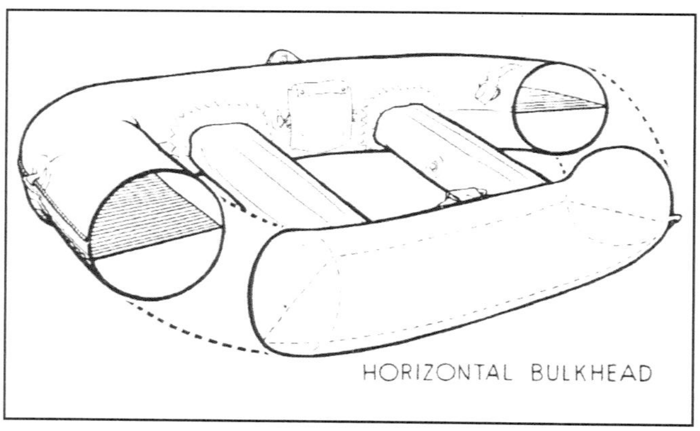


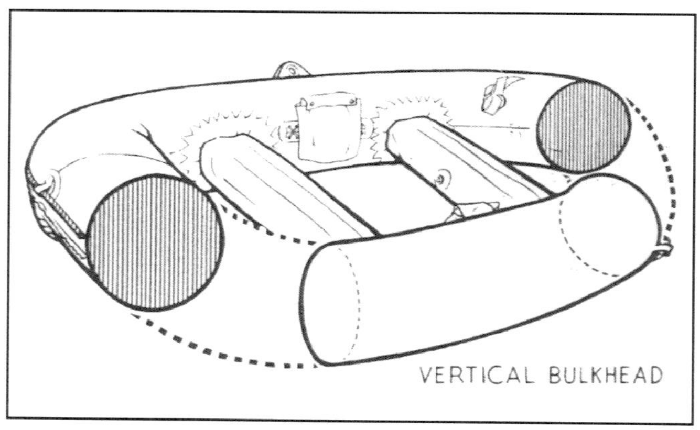


*(Above) Details of bulkheads in life raft bladders relevant to both the BuAer and USAAF. As part of the original design concept, life raft bladders in the M-3 series included the vertical type located at the bow and stern. The purpose was that in the event of damage to one side, the opposing side would stay inflated. As part of Supplement-1 of Amendment-M dated 16 September 1938, MK-I & MK-II rafts would have vertical bulkheads and the MK-IV & MK-VII would be converted to horizontal bulkheads. The horizontal method encompassed the complete circumference of the bladder. The rafts illustrated are USAAF, but the same principle applied to the BuAer M-3.*

*(Left) This USAAF Type A-2 life raft demonstrates the bulkhead principle. One side stays inflated, still rendering lifesaving aid and, once repaired, the other side could be re-inflated via the manual inflation valves. (National Archives)*

*(Right) To continue with bulkhead concepts, this image illustrates a further adjustment integrated into BuAer life rafts as part of Amendment-R. Horizontal types were discontinued in favor of the vertical method, but displaced to the mid-section of the port and starboard bladders.*

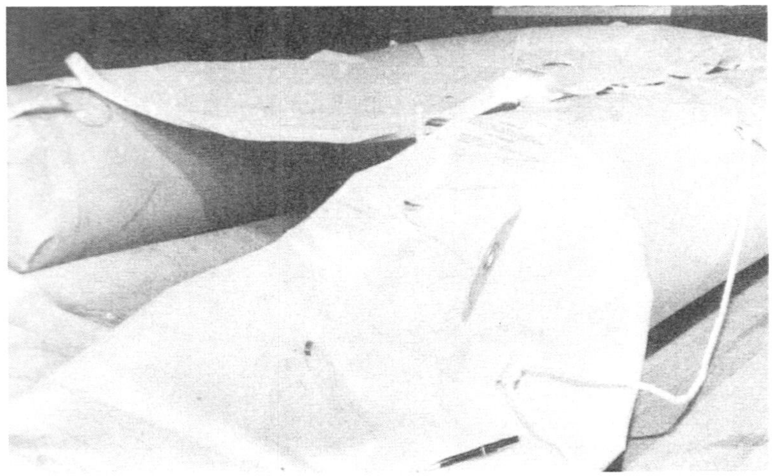

*(Below and top of following page) Life rafts were folded in a specific order before being slipped into their carrier. October 1937. (National Archives)*

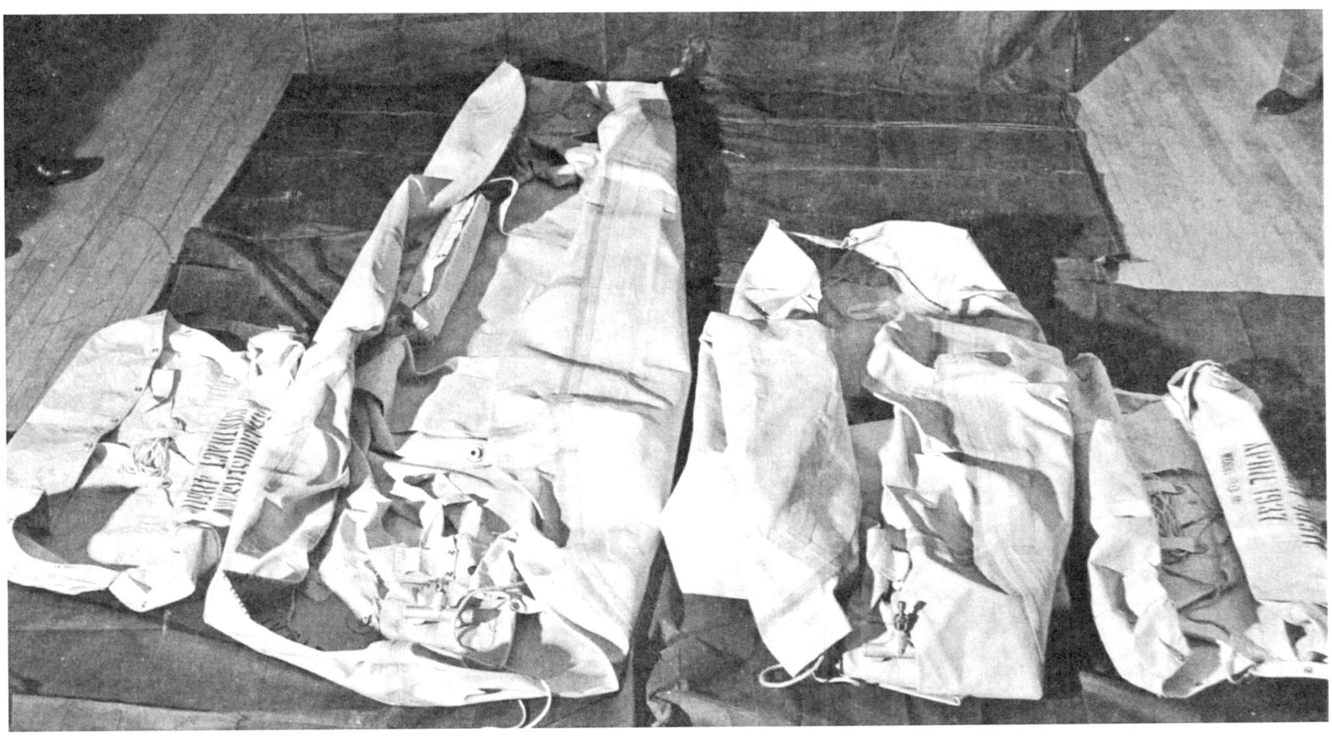

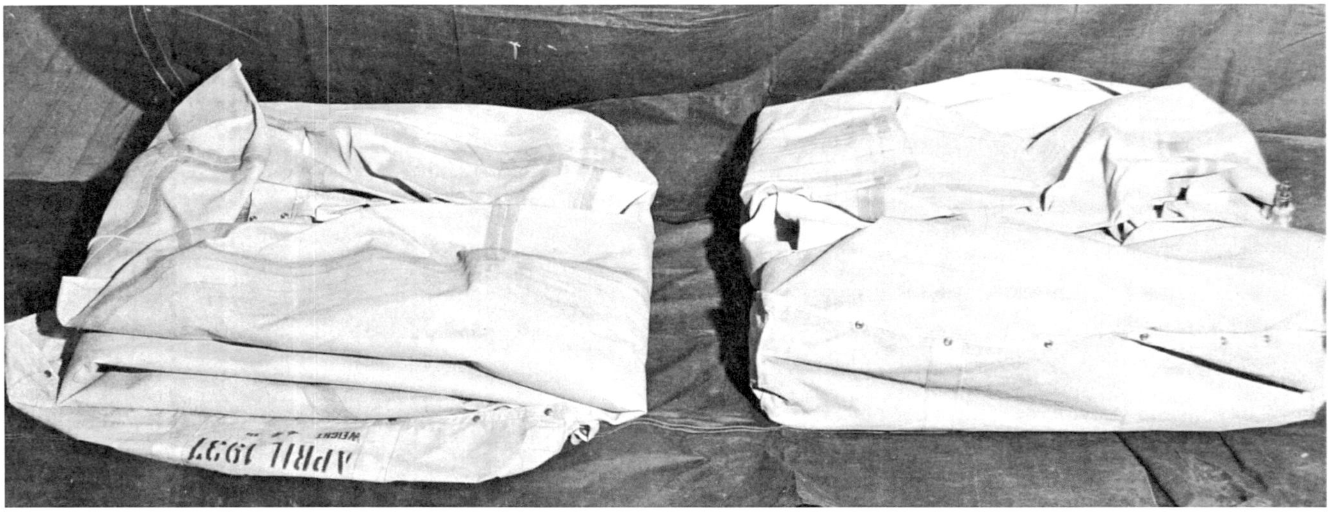

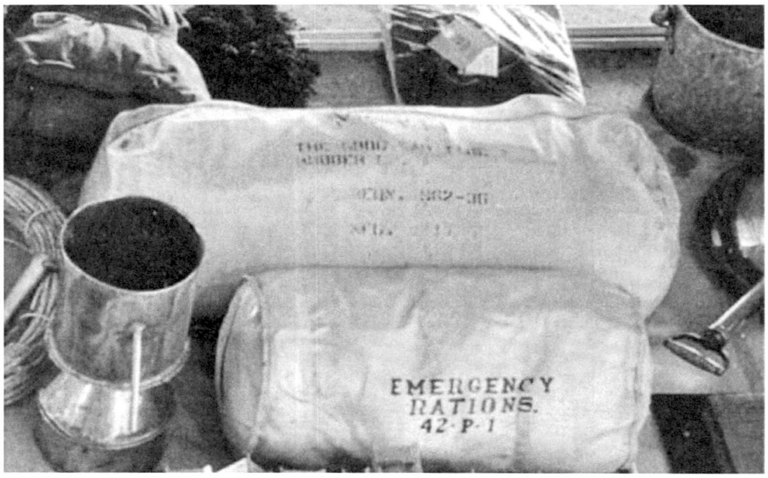


*(Left) A MK-VII as part of a display of equipment aboard a PBY in December 1939. The emergency ration bundle would have been a locally assembled unit and not a standard compliment for M-3 life rafts at this date. (National Archives)*

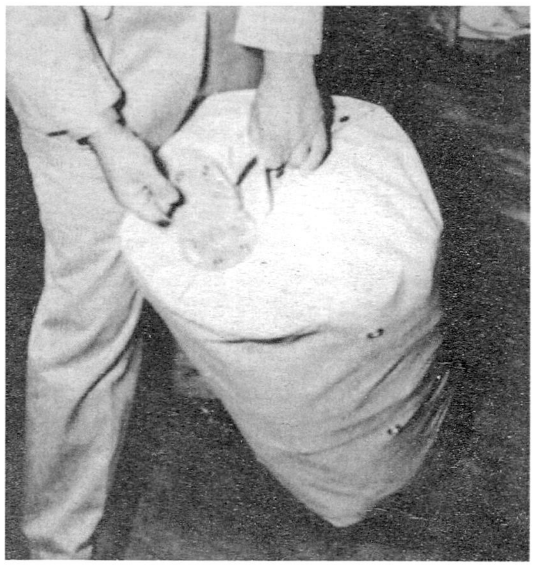

*(Right & Below) Further revisions as detailed in Amendment-N of 27 November 1939 that separated the rafts in two designations, the Type-S (Standard) and Type-D (Droppable). The same raft is used for both types with the differences being the inflation device, carrying case and packing. The Type-S would utilize a zippered case and to inflate the raft it would be completely removed and actuated by pulling a wooden toggle as seen below. The Type-D, designed for air dropping purposes, was contained in a case with snap fasteners and packed so that the hand grip, or rip cord, for inflation, was outside the container. This method did not require the raft to be unpacked, and as the raft inflated it burst the case open. To air drop the package the user slipped it from the plane while holding the rip cord. Then, as the raft descended, it would inflate.*

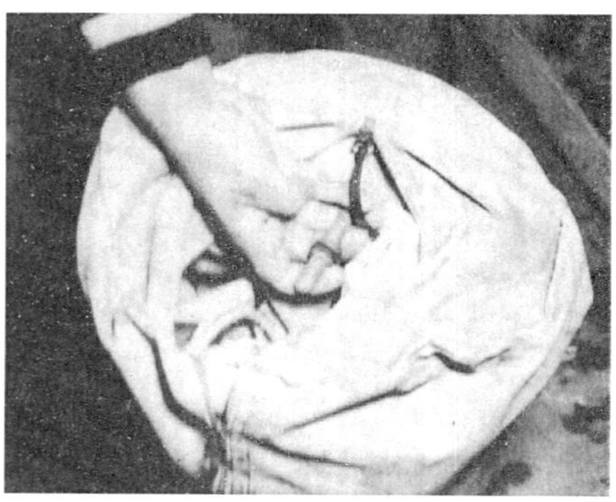

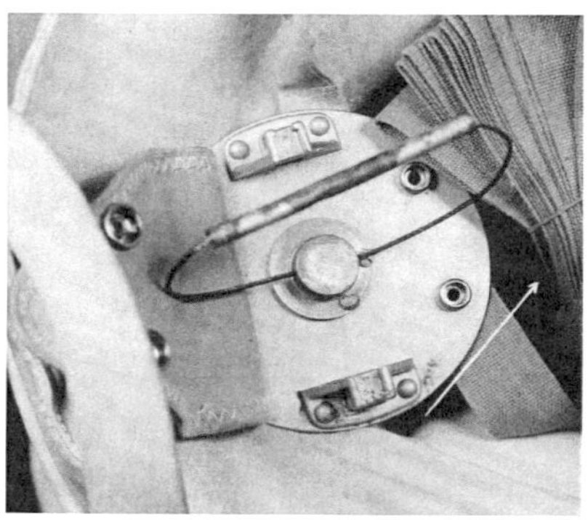

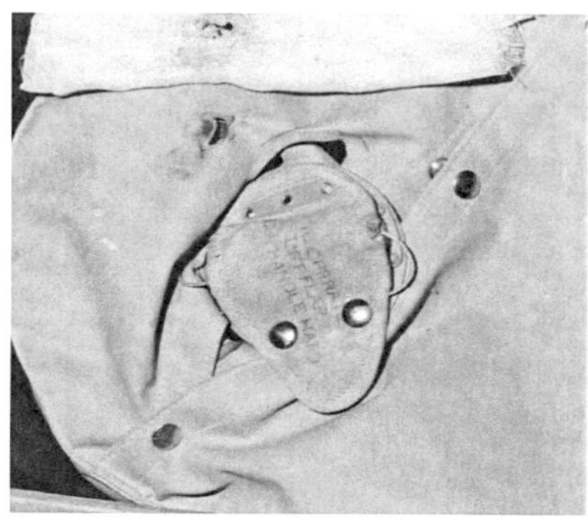

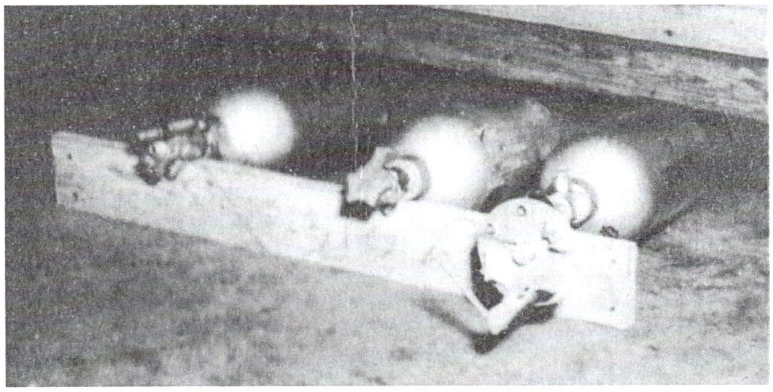

(Left) In June of 1941 the Type-A (Automatic) designation was officially added to specification. It closely resembled the Type-D but the inflation system was operable from a remote source. The typical application was the stowage of the raft in a compartment in the fuselage, attached to sensors that would inflate upon contact with water. Each designation utilized the standard CO2 bottle NavAer Specification M-108 but utilized different valves. Pictured are the three variants, from L-R; Type-S, Type-A and Type-D.

(Right and below) Upon standardization in 1929 M-3 rafts were equipped with a basic compliment of accessories to include oars and a hand pump. The original pump to be included was to be made of brass, but this added excess weight. A new aluminum lightweight pump was adopted 27 December 1933, under NavAer Specification M-197 titled as Pump, Hand, Air (Aircraft). This new pump was 14" long x 2.5" in diameter with a 12" rubber hose. As part of original details, the handle on the plunger was to be a wooden ball, but this feature was eliminated for wartime production and replaced with a domed plate.

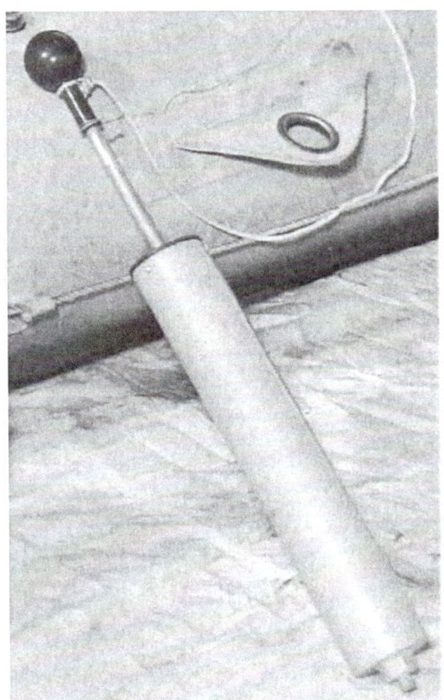

*(Below) For conservation of critical materials the BuAer adopted a new plastic pump 19 August 1942 under NavAer Specification M-550 titled as Pump, Hand, Air, Plastic. This new pump greatly reduced the use of metal parts with the bulk of fabrication being of a phenolic laminated material. The pump measures 10-1/2"x 3-1/2" in diameter and would include a rubber hose. The USAAF did not draft specifications for hand pumps and utilized M-197 for contractors. As part of the consolidation for production efficiency the M-550 was approved for Army-Navy standardization re-designated under Specification AN-P-48 7 July 1943 titled as Pumps; Plastic Hand Air. As detailed in AN-P-48 it calls for two lengths of rubber hoses 13-1/2" and a 9-1/2". Illustrated is an existing example of an M-550 (Left) and AN-P-48 (Right) hand pumps and the two hose lengths.*

*(Below) The oars originally provided were to be spruce in the lengths of 4 and 5 feet. 1& three-man rafts would be provided with 4' oars and 5' for the five-man. Further amendments add the material fir as an approved type. With the re-designation of the capacity the one-man would include a 4' length, 2 & four-man with a 4'-6" length and a 5' length for the seven-man. Oars were to be provided one per oar lock with the seven-man, the only raft that included more than two. To compensate for weight, the BuAer drafted and adopted as standard sectional aluminum oars under NavAer Specification M-162A 22 December 1933 titled as Oars, Metal, Sectional (Aircraft Use). Preliminary specifications M-162 were included in Amendment-J of November 1933. As detailed the new three sectional aluminum oars were to be provided in three overall lengths of 4', 5' and 6' with an exterior finish of an aluminized varnish. The 4' was originally slated for use with the one-man but was soon changed to include the 5' as with the 2 & four-man rafts, the seven-man would include the 6'oar.*

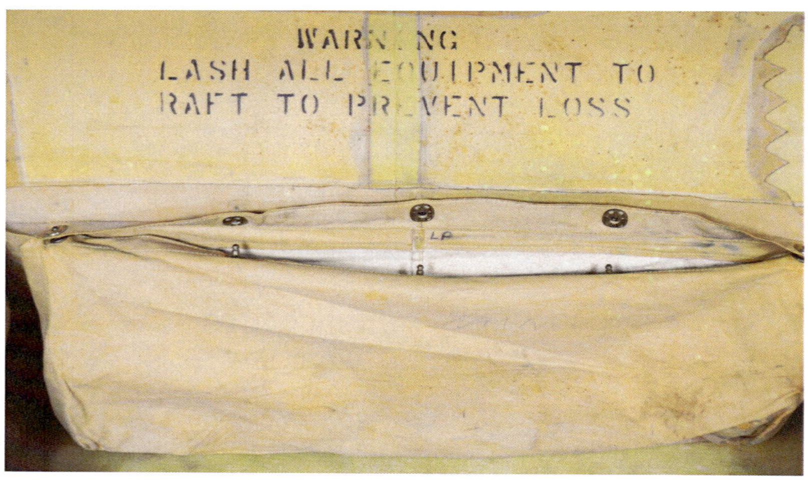


*(Left) Four aluminum oars included in the MK-VII*
*(Above) The oar pocket in a MK-IV life raft.*

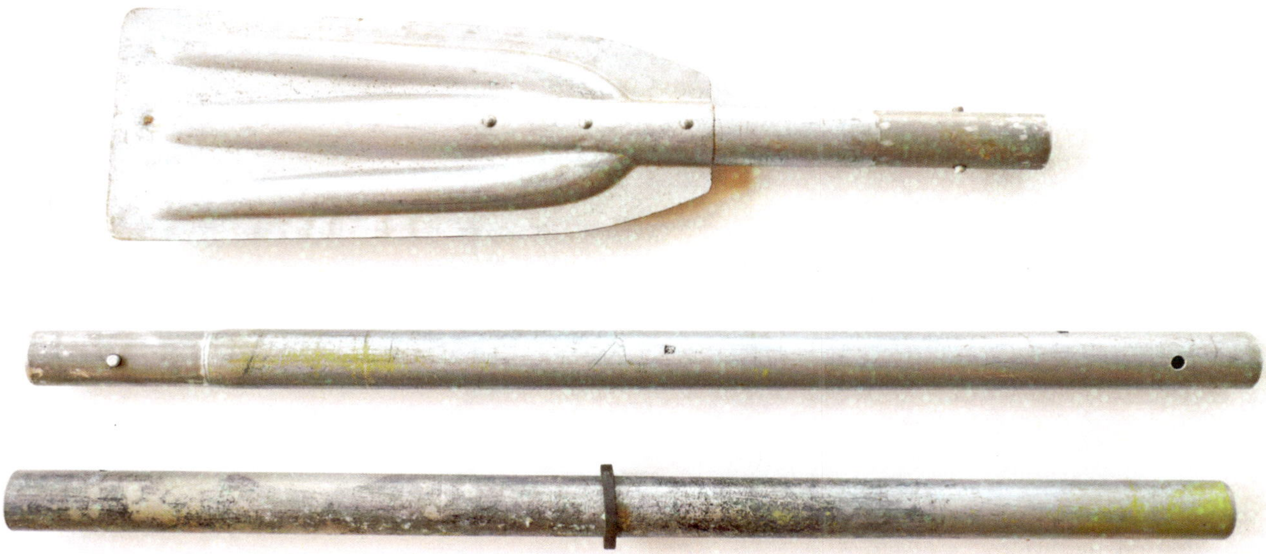

*(Above) Example of U.S. Navy aluminum painted sectional oar. An oxide primer, which looked green, was to be applied prior to application of the exterior finish. Each oar section was not to exceed 25" in length, this assembly is 5' when assembled. The rubber washer on one section prevented loss when is use.*

It was the belief that the aluminum colored oars had the potential to attract sharks so it was directed that oars be painted a dark blue color, formally addressed in Supplement-2 of M-162 Amendment-C 17 September 1942. After this date oars would be painted in a dark blue. Further revisions would occur with an adoption and integration of a fishing spear and the application of reflective material for night rescue, known commercially as Scotchlite. Fish spears and Scotchlite were retrofitted to the oars in service, and factory applied after acceptance of Amendment-D on 1 May 1944. This amendment would remain the standard for the duration of conflict.

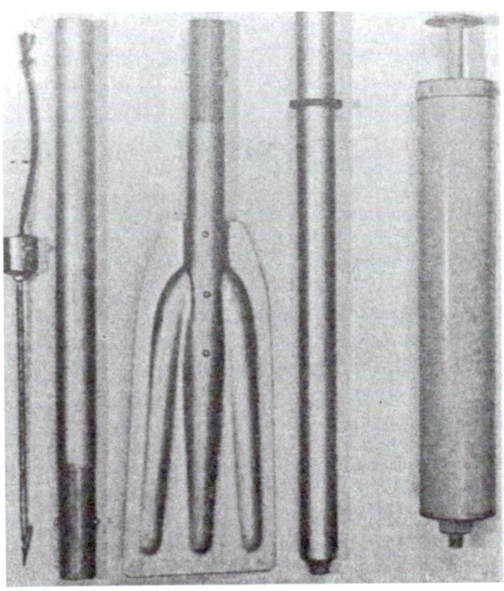

*(Left) A three-sectional aluminum oar that has been retrofitted with the fishing spear. It also includes the M-197 hand pump. The oars and hand pump would be stored together in the oar pocket.*

*(Below) As per Amendment-Q, all life rafts would be equipped with a compliment of three oars either 5' or 6' in length, relative to the seating capacity. The oars illustrated here conform to Amendment-D and have the factory applied Scotchlite and fishing spear on the end of one oar. As per Amendment-D, two grommets were provided on the end of the paddles for lashing them together to form the sailing mast.*

*(Right) Rubber washers (right) were provided to prevent loss when in the oarlock. This example is a friction fitted type.*

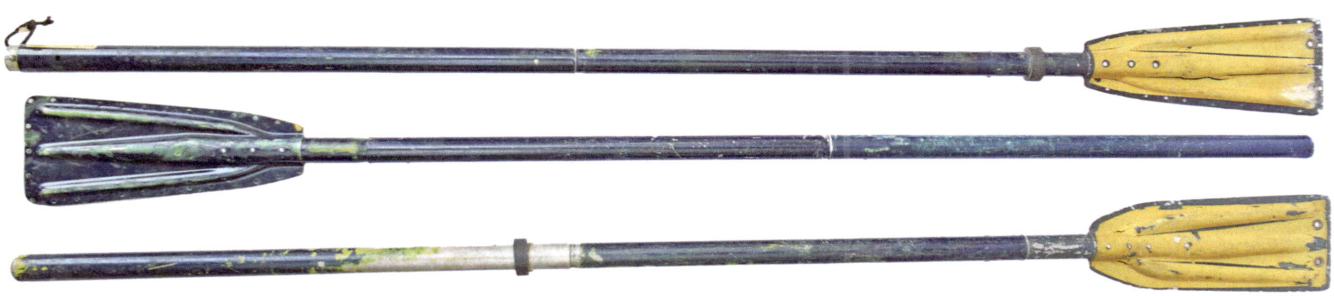

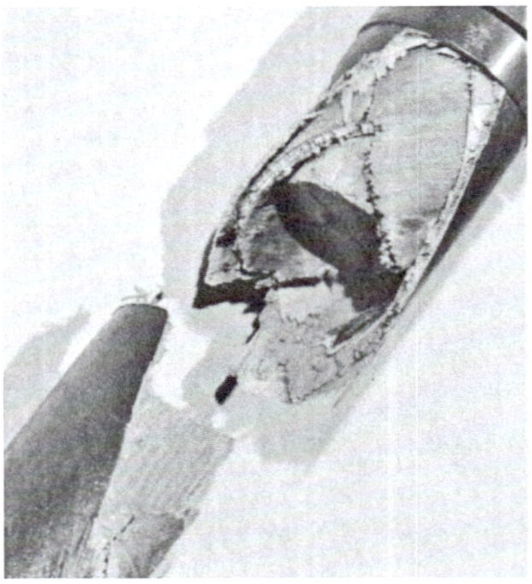

(Left and above) In 1943 plastic oars were developed and placed into service for service testing. They were manufactured of a laminated phenolic material. They soon proved to be inferior and were ordered to be removed from service in May 1944.

Added to the compliment of the M-3 series was a basic repair kit, originally provided with two pieces of aircraft fabric 4" x 12" and one tube of rubber cement, contained in a sealed pouch cemented to the bladder of the raft. As further amendments occurred, items were added and revised. Amendment-E of June 1931 detailed the inclusion of 4 pieces of rubberized cloth, 6" x 12", 4 pieces 4" x 4", one tube of rubber cement, a suitable roughing tool, pliers to fit boat fittings and cutting lines, wires etc. and 40' of line provided for lashing. The pliers were later amended to be specifically cadmium plated and have a screw driver end. Scissors were soon added for cutting the fabric. This would remain the standard complement until 6 wooden leak plugs were added in 1941. These were of a solid core type. Two 5" plugs and four 3" plugs were provided. By Amendment-Q of November 1942, the repair kit was to consist of a rubberized patching material, 6" x 24", five 2" diameter Patches, 2 ounces of rubber cement, 5" adjustable pliers with screw driver end, blunt scissors and 4 sets of nested wooden bullet hole or leak plugs.

(Left) Repair kits were provided as Contractor Furnished Equipment, so variations could occur and did not always strictly adhere to specifications. Illustrated on a display board is a M-3 series repair kit, c.-1945. The rubber cement is protected in the fiber container with a metal screw cap. Under the scissors is the patching material. A sandpaper roughing tool would also have been included.

(Right and below) Typical repair kit components, issued as per M-3Q. Not pictured is the sand paper roughing tool which would have been replaced by a metal type starting in 1943.

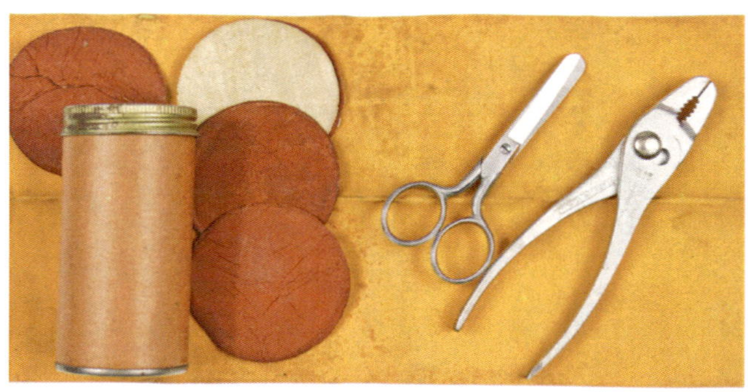

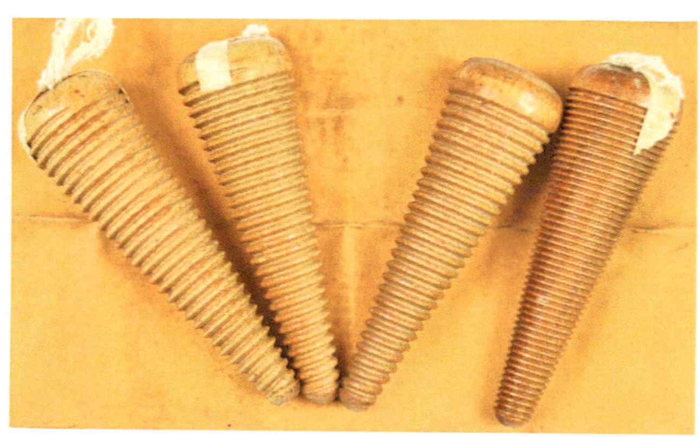

*(Right) The repair kit pocket, cemented on the bladder of a MK-IV life raft. This pocket was originally sealed but later revised to include a slide fastener for access.*

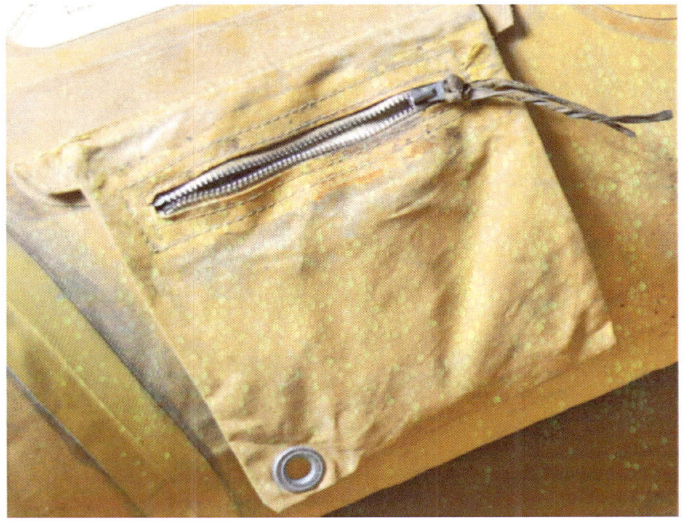

*(Below) The supply pocket was intended to provide additional stowage if needed, and served as a bailing bucket. Unlike the repair and oar pocket which were cemented to the bladder, the supply pocket was detachable. A Naval aircrew member is shown undergoing training exercises and is utilizing the supply pocket for bailing the raft, c.1942.*

*(Below right) The pocket affixed to the bladder of a MK-IV raft. Both the repair and supply pockets were attached at the bow. (National Archives)*

*(Below left) As part of Amendment-N 27 November 1939, a scout knife, police whistle and reflector were added to the accessories. The supply pocket was then repurposed for the stowage of these items and the line for lashing. This is representative of the accessories included as part of the standard compliment to M-3 life rafts in the immediate pre-and inter-war years.*

*(Below right) Accessories as detailed in Amendment-Q of November 1942 and Technical Note 54-42. With the standardization of new items in 1942 such as the M-554 match safe compass, M-552 fishing kit, and the introduction of the smoke grenade clamp, these would be integrated for stowage in the supply pocket. (National Archives)*

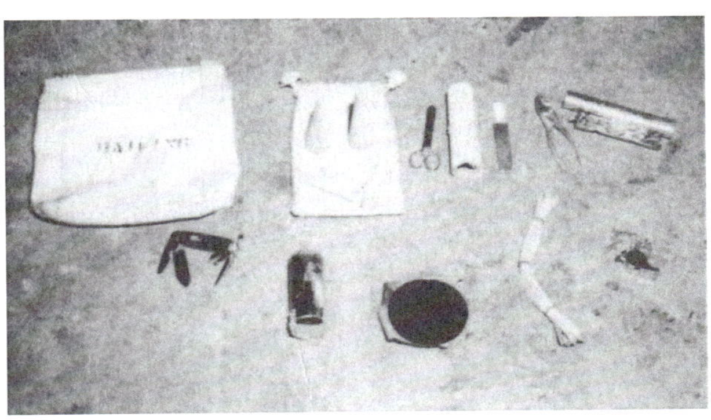

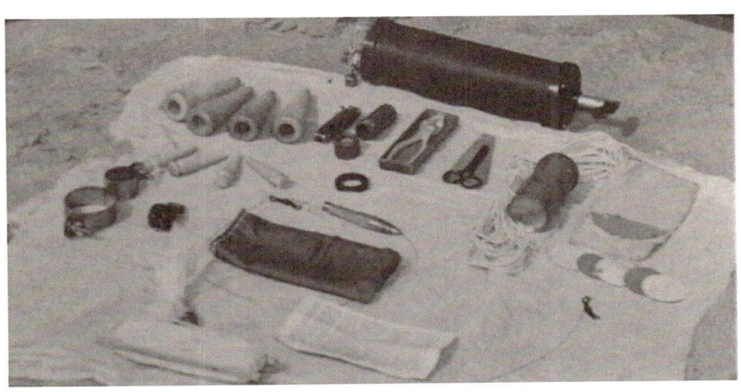

In Amendment-N, as detailed in Supplement-3 10 April 1941, an Emergency Equipment Container was added. It was felt that emergency provisions should be provided for survivors. This was in part influenced by war activities in Europe, and the prospect of being engaged in the conflict across the world in vast uninhabited regions. Signaling devices were implemented such as the whistle and signaling mirror, and further steps were taken to include an aluminum sea marker and two H.C. M-8 smoke grenades. To utilize the grenades, a special clamp was devised that would attach to an oar handle and would be stored in the supply pocket. To assist in basic needs that may be required by a marine survivor, it was decided to add a first aid kit and emergency rations. In June 1941, under Amendment-O, drinking water was added as detailed under NavAer Specification M-504. There continued to be a gradual integration of items through 1942, with the aluminum sea marker being replaced by the fluorescein Type M-528 in January.

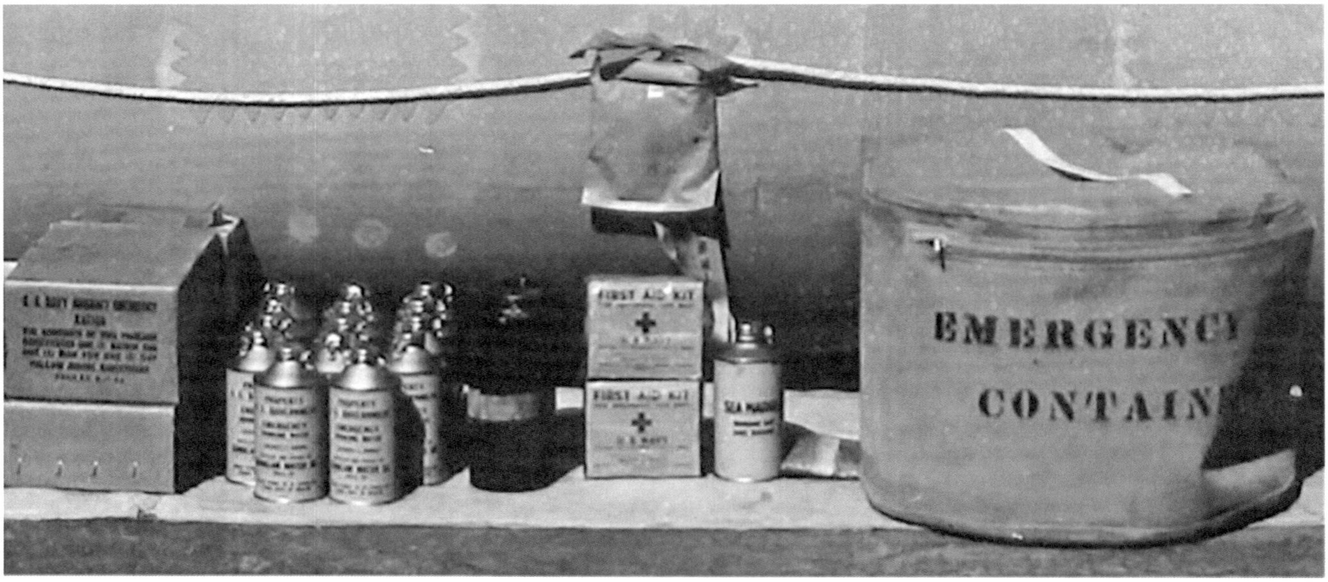


*(Above) When the emergency container was adopted, the rations and first aid kit had not been standardized, but were formally adopted in 1942. Stowed in the container would be the standardized items of one first aid kit S2-1055, one ration per man M-539, two cans water per man M-504/AN-W-5, one can sea marker M-528 and two smoke grenades H.C. M-8/AN-M8. Illustrated above is a complement representative of late 1942 packing for a MK-VII life raft. Two first aid kits were included, but only one was detailed to be provided. This raft was being utilized in field tests, and as part of those tests the life preserver sea marker packet (shown tied to the lifeline above the first aid units) was being evaluated. Of particular note is the first-generation equipment container provided in varying sizes, corresponding to the capacity of the raft, to accommodate the volume of rations and water. (National Archives)*

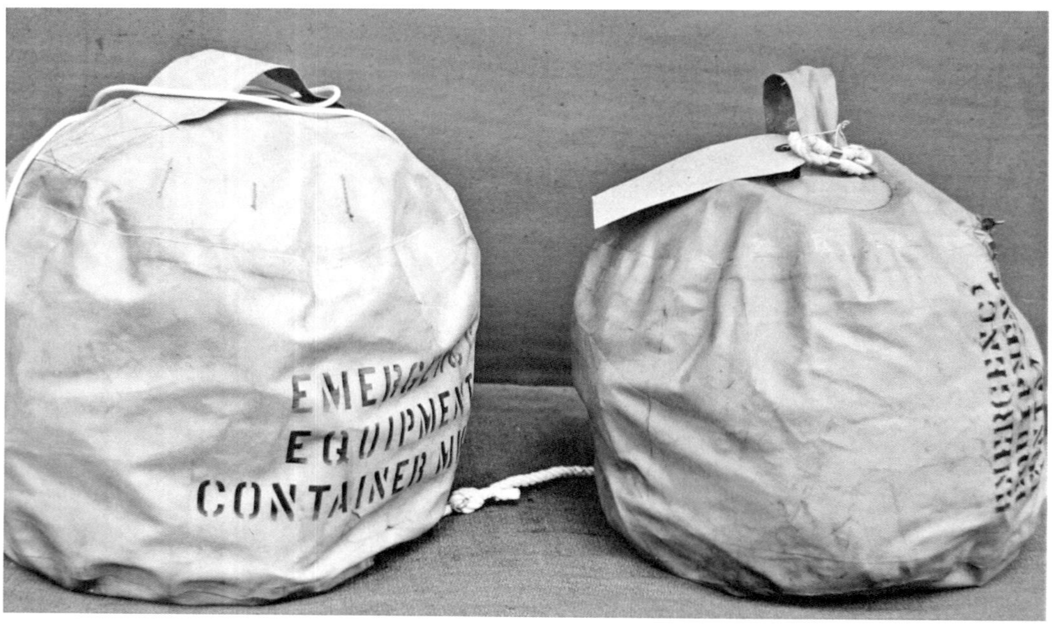

*(Left) The new equipment container was provided as detailed under NAF Dwg. No.214715. Under this drawing four different sizes would be available for each capacity rated raft: 8-1/2" long x 12 inches in diameter for the MK-II, 9-1/2" long x 14 inches in diameter for the MK-IV and 13 inches long x 15 inches in diameter for the MK-VII. For distinction, a dash number would be provided after the drawing number for example 214715-4 would be for the MK-IV life raft. (National Archives)*

*(Above and right) Items were loosely stored in the container with no particular order, and a slide fastener provided for access. The emergency container was stowed separately from the raft but tethered by a lanyard. When the raft was inflated, the emergency provisions would remain in the water temporarily, acting as a sea anchor with the carrying case. The 10' lanyard from the container was threaded through a grommet in the carrying case and tied to the life line on the raft as illustrated to the right. (National Archives)*

*(Below) Emergency Equipment Container for a MK-IV M-3 series life raft. They were provided with a grab handle on the top and a zipper that spans one side and the bottom. The 10' lanyard tied to the grab handle and had a spring clip for attachment to the life line of the raft bladder.*

*(Bottom) The Container shown tethered to the life raft case.*

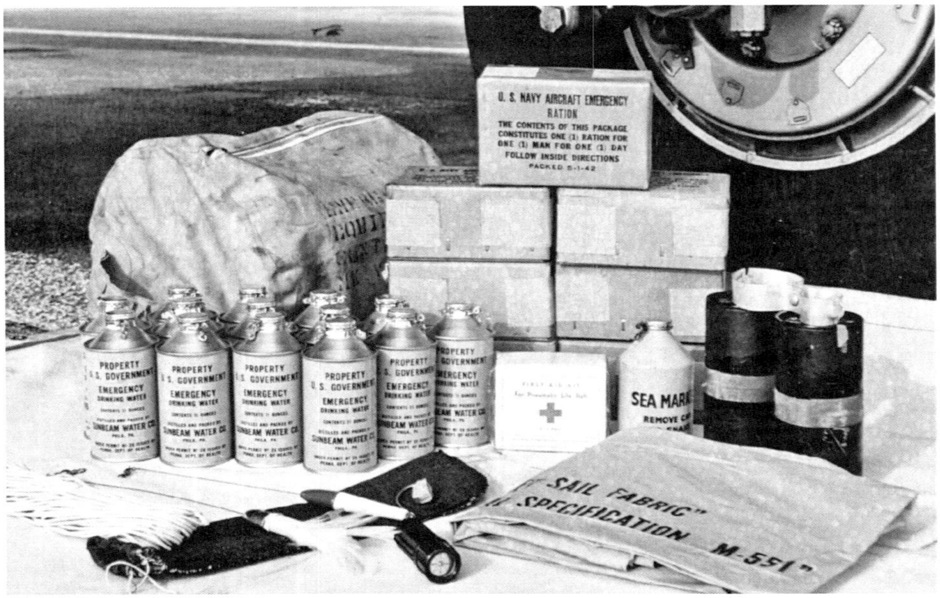


*(Left) Another look at the components contained in the emergency container. The grenade clamp, match safe, and fishing kit would be stowed in the supply pocket with the sail stowed in the oar pocket. (National Archives)*

*(Right) Technical Note 6-43 of February 1943 lists the addition of the projector kit and mosquito headnet M-565. Shown here is the complement of emergency provisions. Rations, water and headnet would be provided in quantities as per the capacity of the raft.*

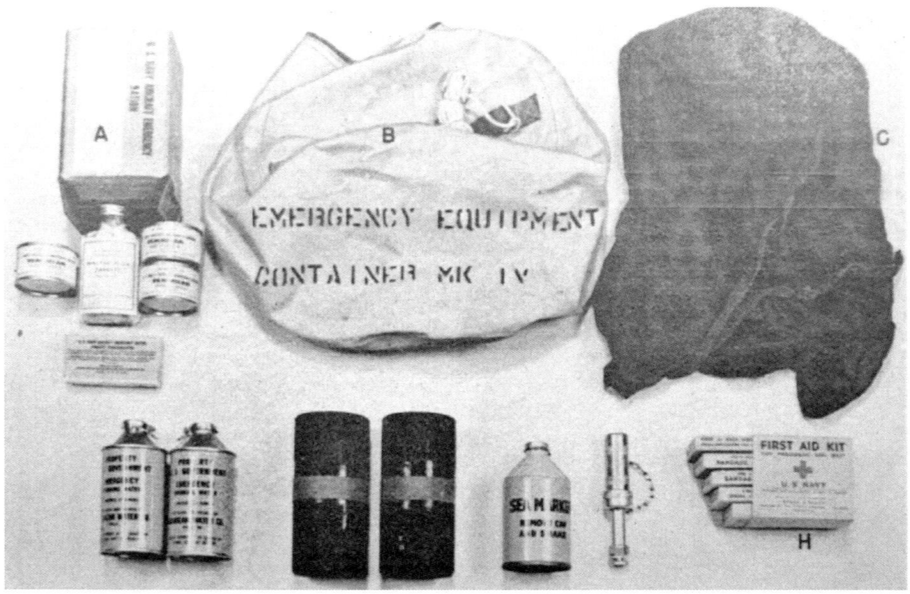

*(Left) Naval aircrew disassemble a life raft assembly for inspection. Laid out are the components of the Emergency Equipment Container. Three pockets for the repair kit, supply and oars can be seen. (National Archives)*

*(Below) Upon declaration and entry into the war, life rafts in service comprised accessories from Amendment-N. The Emergency Equipment Container and its provisions were a new item and not yet available in quantity. Through 1942, new items were adopted and integrated. Listed below are the consolidated contents as detailed in Amendment-Q of 1 November 1942, and Technical Note 54-42 drafted several months prior. From this date, life rafts would have been equipped with these items from the contractor and supplied for rafts in service as supply permitted.*

** Stored in oar pocket*

*** Stored in repair pocket*

**** Stored in supply pocket*

***** Stored in equipment container*

***** The jackknife provided is that as detailed in Specification M-3, as of March 1943 it would be referenced under NavAer Specification M-575.*

| ITEM | STOCK NO. | SPECIFICATION NO. |
|---|---|---|
| Raft, Pneumatic, Droppable, Complete | MK-II R83-R-15510, MK-IV R83-R-15530, MK-VII R-83-15570 | M-3Q |
| Carrying Case | MK-II R83-C-8510, MK-IV R83-C-8530, MK-VII R83-C-8550 | M-3Q |
| Sea Anchor | MK-II R6-A-1960, MK-IV R6-A-1970, MK-VII R6-A-1980 | M-3Q |
| Cylinder Valve Assembly | MK-II R83-C-8165, MK-IV R83-C-8175, MK-VII R83-8185 | M-108 |
| Container Emergency Equipment | MK-II R83-C-66755, MK-IV R83-C-66765, MK-VII R83-C-66775 | M-3Q |
| Oars (3-per raft)* | MK-II & MK-IV 5 ft. R83-O-4010, MK-VII 6 ft. R83-O-4020 | M-162 |
| Leak Plugs 4-sets** | R83-P-408550 5" | M-3Q |
| Repair Kit** Kit contains;<br>Patching material 6"x 24"<br>Roughing tool<br>2 oz. cement<br>5-2" dia. patches<br>1-pair 5" adjustable pliers<br>scissors | R83-K-710150 | M-3Q |
| Hand Pump* | R-83-P-408500 | M-197 or AN-P-48 |
| Sail* | R83-C-76950 (1-per MK-II, 2-per MK-IV & MK-VII) | M-551 |
| Fishing Kit*** | R83-K-300 | M-552 |
| Match Box Compass*** | R83-C-2500 | M-554 |
| Smoke Grenade Holding Clamp*** | R83-C-44600 | M-3Q |
| Signal Mirror Reflector*** | R83-R-20500 | M-3Q |
| Jackknife*** | R41-K-365 | M-3Q***** |
| Whistle*** | R42-W-24000 | M-3Q |
| Line 25'-75# test*** | 21-R-150 | M-3Q |
| First Aid Kit**** | R57-K-8525 | 57-K-0366 BuMed |
| Ration**** | R56-R-6125 (1-Per Man) | M-539 |
| Emergency Drinking Water**** | R51-W-135 (2-Per Man) | AN-W-5 |
| Sea Marker**** | R83-M-160500 | M-528 |
| AN-M8 Smoke Grenade (2-per raft)**** | R83-G-770100 | BuOrd |

*(Below) Items were continually developed, and added most often in response to service activities from those operating in the combat zone, based on actual survival experiences. As part of the continuing effort, new articles were added to the complement of the M-3 life raft detailed in Technical Note 6-43 of February 1943, which superseded Technical Note 54-42 for the inspection and maintenance of multi person life rafts. The new items addressed concerns about signaling, protection from the elements, the gathering of food and navigation. Accumulatively these articles would remain standard and were utilized for the duration of hostilities. New items, such as the solar still, would be adopted in 1944, superseding some articles, but would have not become available until 1945.*

| ITEM | STOCK NO. | SPECIFICATION NO. |
|---|---|---|
| Pyrotechnic Signal Kit**** | R83-K-710309 | None |
| Mosquito Headnet (1-Per man)**** | R27-H-261 | M-565 |
| Fish Spear* | R83-S-648150 | S-146 |
| Chart Navigation Set**** | R83-C-22310 | None |

*(Below) MK-IV Type-D NavAer Specification M-3Q life raft. This example was manufactured by Air Cruisers Inc. Other contractors included the U.S. Rubber Co., Rubbercraft Corp. of America, Goodyear Tire & Rubber Co., Firestone Tire & Rubber Co., B.F. Goodrich Co., Hood Rubber Co., New York Rubber Corp. and the Armstrong Rubber Co. Subtle variations can be observed between contractors but were all principally manufactured per specification M-3. An estimated 55,000 M-3Q MK-II, MK-IV and MK-VII life rafts were procured from 1942 to 1944. By the issuance of this amendment, the MK-I would be recommended for discontinuation and replaced by the parachute type. Though the specification still covered the Type-A, Type-D and Type-S, the bulk of procurement was of the Type-D. The Type-A was beginning to be phased out for use in aircraft. The Type-D was more universal than the Type-S and easier to use. Records indicate that all procurement from 1943 on was of the Type-D. Type-S rafts in service were modified by adding Type-D inflation systems. The example shown has been modified at the stern for erection of the MX/138-A radar reflector, the socket for the sailing mast is obscured by the supply pocket. This particular example was procured by the USAAF as evidenced by the contract number. In 1943, the USAAF procured 2,500 MK-IV rafts for expedited shipment to Europe. By the time of the purchase, the Materiel Command was revising the series of life rafts in which they had field tested BuAer models. They concluded that the BuAer design was superior to their Type A-2, with some its features incorporated into the new Type A-3. To temporarily satisfy the concerns expressed by ETO commands regarding the unsatisfactory nature of the Type A-2, Materiel Command approved procurement of the MK-IV until the A-3 became available.*

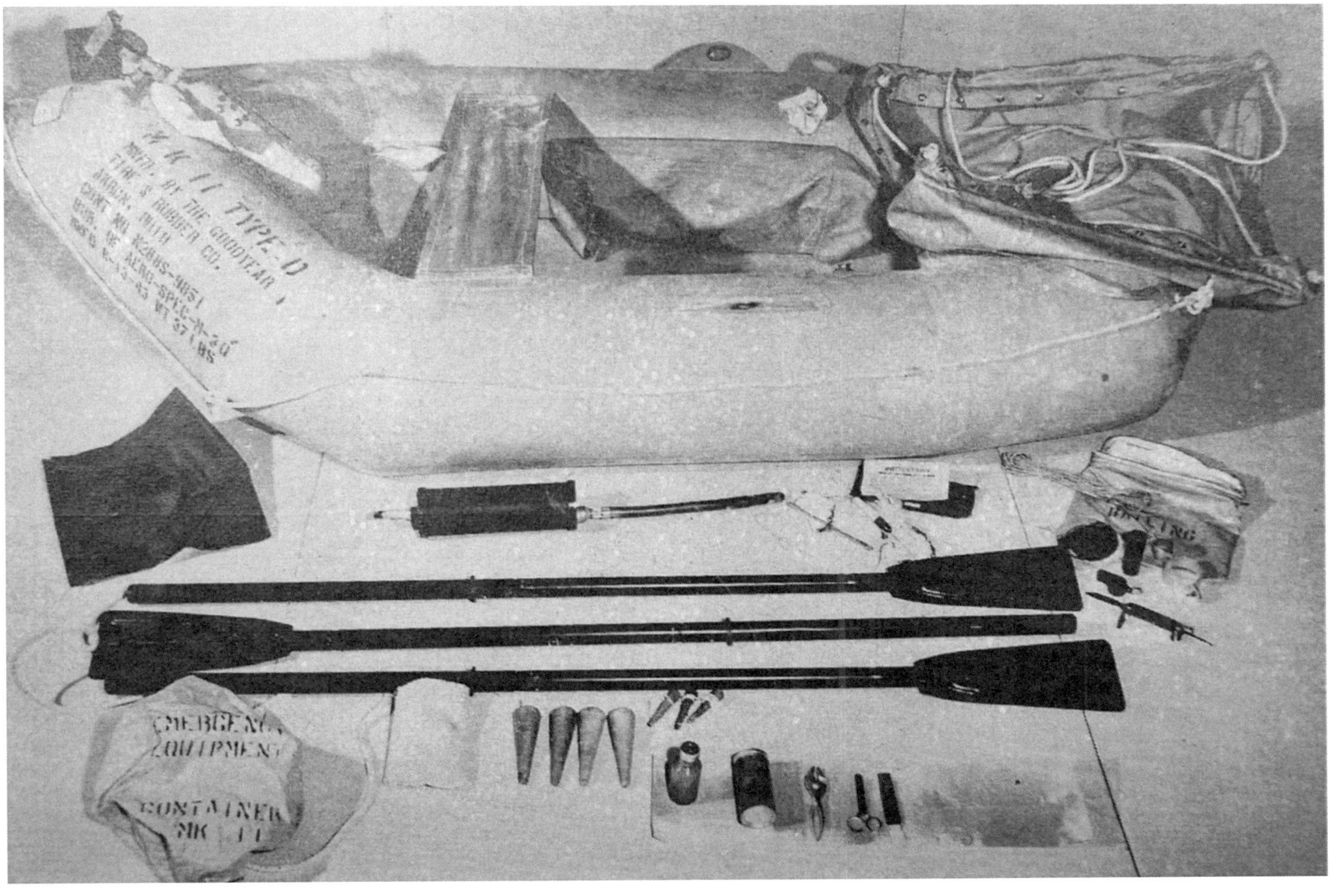


(Above) Type-D MK-II. Rafts manufactured prior to Amendment-Q would have been updated in assembly and repair facilities to incorporate modifications outlined in the new requirements, such as adding of the third oarlock at the stern and addition of the equipment container. This display of the MK-II and its accessories is as detailed in the lists previcusly outlined. The supply of items was not always as consistent as the component lists. A. MK-II M-3Q Life Raft, B. CO2 Inflation Cylinder, C. Repair Pocket, D. Oar Pocket, E. Carrying Case, F. Sail, G. Hand Pump, H. Fishing Kit, I. Oars, J. Signaling Mirror, K. Matchbox Compass, L. Whistle, M. Jackknife, N. Supply Pocket, O. 25' Line, P. Equipment Container, Q. Projector kit, R. Leak Plugs, S. Repair Kit.

(Below) Similarly updated Type-S MK-II.

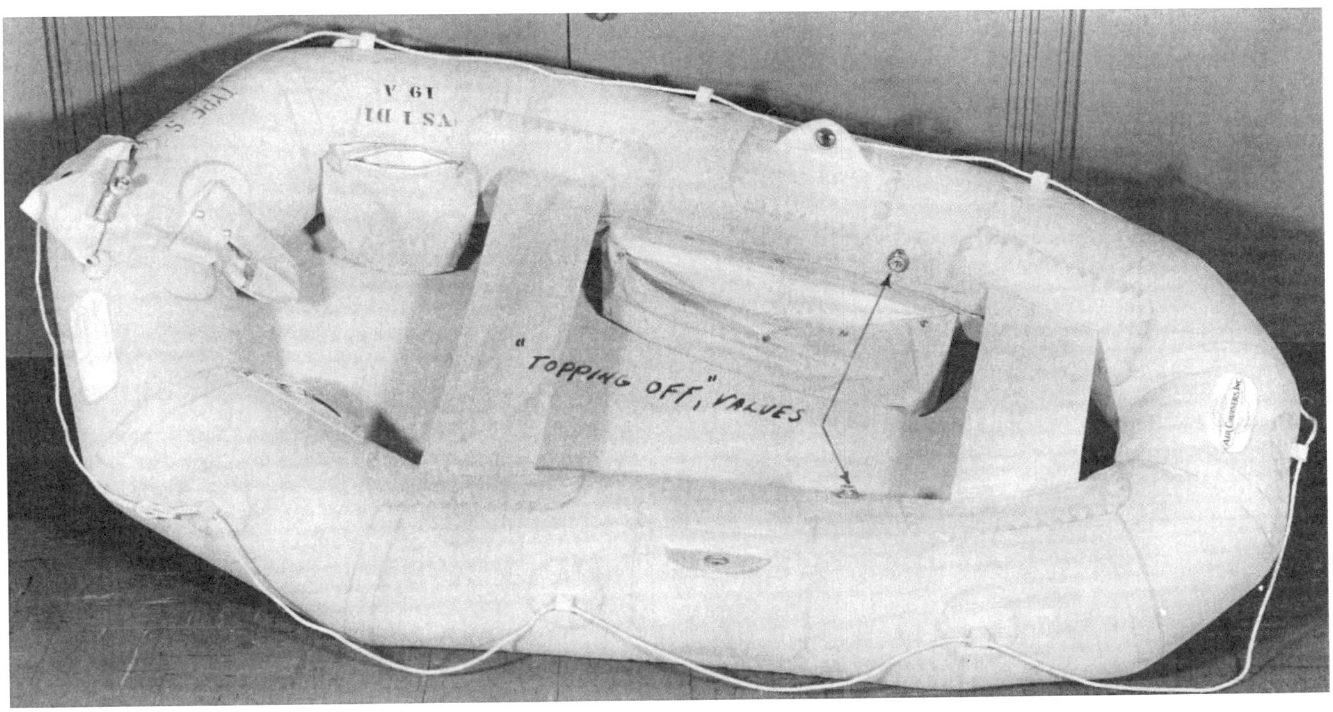

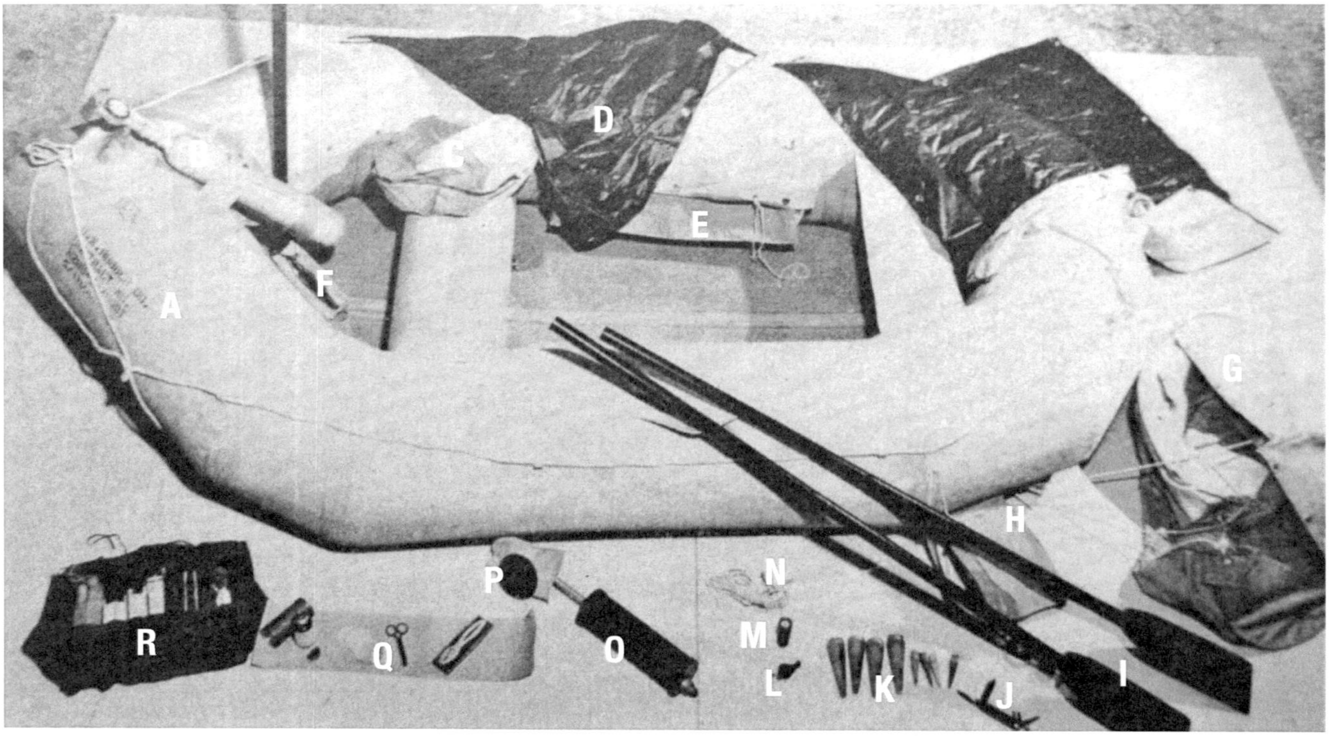


*(Above) A distinguishing detail between the 2 and four-man rafts was that the MK-IV included an inflatable cross member near the bow, due to its width. Several features to note in this image are the inclusion of the updated model fishing kit M-552A or AN-K-2, a metal roughing tool with the repair kit, one oar is in the sailing socket. Also shown is its complement of two sails. A. MK-IV M-3Q Life Raft, B. CO2 Inflation System, C. Equipment Container, D. Sail, E. Oar Pocket, F. Repair Pocket, G. Carrying Case, H. Supply Pocket, I. Oars, J. Jackknife, K. Leak Plugs, L. Whistle, M. Matchbox Compass, N. 25' Line, O. Hand Pump, P. Signaling Mirror, Q. Repair Kit, R. Fishing Kit.*

*(Right) This sterile view of the MK-IV Type-S gives further detail on a pre-M-3Q life raft. It is void of the socket for the erection of a sail but has been modified with an oar lock on the stern, and socket with support strap for the MX/138-A radar reflector.*

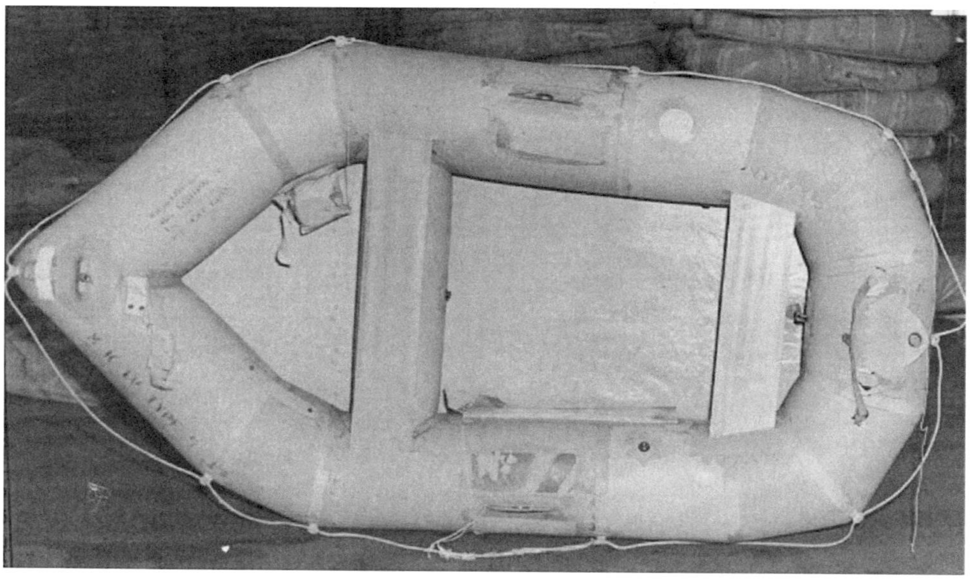

*(Below) All three capacity ratings of rafts were similar in appearance other than actual size. These inflated MK-II and MK-IV rafts offer a perspective on size differences between the two. Both contained the same accessories. (National Archives)*

(Left) The MK-II, 2-man raft inflated and occupied. (National Archives)

(Left) The MK-IV raft fully occupied for testing. (National Archives)

(Below) View of a MK-IV M-3P demonstrating its basic features. It did not include a stern oar lock and would not have a socket for the sailing mast near the CO2 cylinder. (National Archives)

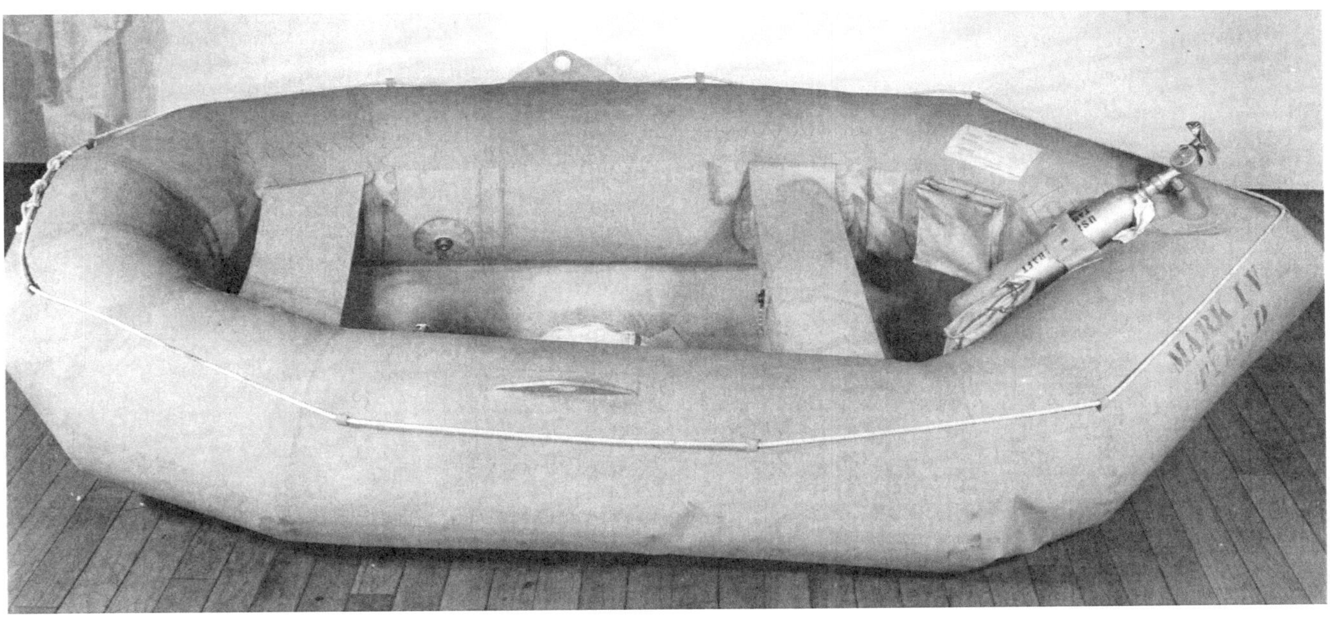

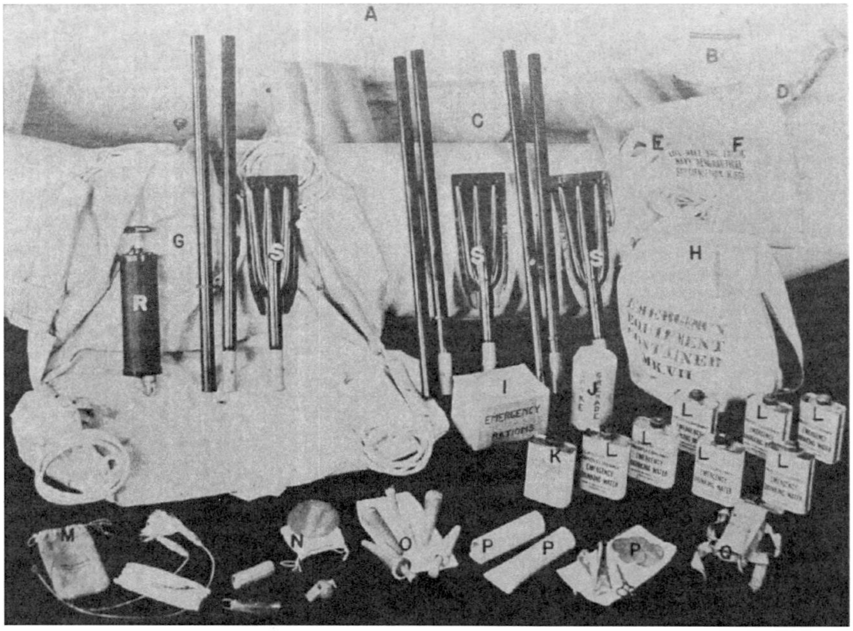


*(Left) A display of provisions included with the MK-VII for 1943 packing. The most peculiar item included is the USMD parachute first aid packet. This and the other equipment layouts are representatives of what would have been included as per specification and technical notes, but would vary according to availability and service activities. In 1943, the revised AN-W-5 flat water can become available and is illustrated here. A. MK-VII M-3Q Life Raft, B. Carrying Case, C. Oars, D. Sail, E. Repair Pocket, F. CO2 Inflation Cylinder, G. Equipment Container, H. Hand Pump, I. Fishing Kit, J. Matchbox Compass, K. Jackknife, L. Signaling Mirror, M. Whistle, N. Leak Plugs, O. Repair Kit, P. Rations, Q. Smoke Grenade, R. Sea Marker, S. Drinking Water, T. First Aid Kit.*

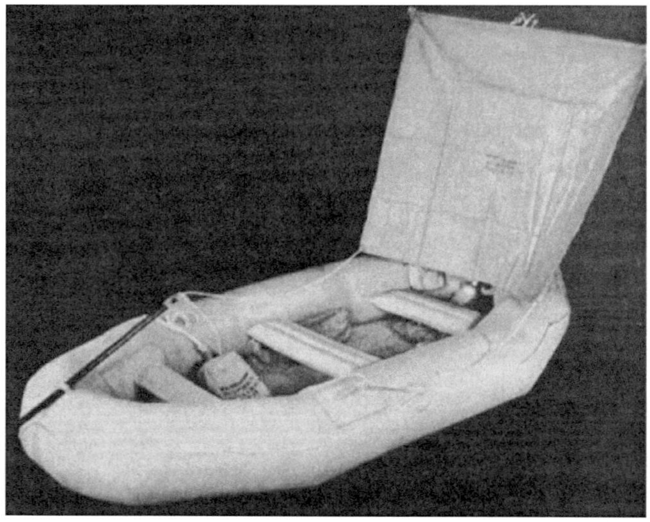

*(Right and below) MK-VII M-3Q life raft. Only three oarlocks were provided, versus four on previous models. The complement of three oars allowed two to be used as a sailing mast with the third in the stern lock acting as a rudder. The 25' of line stowed in the supply pocket would be used to rig the sailing system. Variation between contractors can be seen in the inflatable seats or cross members. The illustration below shows the features of the seven-man raft.*

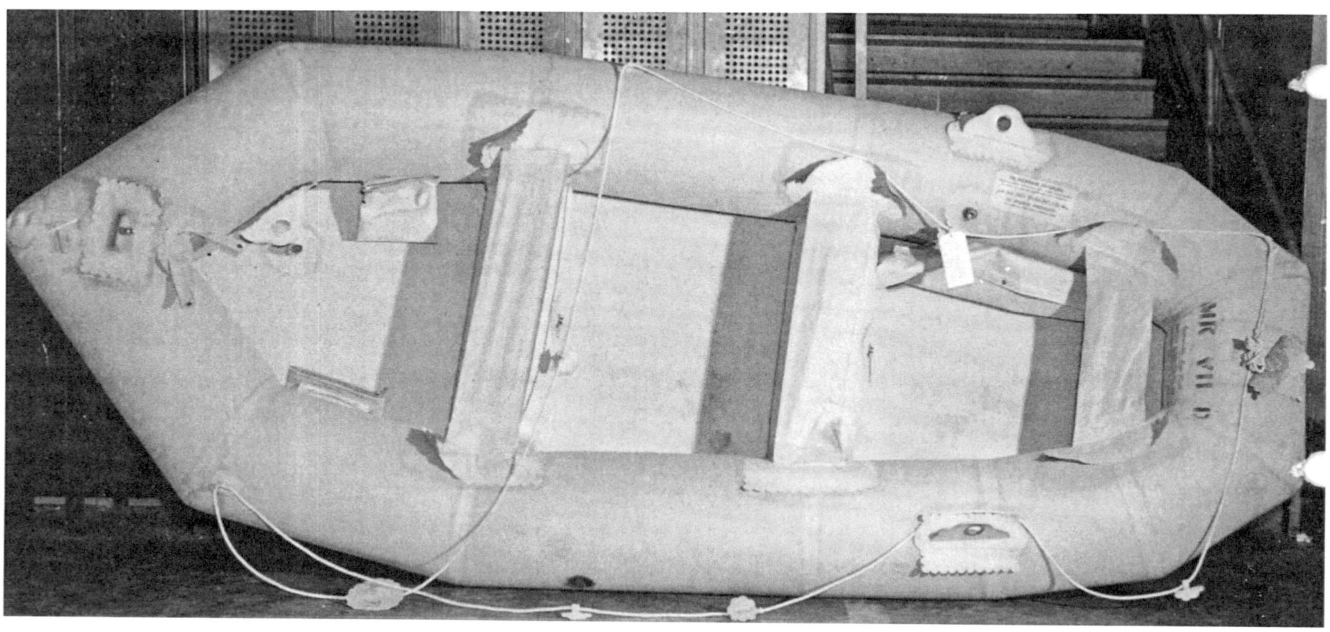

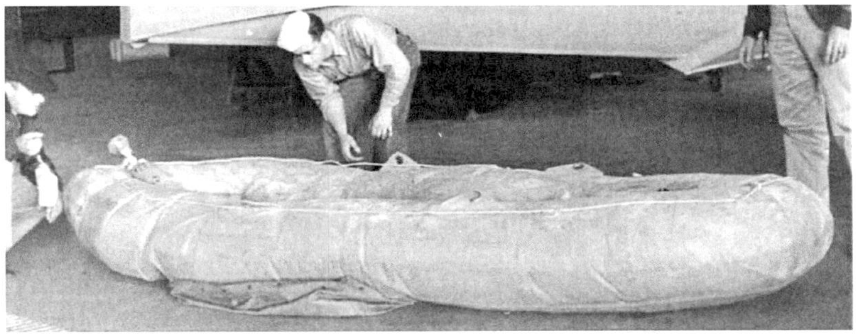

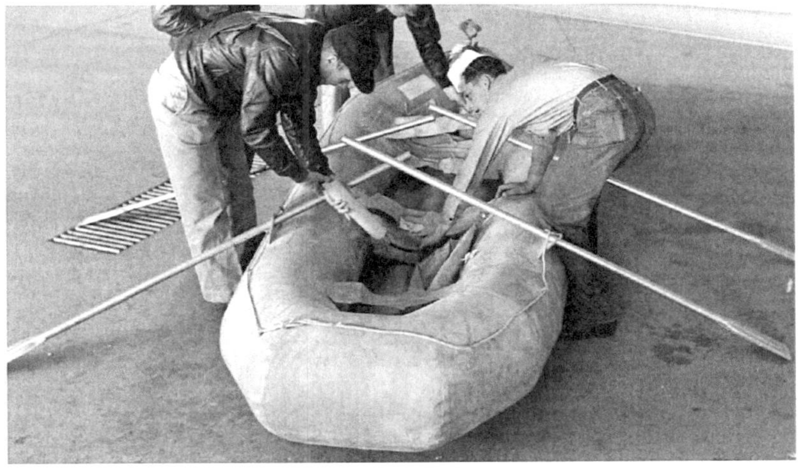

*(Above and top) Three images illustrating the packing and inflation of a Navy multi-person life raft. Once inflated, the cross members would be manually filled by the hand pump. (National Archives)*

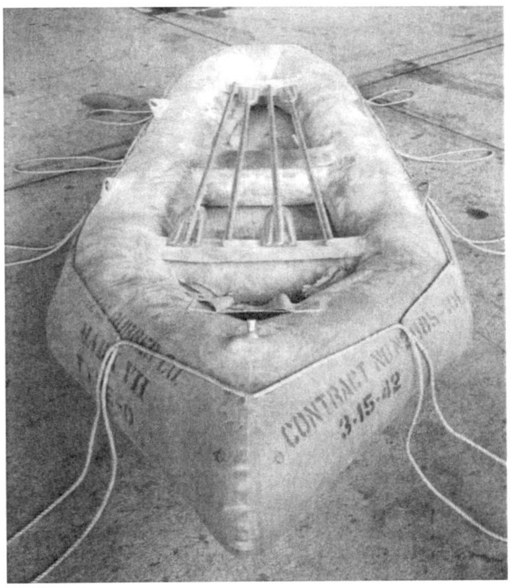


*(Left and below) MK-VII life raft that would have been manufactured under the active Amendment-P of 29 January 1942. Though the equipment container would had been detailed in preceding amendments, it may not have been available with the delivery of this raft. This image also lends a perspective to the meager provisions survivors would have had access to during episodes through 1942, as compared to those items detailed in M-3Q. (National Archives)*

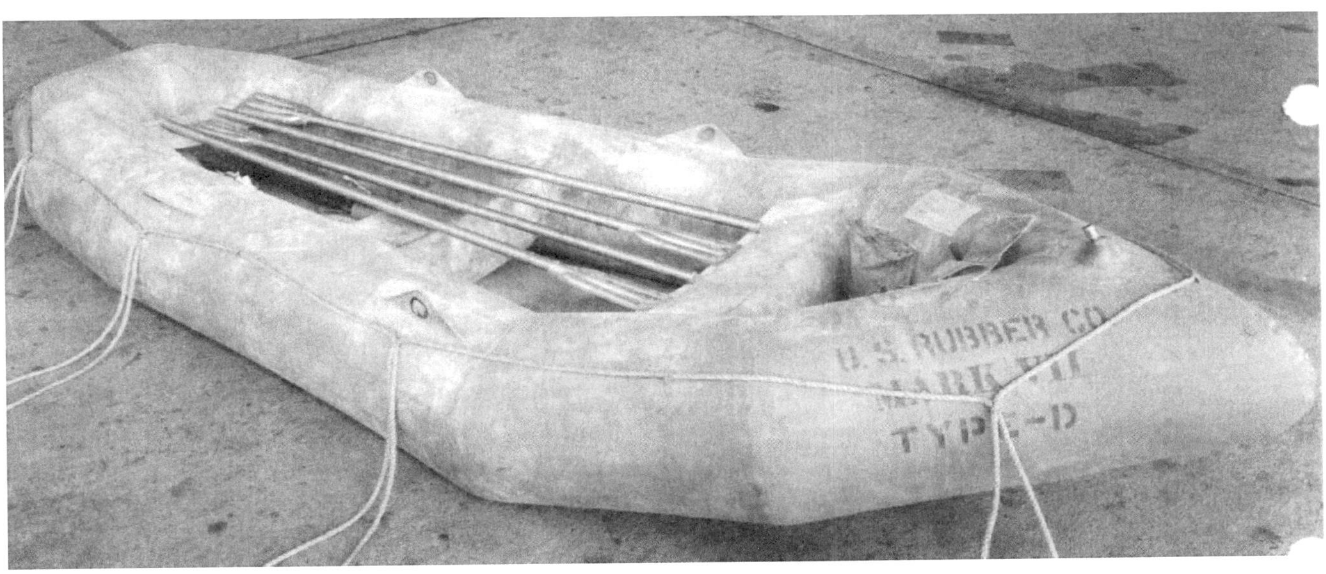

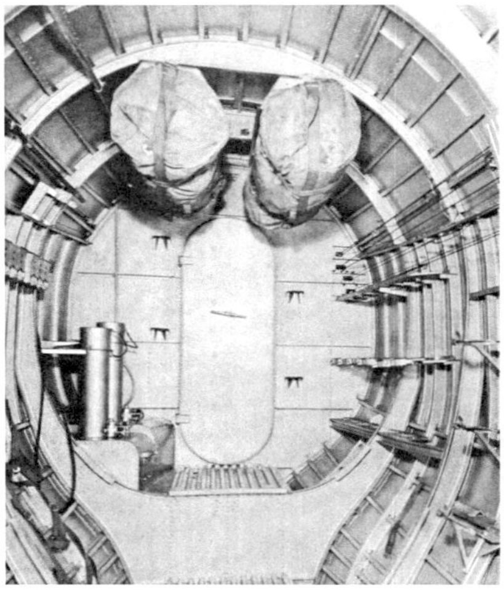

(Above) In PBY or PBM aircraft, life rafts were stowed in a location where they were easily accessible and visible in the event of an emergency. Typically, near an emergency exit. Both the Type-S and Type-D can be observed. Two rafts and SCR-578A emergency transmitter were stored in a convenient location aboard transport aircraft, with the emergency equipment containers in a rack above. Airplanes delivered to the US Navy from the USAAF such as the PBJ (B-25) and PB4Y (B-24) would most often be equipped with USAAF life rafts in the compartments provided on the aircraft. (National Archives)

(Left) Life rafts in airplanes rated for three persons or less had specific locations integrated into the fuselage. A MK-II is stowed on a sliding tray in this land based OS2U Kingfisher near the gunner's compartment. A principle reason for the primary procurement of the Type-D can be illustrated in these images, especially with regard to patrol and scouting aircraft. With the Type-D as standard they could be more efficiently dropped to survivors. (National Archives)

(Below) A MK-II Type-D life raft can be seen in the life raft compartment of this SB2C Helldiver aircraft behind the gunner, Joe Genne. When pilots were forced to bailout or ditch it became a common procedure for one or more planes to circle over survivors until rescue arrived. If it were evident they were without a life raft, squadron members could drop their provisions to assist. (National Archives)

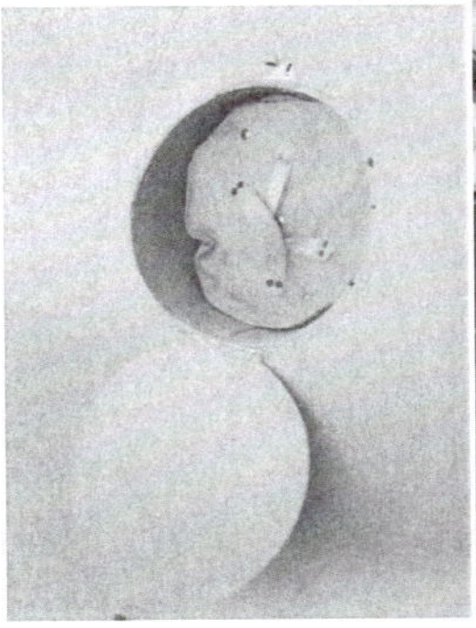

*(Left and inset) Not all aircraft had the life raft located in an easily accessible location while in flight. The compartment for the SBD Dauntless was located on the port side behind the gunner's position. In the event of a successful ditching, the crew would climb out onto the wing and retrieve the raft. If a Type-S, this would require them to unpack it on the wing, then inflate it. With an unstable and fast sinking aircraft this proved to be a daunting task. Gunner John Liska ARM1/c has access to a modification of the life raft compartment with a quick release cable. Oahu, Hawaii, February 1944. (National Archives)*

*(Left and below) The same principle applied to the TBF/TBM Avenger, as illustrated by these aircrew members undergoing training exercises. The Type-D proved more practical and easier to inflate in emergency situations where time was critical while in the water. The size of life raft provided depended on the size of the crew of the aircraft. In the case of the three crew positions on an Avenger, a MK-IV raft would be issued to it. The life raft compartment is just in front the ball turret. Early deliveries of the Avenger included the MK-IV life raft in a standard carrier. At an unknown date a binder replaced the case – a simple cloth bag with open ends, a zippered closure and grab handles on the ends. Once removed from its stowage, the raft could be released by a swift tug of the zipper pull, acting as a quick release. (NNAM)*

Upon entering the war, the BuAer was certainly better prepared for survival at sea than the USAAF. The BuAer was in the initial stages of providing drinking water and rations, but these were not available in quantity until the end of 1942. The most important factor was a coordinated effort for the exchange of information. Based on early reports, new items were developed to help save lives and reduce hardships survivors would encounter. This affected all agencies involved in the war effort and they were not blind to the problems presented. Each respective service or agency was concerned with lifesaving at sea and safety in the event of airplane crashes. As can be evidenced by the list of components detailed for the M-3Q of 1942, the BuAer made serious efforts to assure the safety and the health of survivors. Development continued at a steady pace.

The items were developed in a laboratory setting, then placed into service use and expedited to meet those demands and concerns expressed. With this awareness, institutions were established within the services and coordinated committees were founded to disseminate information. In the spring and summer of 1943, life raft field tests and trials were conducted under a series of controlled conditions. The bulk of these trials took place in the Gulf of Mexico at Eglin Field and Pensacola, and off the coast of North Carolina. In August 1943, the Office of Coordinator of Research & Development, U.S. Navy, held an informal meeting of representatives from all services concerned with the use, design, procurement and efficiency of airborne rubber rafts and sea survival equipment. Within this meeting, all findings of sea trials were shared, concluding with a report of unanimous recommendations. Moving forward with these recommendations, a series of modifications for life rafts and their accessories was considered.

The BuAer and USAAF experimented with sailing characteristics by adding ballast, rudders and keels in various configurations in some of those trials. It was felt that if the life raft could become a reliable sailing apparatus, survivors had the means of self-preservation. However, it proved impractical with the primary deficiency of adding too much extra weight and bulk, making it impractical for stowage on airplanes. Focus was then orientated towards preservation of survivors, comfort and stability of the raft. An accumulative list of modifications was adopted and detailed in Amendment-R 21, April 1944. Upon acceptance of the amendment, the emphasis had swayed towards providing signaling devices.

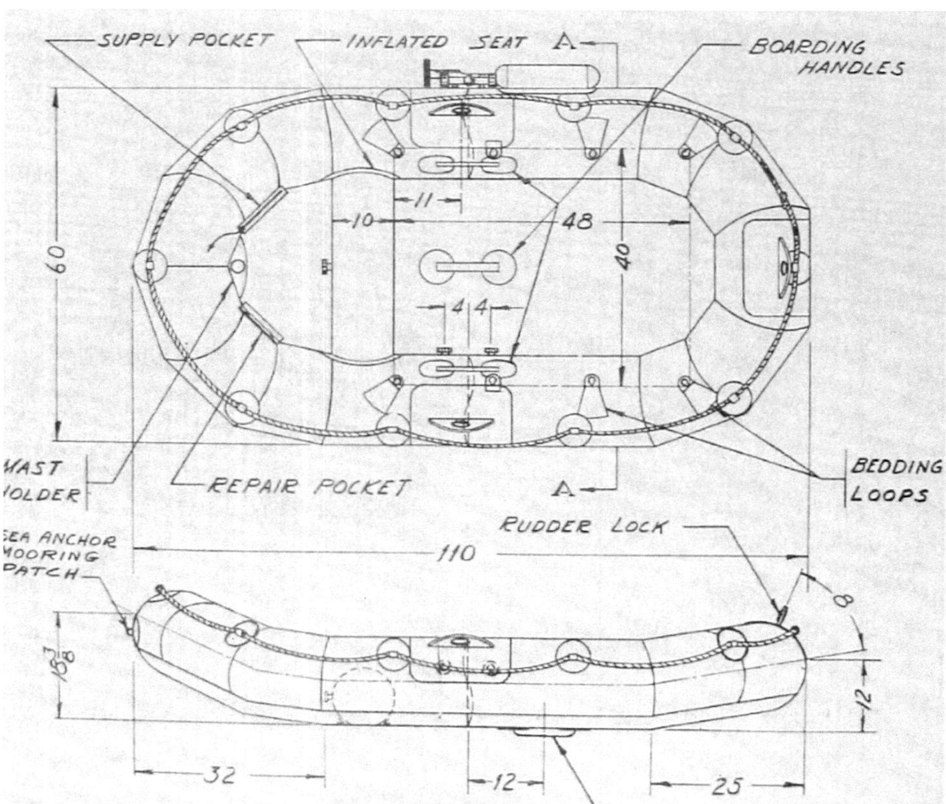


*(Left) The drawing that accompanied Amendment-R. The raft was completely overhauled to an oval, encircling buoyant tube, with vertical bulkheads, mid port and starboard bladders. This shape improved stability, minimizing the risk of capsizing, and created more room for increased comfort. Boarding handles were provided in the raft, and righting handles underneath, if it were to capsize. The CO2 inflation system was moved to the side to accommodate the adjustment of the bulkheads. Seats had been eliminated but one is provided by the support beam. Some of the same features carried over by providing repair, supply and oar pockets, but the oar pocket was no longer cemented to the bladder. A mast socket and three oarlocks were retained. New and updated accessories were adopted and listed in the chart on the following page.*

*(Below) Listed are the contents for the M-3R life raft.*

** Stored in oar pocket*
*** Stored in repair pocket*
**** Stored in supply pocket*
***** Stored in equipment container*

| ITEM | STOCK NO. | SPECIFICATION NO. |
|---|---|---|
| Raft, Pneumatic, Droppable, Complete | MK-II R83-R-15510, MK-IV R83-R-15530, MK-VII R-83-15570 | M-3R |
| Carrying Case | MK-II R83-C-8510, MK-IV R83-C-8530, MK-VII R83-C-8550 | M-3R |
| Sea Anchor | MK-II R6-A-1960, MK-IV R6-A-1970, MK-VII R6-A-1980 | M-3R |
| Cylinder Valve Assembly | MK-II R83-C-8165, MK-IV R83-C-8175, MK-VII R83-8185 | AN-C-73 |
| Container Emergency Equipment | MK-II R83-C-66755, MK-IV R83-C-66765, MK-VII R83-C-66775 | M-3R |
| Oars (3-per raft)* w/Fish Spear | MK-II & MK-IV 5 ft. R83-O-4010, MK-VII 6 ft. R83-O-4020 | M-162D |
| Leak Plugs 4-sets** | R83-P-408550 5” | M-3R |
| Repair Kit** Kit contains;<br>Patching material 6”x 24”<br>Metal roughing tool<br>2 oz. cement<br>5-2” dia. patches<br>1-pair 5” adjustable pliers<br>scissors | R83-K-710150 | M-3R |
| Hand Pump* | R11-P-562 | AN-P-48 |
| Life Raft Paulin* (1-per raft) | MK-II R83-P-92005, MK-IV R83-P-92010, MK-VII R83-P-92020 | M-616 |
| Fishing Kit*** | R83-K-300 | AN-K-2 |
| Match Box Compass*** | R83-C-2500 | AN-C-101 |
| Smoke Grenade Holding Clamp*** | R83-C-44600 | M-3R |
| Signaling Mirror*** | R83-R-20525 | M-580 |
| Jackknife*** | R41-K-365 | M-575 |
| Whistle*** | R42-W-24000 | M-3R/74W1 |
| Line 25’-75# test*** | 21-R-150 | M-3Q |
| First Aid Kit**** | R57-K-8525 | 57-K-1 (INT) BuMed |
| Ration, Tablet**** | R56-R-6300 (3-Per Man) | M-539B |
| Emergency Drinking Water**** | R51-W-135 (1-Per Man) | AN-W-5B |
| Sea Marker**** | R27-P-25 (3-per raft) | M-566 |
| AN-M8 Smoke Grenade (2-per raft)**** | R83-G-770100 | BuOrd |
| Hammock, Bed (2-per MK-VII)* | MK-II R83-H-500, MK-IV R83-H-5001, MK-VII R83-H-5002 | M-3R |
| Bailing Sponge*** | R83-B-648500 (2-per MK-IV & MK-VII) | L-S-626 |
| Drinking Water Kit**** | R83-K-511000 (1-per man) | M-613 |
| Water Stowage Bag**** | R83-B-30175 (1-per man) | BuAer Dwg.8985 |
| Projector Kit**** | R83-K-710309 | None |
| Hat Headnet**** | R83-H-200000 | M-565A |
| Retaining Line | R83-L-332760 (1-per raft) | None |
| Instruction Booklet**** | NavAer Publication 13-1-501 | None |

*(Below) The M-3R had been updated with new items such as the drinking water kit M-613, headnet M-565a and paulin M-616, but were not yet available upon standardization. Development work in refining the raft continued through 1944 to assure that all possible deficiencies were addressed before it was placed into service. In Supplement-1, 8 September 1944, the AN-K-2 fishing kit was directed to be replaced by a new pocket kit M-627, also adding sunburn ointment and a radar reflector to its complement. The chart below lists these three new items as detailed in Supplement-1.*

| ITEM | STOCK NO. | SPECIFICATION NO. |
|---|---|---|
| Fishing Kit*** | R37-K-110 (1-per MK-II, 2-per MK-IV & MK-VII) | M-627 |
| Sunburn Ointment, 4-ounce can**** | R57-O-100 (1-per MK-II, 2-per MK-IV & MK-VII) | Mfg. Pt. No. P-40469 |
| Radar Reflector* | R83-R-AN-MX-138/A | M-619 |

*(Right) An additional feature added was the retaining line. One end was secured to the neck of the CO2 bottle for the purpose of securing to the user before inflation. It consisted of 10' of 1/4" cotton line, with one end having a snap hook for securing to the life preserver. An instruction tag was provided.*

*(Below) In M-3R the oar pocket was now a separate assembly, secured to the raft by 5' of line to a grommet on the container. It still retained the same function, with the additional inclusion of the hammock and later the radar reflector. Illustrated below are two variations of the segregated oar pocket.*

*(Right) The sea anchor utilized in all M-3 series life rafts remained consistent, differing only in diameter relevant to the number of people in the raft. The funnel shaped anchor was made of a cotton cloth with 25' of cotton cord, later revised to nylon. The MK-II anchor was 10" in diameter and 9" long, the MK-IV was 12" in diameter and 12" long, and for the MK-VII it was 14" in diameter and 14" long.*

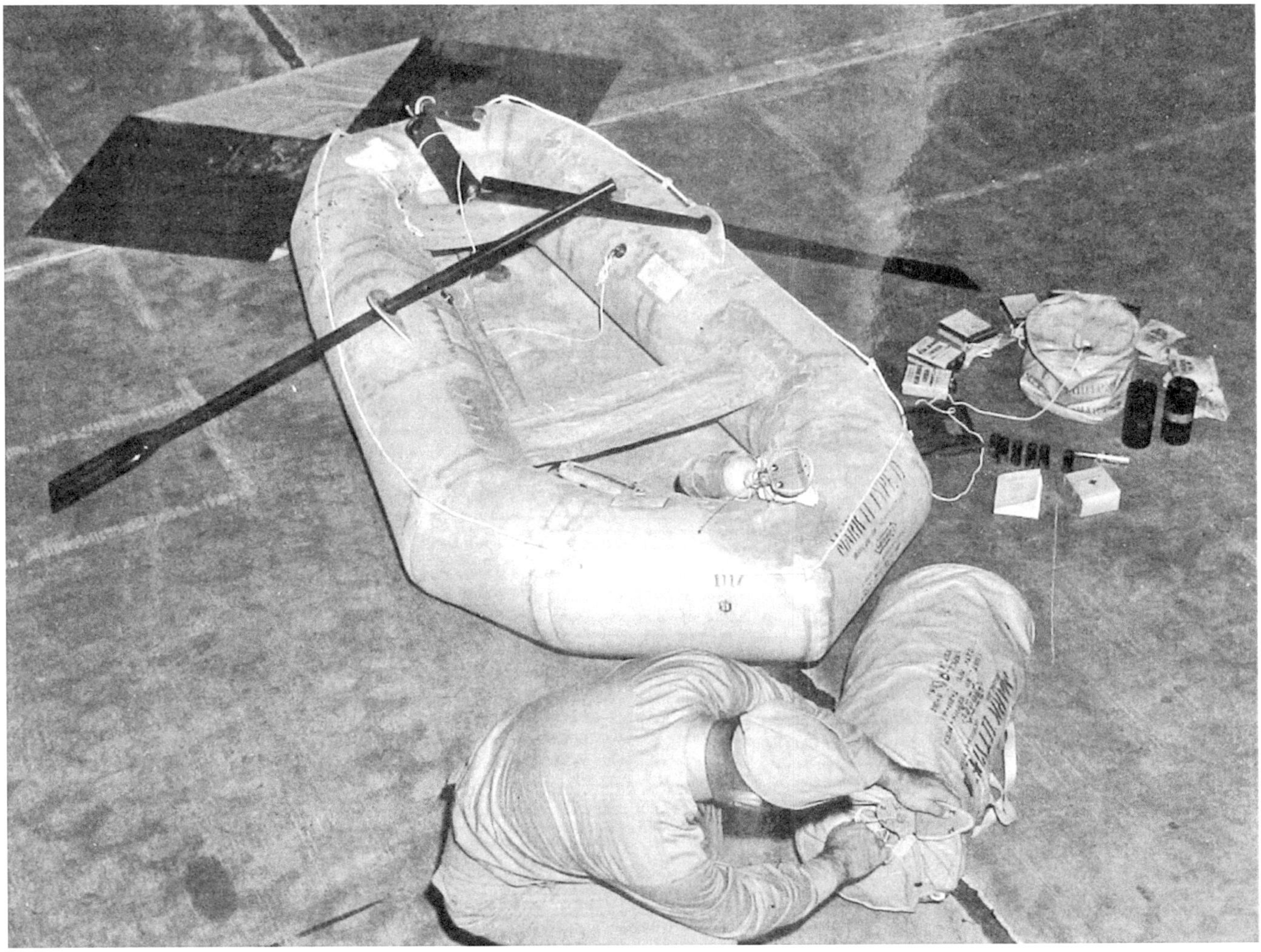

(Above) Upon standardization of M-3R, contractors were directed to convert undelivered M-3Q rafts to Amendment-R specifications regarding accessories. Illustrated above is an M-3Q MK-II that has the additional components displayed with it principally the emergency equipment container, NAS Lakehurst September 1944. All contracts awarded after April would conform to drawings included in M-3R. Type-S and Type-A rafts were officially cancelled for procurement.

(Below) MK-II as per drawing M-3R. The principle dimensions remained the same at the longest and widest points, with the MK-II being 7' 6" x 4'. The original drawing detailed one oar socket, in Supplement-2 3 February 1945, an additional socket was added at the stern. All three sizes included one seat/support beam but only a flap style in the MK-II. The seat on this example has been detached from the opposite bladder. The port and starboard bladders have 4 each web loops, these are for the attachment of a bedding hammock. (National Archives)

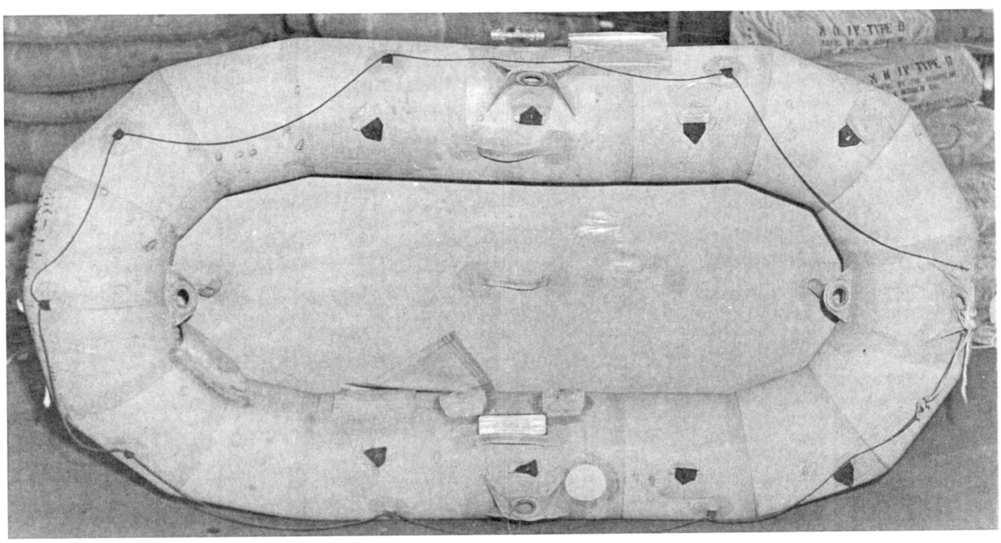

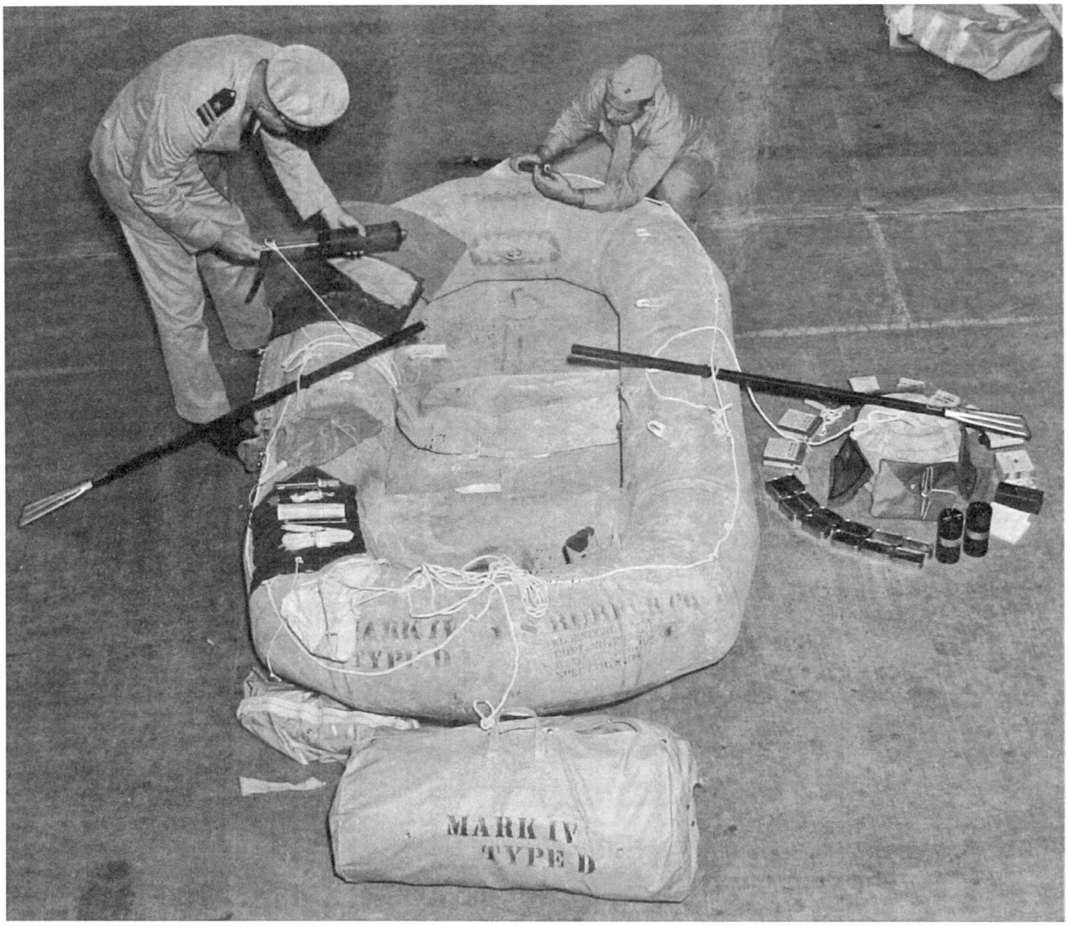


*(This page) Pictured above is the complete assembly of a MK-IV M-3R life raft, NAS Lakehurst 1944. Due to the width, an inflatable cross member was provided for lateral support. Inflated, the MK-IV measured 9' 2" x 5'. A typical finding in sea trials conducted in 1943 was the inability to lay out comfortably, where seats proved more of a hindrance. It was also reported in survival accounts that seats were useless and often cutout. By eliminating the seats, and the taper of the bladder towards the stern, more room became available on the floor of the raft, providing survivors more comfort. In continuation, the raft floor was consistently moist or retained standing water leading to immersion wounds and deteriorating physical health. To help with this issue, a bailing sponge was adopted to keep the floor dry. To further improve comfort and allow occupants to lay out flat, a hammock bed that secured to the bladder of the raft was integrated into the design, as illustrated below right. (National Archives)*

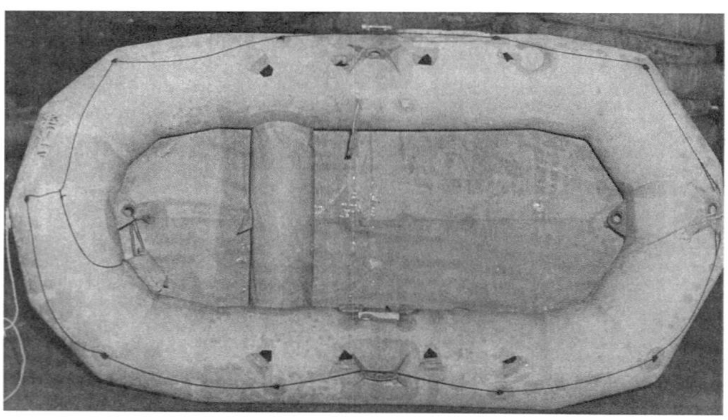

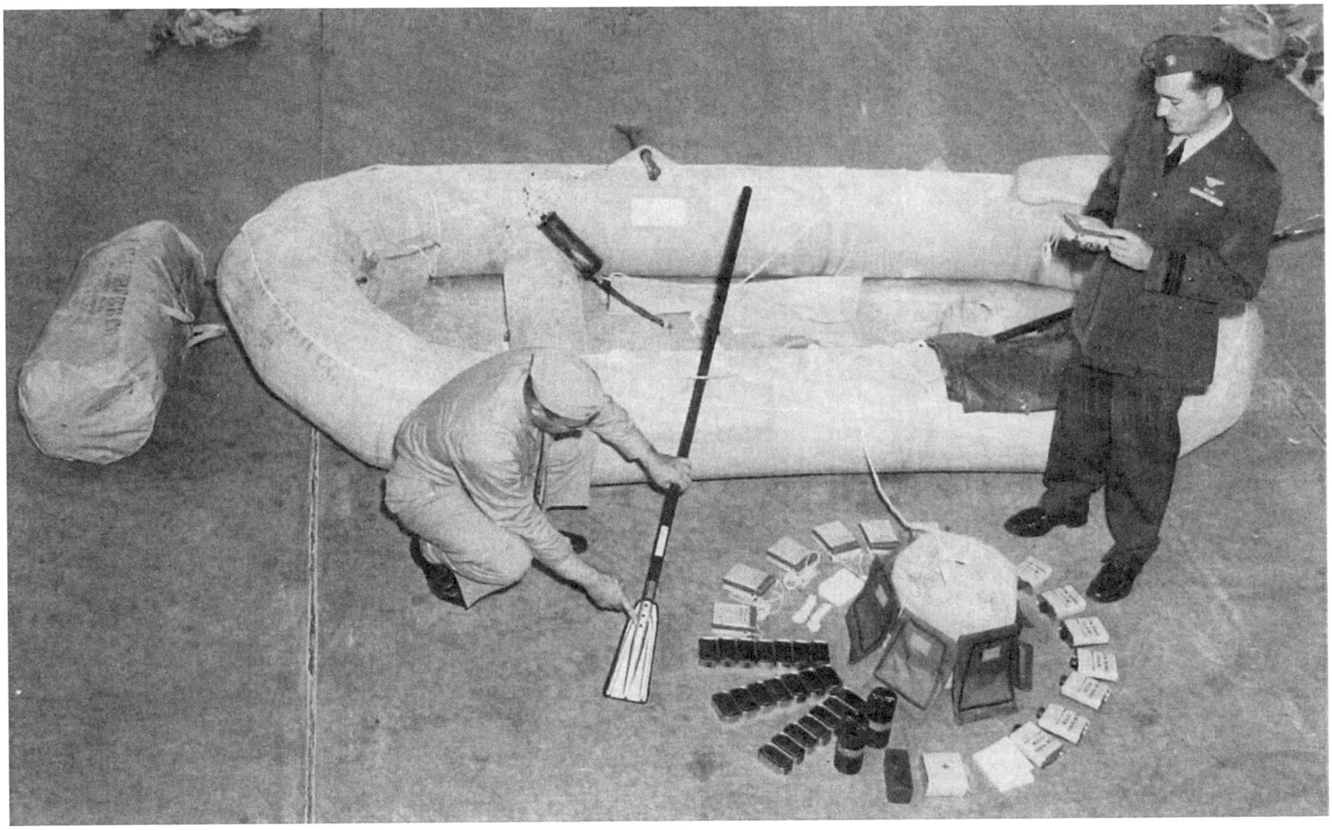

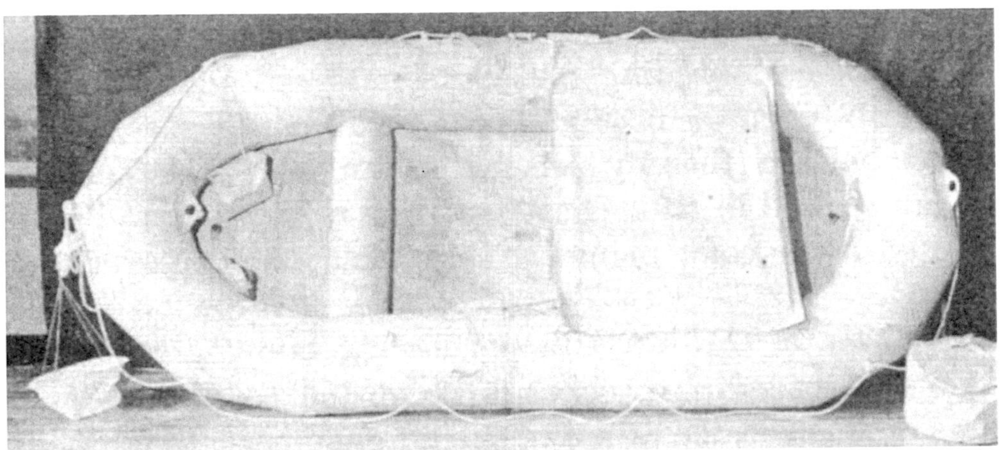

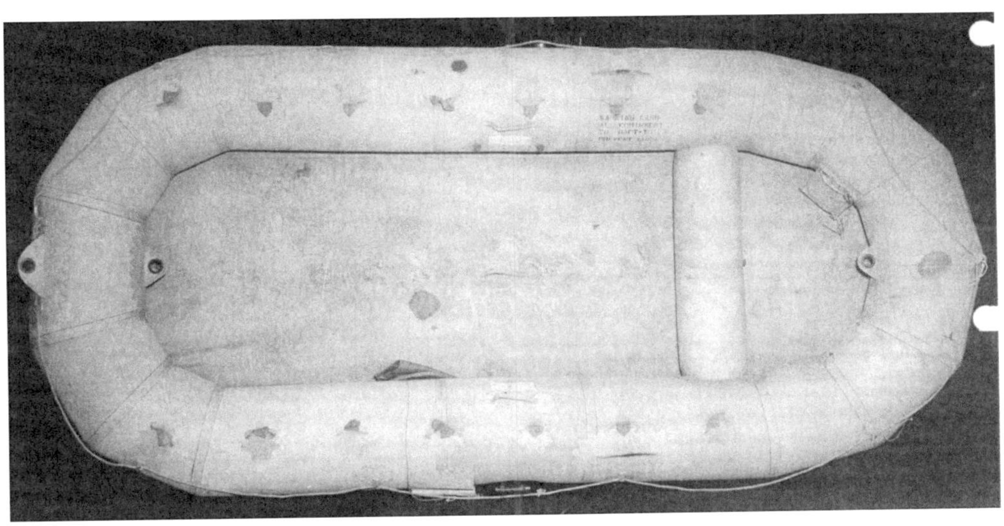

*(Above) MK-VII M-3R life raft which measures 12' x 5' 8". The volume of provisions in the emergency equipment container can be seen. The individual kneeling is pointing to the Scotchlite on the oar paddle M-162D. (National Archives)*

*(Right) Two hammocks are provided to accommodate the volume of potential occupants. (National Archives)*

*(Below right) With contents removed to show the space for laying out flat. (National Archives)*

*(Left) Components for the MK-VII M-3R life raft upon standardization. M-3Q rafts under procurement were to be updated with these components. Later packaging, as per amendment, would replace the AN-K-2 with the fishing kit M-627. In September 1944, Technical Order 119-44 was distributed detailing pertinent data relative to multi-person life rafts, superseding Technical Note 6-43. In this order it was officially stated that the Type-D was the only type of life raft to be provided for use in naval aircraft. It also updated the list of emergency equipment that was to be provided in all M-3 life rafts. When available, the fishing kit M-627 was to replace the AN-K-2. Match safe compass AN-C-101 was to be replaced by the Magnetic Card Type pocket compass. Tablet rations M-539B were to replace the M-539. 3 sea markers M-566 were to replace the can of sea marker M-528/AN-S-10. M-565 was exchanged for the M-565A. Additions also included one drinking water cup, one waterproof flashlight and replacement of the smoke grenades with the MK-I MOD-0 hand held smoke signal (4 per grenade). (National Archives)*

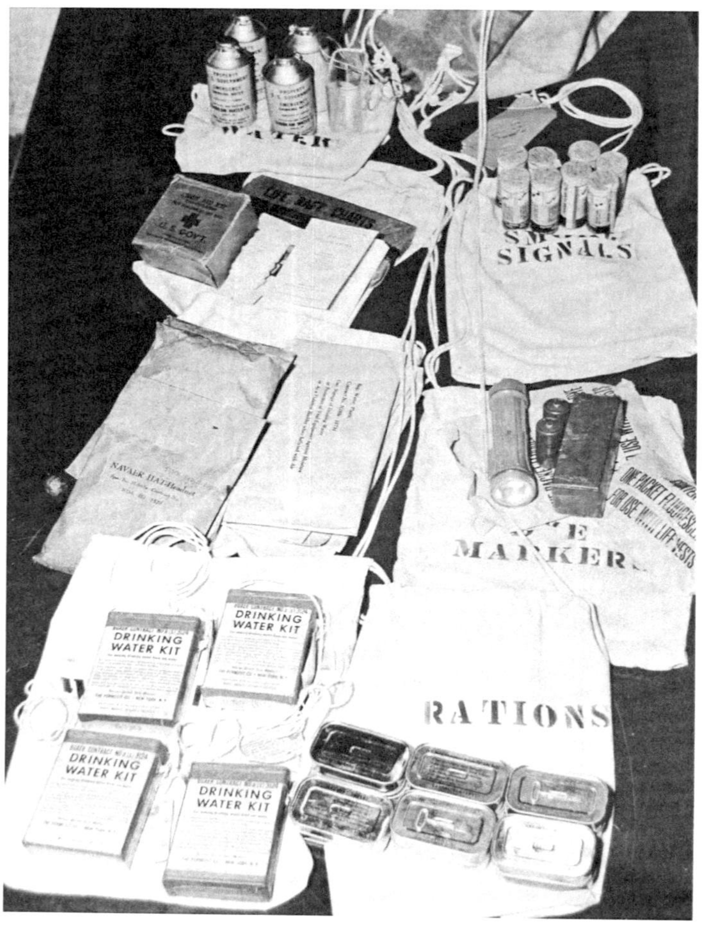


*(Left) Updates included the drinking water cup, hand held smoke signals, waterproof flashlight, M-565A head-net and the new revised S2-1061 first aid kit. The bags shown under these components were part of a unit level modification to separate the accessories in the equipment container. (National Archives)*

*(Below) Amendment-S, 10 May 1945, was the last revision before VJ-day. The differences in this revision were minimal from M-3R and included all updated items as per Technical Order 119-44. Notable additions included a mast clamp, as pictured, making the erection of the sailing assembly easier. The supply and oar pockets were to be screen printed with their corresponding components. M-3S was still active as of September 1947.*

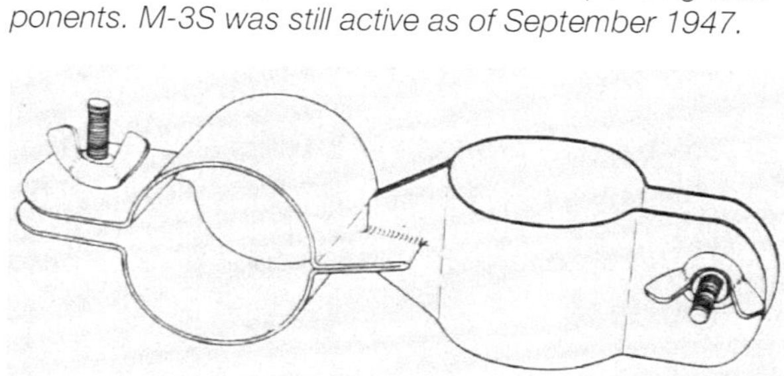

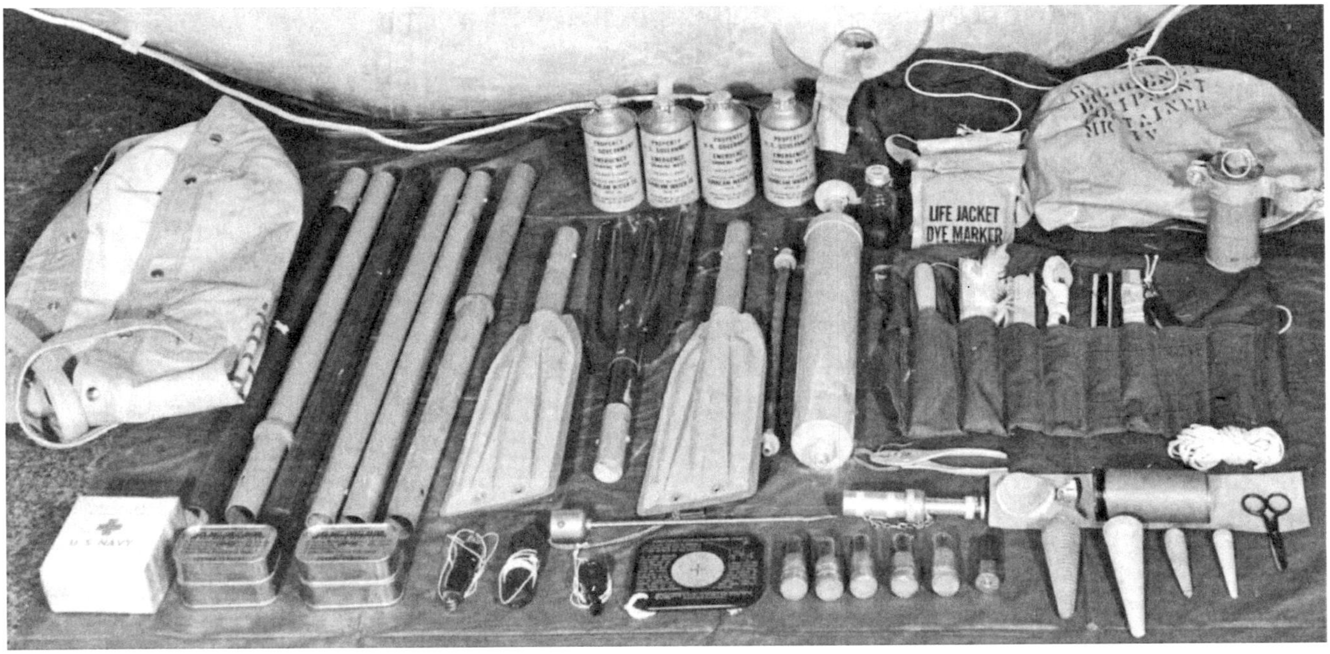


(Above) Deliveries of the M-3R were in limited quantities before the war came to a close, seeing little if any use. The components as detailed in Technical Order 119-44 would integrate into M-3Q or previous models during hostilities. The equipment displayed with this MK-IV M-3Q illustrates such a progression, c.-1945, utilizing some of the older components such as the M-197 hand pump, smoke grenades and round water cans, but principally compliant with Technical Order 119-44. (NNAM)

(Below) Equipment layout demonstrating components for M-3 life rafts at NAS Norfolk c.1945. (NNAM)

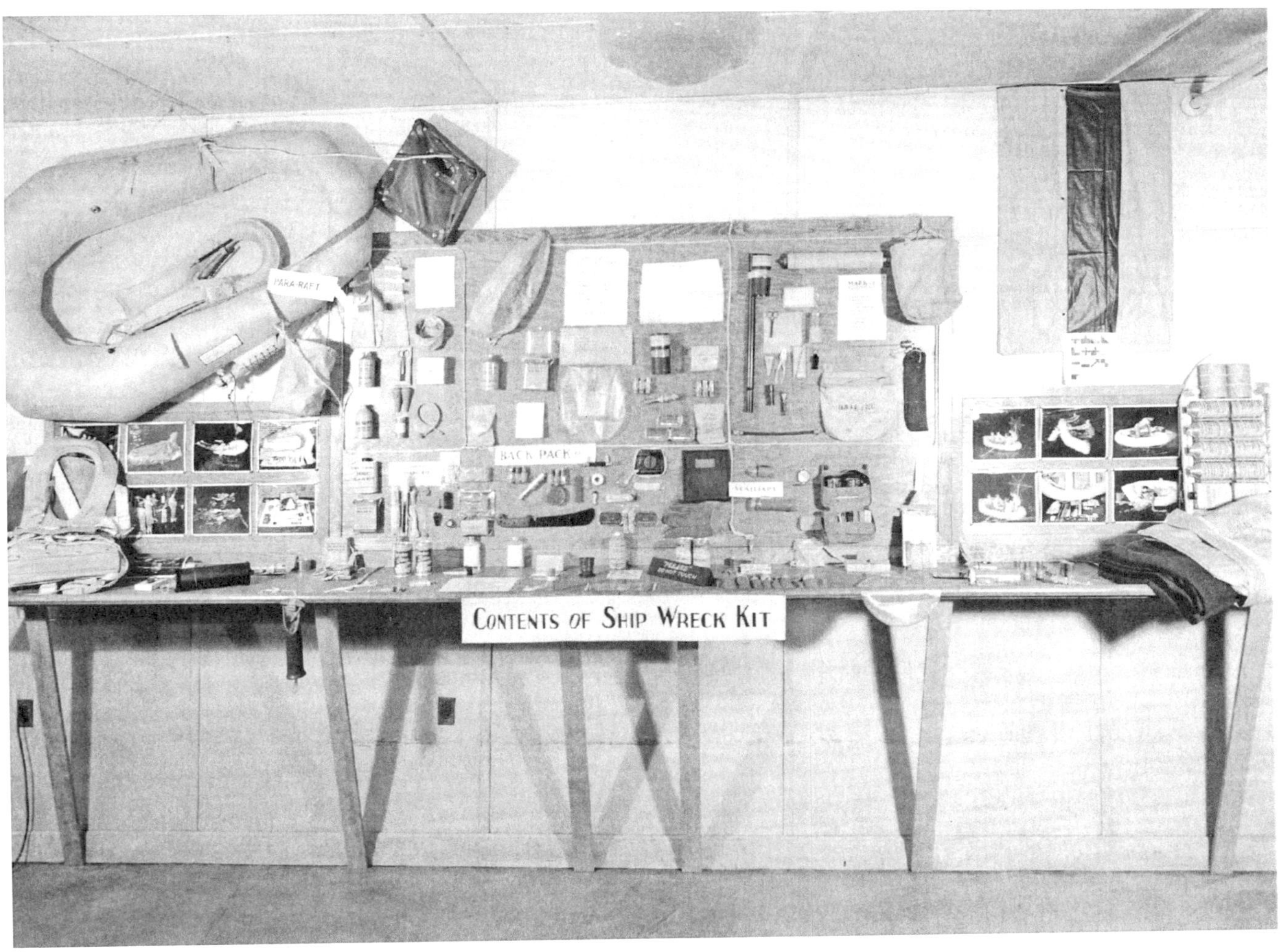

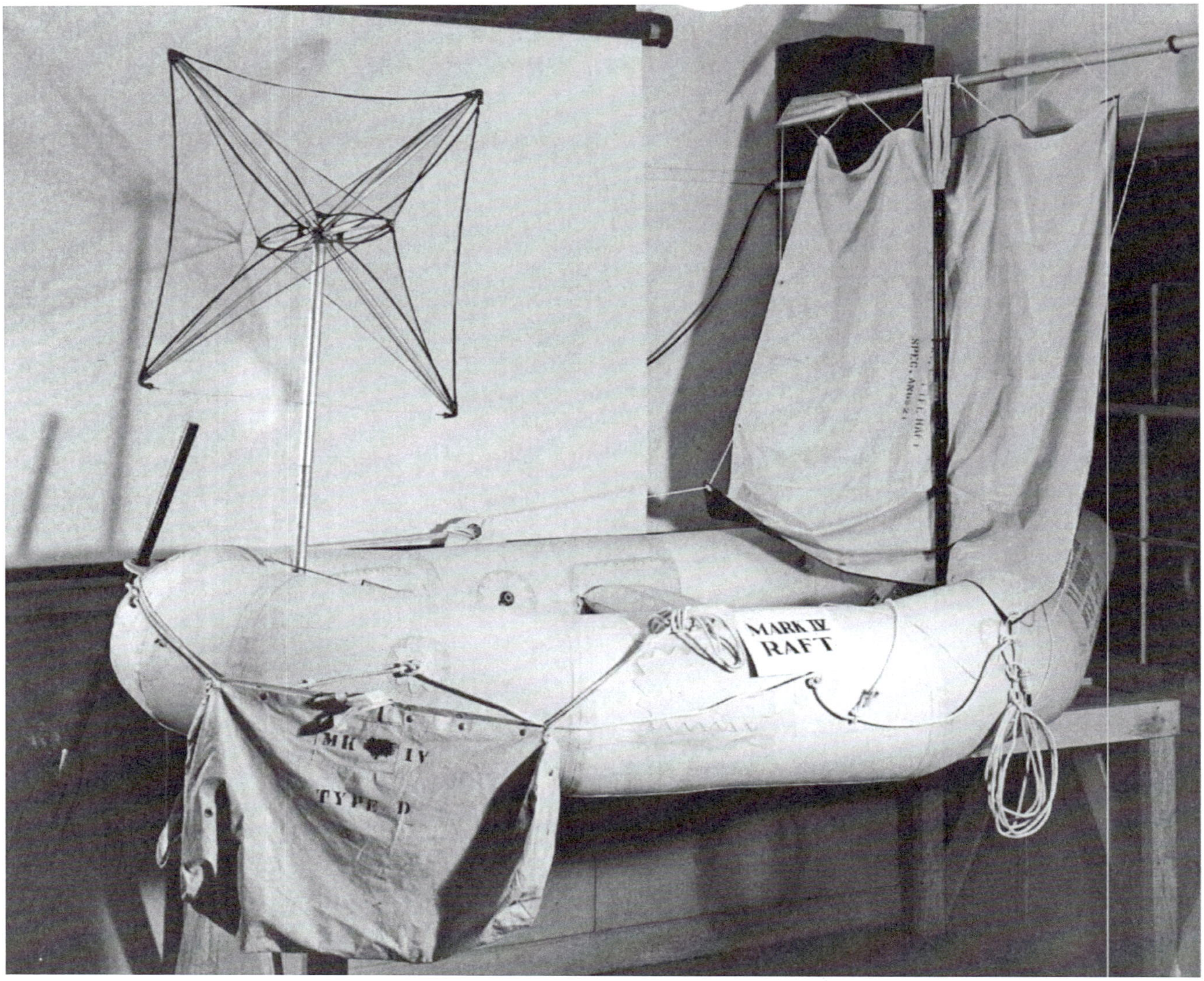


(Above) M-3Q life raft with erected sail and radar reflector. Rescuers could locate the raft via radar and spot survivors with the assistance of the sail.

(Below) Two types of support sockets for the sail mast or radar reflector pole – both types of socket were universal for either item. The MX-138/A did not come with a telescoping pole, two sections of oars were utilized which slipped into a socket on the bottom of the reflector.

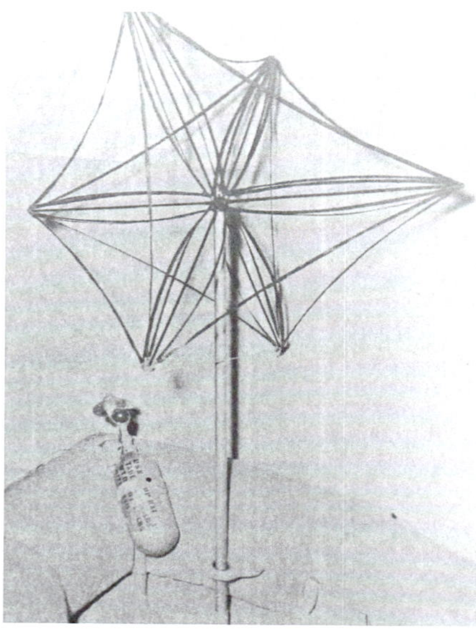

*(Below) This chart comprises of BuAer Technical Notes relative to operation, inspection and maintenance of multi-person life rafts.*

| | | |
|---|---|---|
| 39-38 | Life Rafts & Floatation Bags-Patching Of | 25 October 1938 |
| 11-36 | Life Rafts | 9 April 1936 |
| 3-40 | Life Rafts | 15 January 1940 |
| 54-42 | Standard, Droppable & Automatic Life Rafts-Description, Insp. & Maint. of | June 1942 |
| 6-43 | Standard, Droppable & Automatic Life Rafts-Description, Insp. & Maint. of | 15 February 1943 |
| 77-43 | Type-D Life Raft Inflation Gear | 13 November 1943 |
| 15-44 | Installation of Righting Lines on MK-II, MK-IV, & MK-VII Life Rafts | 12 February 1944 |
| 69-44 | Sealing of the Life Raft Emergency Equipment Containers | 1 August 1944 |
| 104-44 | Life Rafts-NAVAER M-3R -Folding & Packing into Carrying Case-Procedure for | 21 December 1944 |

*(Below) This chart comprises of BuAer Technical Orders relative to operation, inspection and maintenance of multi-person life rafts.*

| | | |
|---|---|---|
| 19-42 | Life Rafts– Type-S & Type-D –Modification of | 7 March 1942 |
| 126-43 | Pneumatic Life Raft Oars-Reflectorizing of | 10 December 1943 |
| 76-44 | Oars, Plastic or Plywood– Removal of from Pneumatic Life Rafts | 27 May 1944 |
| 119-44 | Multi-Place Life Rafts & Life Raft Equipment– Description, Insp. & Maint. Of | 2 September 1944 |
| 132-44 | Life Raft Corner Reflectors– Installation of | 13 November 1944 |
| 16-45 | Life Rafts-NavAer Specification M-3Q and Prior– Revision of | 31 January 1945 |
| 62-45 | Multi-Place Rafts-Inflation & Deflation Hand Pumps | 9 July 1945 |
| 84-45 | Multi-Place Life Rafts– Repair & Survey of | 14 September 145 |

In Amendment-N, as detailed in Supplement-3, dated 10 April 1941, an Emergency Equipment Container was added. It was felt that emergency provisions should be provided for survivors. This was in part influenced by war activities in Europe and the prospect of being engaged in the conflict across the world in vast uninhabited regions. Signaling devices were implemented, including the whistle and signaling mirror, and further steps were taken to include an aluminum sea marker and two H.C. M-8 smoke grenades. To utilize the grenades, a special clamp was devised that would attach to an oar handle, and was stored in the supply pocket. To assist in basic needs that may be required by a marine survivor, a first aid kit and emergency rations were added. In June of the same year, under Amendment-O, drinking water was added, as detailed under NavAer Specification M-504. A gradual integration of items continued through the year.

The BuAer spent a great deal of effort supplying emergency provisions for aircrew in the wartime environment. Initial steps were taken to equip the aircraft with emergency supplies as part of standard equipment, often stowed in specific locations. Emphasis then shifted to equip the individual with emergency back pad kits and parachute life rafts. In some respects, there was a redundancy to assure the safety of individuals.

There was a period when ditching or bailing out was a very precarious proposition. In the early stages of the war, aircrew typically only had a life preserver at their disposal if they bailed out, and this remained the case until early 1943, when the pararaft became available in quantity. Air-sea rescue indoctrination was in its infancy, with very little dedication to bringing aid. Ditching at least presented the possibility of retaining the emergency equipment aboard, but this was dependent on the physical condition of the individuals and the integrity of the aircraft. Survivors were often disorientated or possibly injured as a result of the violent crash, and may only have had minutes to escape from a fast sinking plane. It had been consistently reported that the emergency provisions were left behind, even when the life raft was retained. These scenarios influenced the procurement of the Type-D life raft, enabling carrier based scouting and patrol planes to air drop provisions to survivors. With operations being conducted far out at sea, carrier task forces were heavily relied upon for rescue work. It became necessary to develop airborne emergency equipment to expedite rescue operations from carrier based aircraft. Operational experience had shown that downed airmen were, at times, near enemy waters or in regions where rescues by surface vessels and amphibious aircraft could not be carried out safely. Other factors included waters that were too rough to land seaplanes, or places too distant to be reached by surface craft.

In May 1943, the British made the first successful operational deployment of an airborne lifeboat off the Dutch coast, followed by another deployment in July for the rescue of 6 Wellington bomber crew members off the French coast. Their airborne lifeboat was developed by an internationally known British boatman, Uffa Fox, a well-known author and authority on small boats, having crossed the Atlantic in 60', 40' and 36' boats. He presented the idea in 1941, and working in coordination with the British Deputy Director of Air Sea Rescue and Minister of Aircraft Production, they designed a lifeboat that could be dropped by parachute.

The operational model was designated Airborne Lifeboat Mark-I, consisting of double skin mahogany, 23' long and weighing 1,600 pounds, with the capacity to carry seven men. The boat was provided with two gasoline engines and had the potential to travel 100 miles under power. The boat could also be operated using the sails which were provided, along with supplies of gear stowed aboard for the crew. The first of these lifeboats was air dropped by a Lockheed Hudson, with the updated model (designated Mark-II, having a capacity for 10 men and 30' long) being designed for stowage under the Vickers Warwick rescue aircraft. Influenced by its success, both the BuAer and USAAF entered into designing similar concepts.

To meet requirements for the carrier fleet, the BuAer first turned to the development of a small nonrigid boat that could be carried by a carrier based airplane. Evaluation of aircraft in their arsenal showed the TBF/TBM Avenger to be the most practical for deploying an air droppable assembly. By July 1944, a suitable model had been developed, designated as the AR-10 Rescue Assembly and designed to be dropped from the bomb bay of an Avenger. 500 complete assemblies were ordered and projected to be available in September 1944, available from the ASO under stock No. R83-R-350800, and provided at a ratio of ten per carrier.

*(Below left) The complete AR-10 assembly was provided in five containers, which included a ten man boat, two shipwreck kits, one outboard motor and a fuel container with eight gallons of fuel weighing 700 pounds. The gear was dropped in a train, tethered together by 35 yards of buoyant rope. The five containers are shown in the bomb bay of an Avenger aircraft. (National Archives)*

*(Below right) The assembly was air dropped approximately 100' upwind of survivors, and when it reached the surface it extended for a distance of 280'. When a survivor reached the assembly at any point he could pull the rest of the gear to him. (National Archives)*

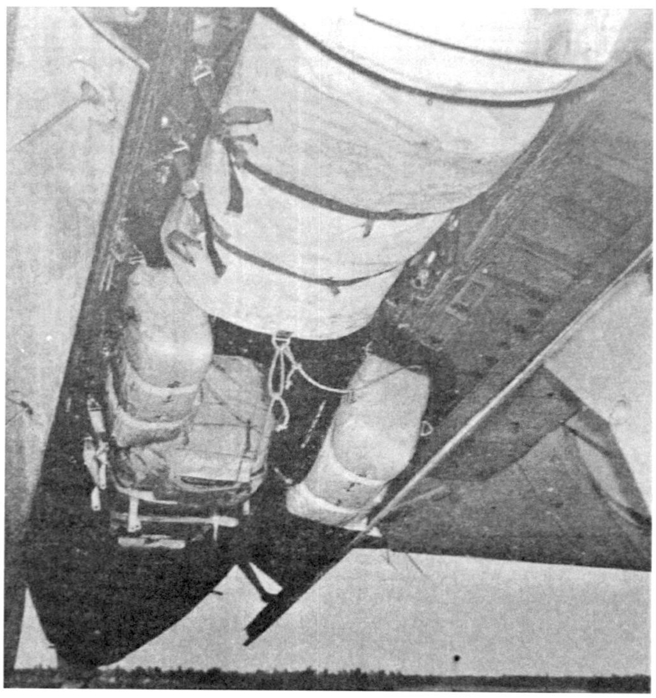

*(Above and below) Once survivors assembled, the boat was inflated by pulling a release cable. Once fully inflated it was boarded at the stern were a ladder was provided. Fully-inflated the boat measured 15' x 8' and comprised two vertical bulkheads, an inflatable floor and 2 cross tube seats with a top rail or splash tube. A 9.7 H.P. outboard motor was provided, as with a sailing rig. Aside from the two shipwreck kits, the raft contained other provisions such as oars, hand pump, hammock beds, repair material and paulins for sun shade. Attachment, description, operation and maintenance were detailed in Technical Note 99-44 of 29 November 1944 and Technical Note 107-44 of 28 December 1944. (National Archives)*

In September 1944, the BuAer field tested other airborne assemblies to accommodate specific needs that did not require the use of the AR-10. The same concept was applied, but orientated towards a design that utilized standard equipment in order to make them immediately available. These new airborne assemblies were designated as the AR-1, AR-2, AR-4 and AR-7. The standard M-3 life raft was to be used in conjunction with the shipwreck kit, eliminating the emergency equipment container.

As detailed in Technical Note 103-44, dated 20 December 1944, for description and fabrication in the field, the AR-2 consisted of one MK-II raft with two shipwreck kits. The AR-4 consisted of one MK-IV raft with two shipwreck kits, and the AR-7 consisted of the MK-VII raft with four shipwreck kits.

*(Below) As part of the procedure for dropping the assembly, a smoke signal would be deployed to determine wind and current direction, and a sea marker attached to the assembly for location spotting. A MK-5 drift signal is illustrated. Shipwreck kits M-594 and M-594A required modification for the attachment of a rope. Securing straps were fabricated for securing to the bomb bay. The shipwreck kit M-594B was factory converted for inclusion with the AR-10. (National Archives)*

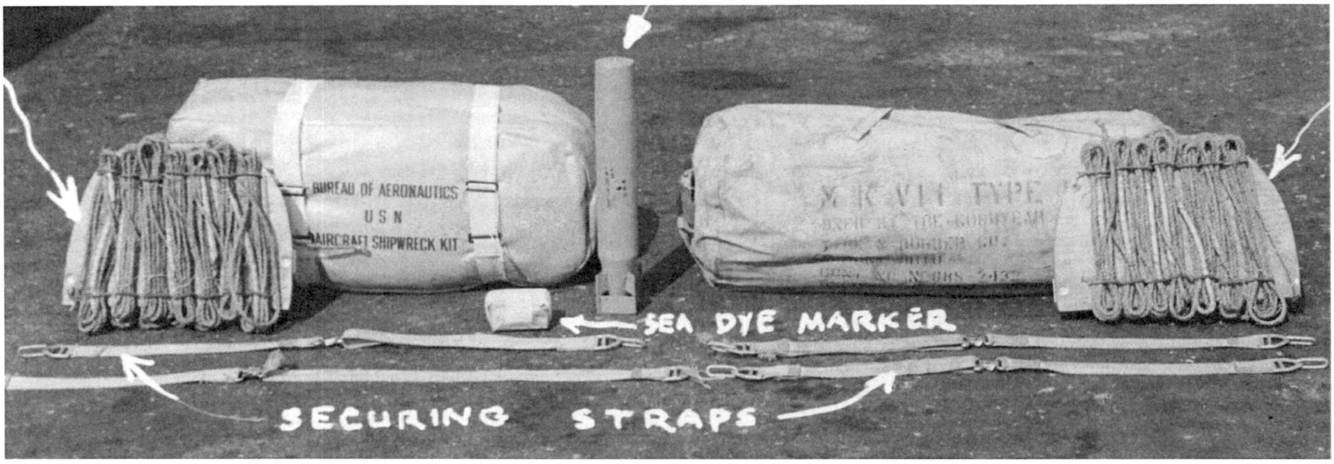

*(Above and left) An AR-4 assembly being demonstrated at NAS Lakehurst, October 1944. Two shipwreck kits can be seen floating next to the M-3R life raft. The AR-2, 4 and 7 were adaptable for air dropping from most aircraft, but intended for use by carrier types, principally the Avenger. (National Archives)*

*In 1944 the BuAer began development on a rigid airborne lifeboat designated as the AR-8. This boat was designed to be carried under the wing of a PBY. The AR-8 offered benefits over the AR-10. Primarily, since it was dropped ready to use, it required no inflation, had better seaworthiness and cruising qualities, and carried provisions for eight men for 12 days. The 18' x 6' boat was built out of five-ply mahogany plywood, and if capsized it could be righted by one man. It was provided with a one cylinder, four cycle engine and a sailing rig capable of traveling, under power, for 250 miles. The AR-8 became operational and available in the summer of 1945. It came complete with survival equipment, including the SCR-578B emergency transmitter. Technical Note 81-45, dated 6 September 1945, was distributed with instructions on dropping from a PBY and Technical Note 76-45 for the description and instructions for rigging to a PBY. (National Archives)*

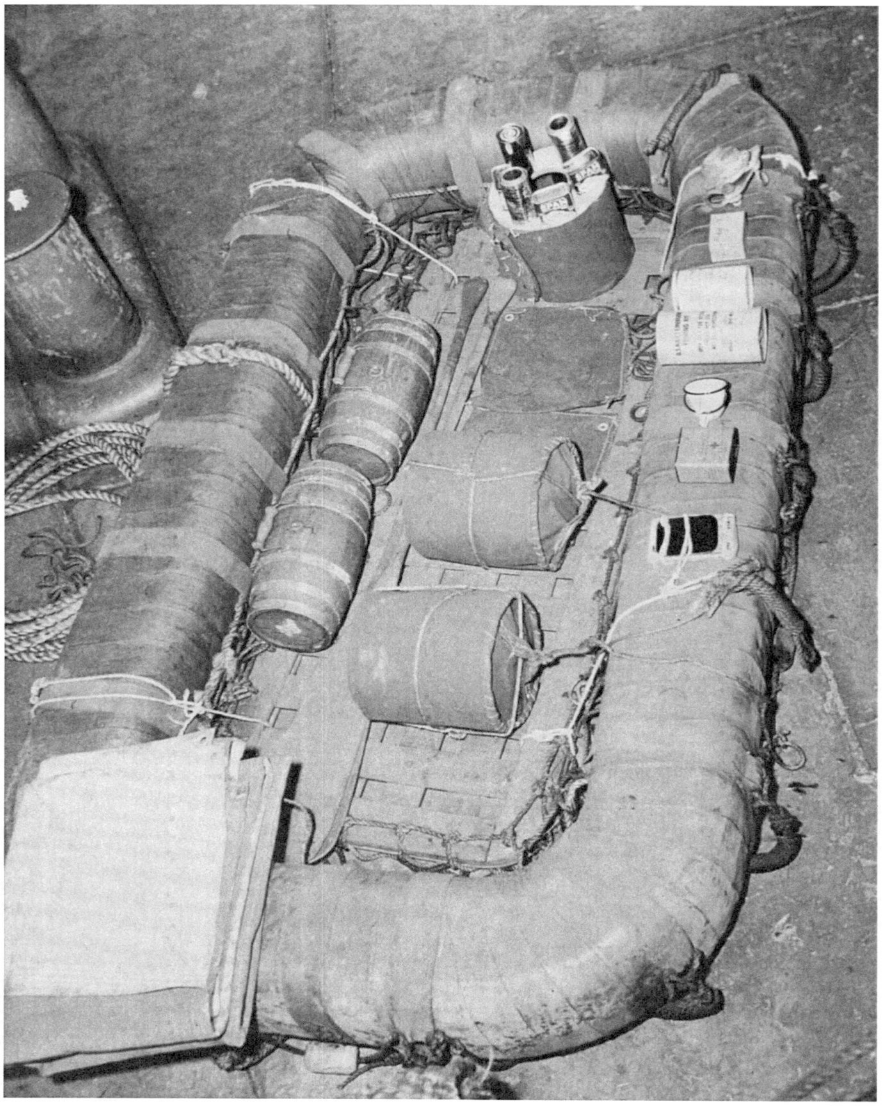

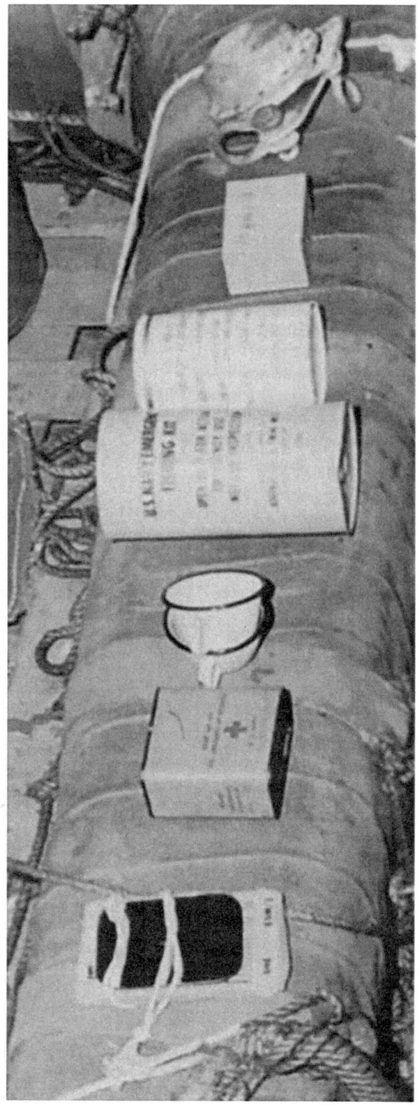

*(Above) For comparison purposes, a US Navy Bureau of Ships life raft, which includes many of the same items, at least as described, as the BuAer type. Secured in the ring are two water breakers and three ration tins. Displayed along the rim of the ring is an M-580 signal mirror, S2-1062 first aid kit, drinking cups, emergency fishing kit 37K1, Abandon Ship Outfit (pyrotechnic kit) and an unidentifiable article. Also included are oars and shade cloth. (National Archives)*

Life rafts for the USAAC were not as a critical as with Naval aviation units as the Air Corps conducted very little flying over water during the 1920s and 1930s. Surprisingly, the need for life rafts was less for aircraft in foreign service than for operations in the continental United States. Flight paths, particularly with larger aircraft, were conducted along both the eastern and western seaboards. From 1925-27, commercial life rafts were procured for experimentation purposes to evaluate a suitable design to equip aircraft when their flight path was over water. Rafts from the New York Rubber Co., Air Cruisers Inc. and Rubbercraft Corp. were field tested but did not meet the rigid requirements established for emergency life-saving devices by the USAAC. They were determined to be too flimsy and not seaworthy. Utilizing these designs as a base, the USAAC Materiel Division drafted specifications for a suitable life raft to meet their needs. What they designed was an 8' x 4'4" oblong raft with two seats, constructed of a rubberized heavy duck fabric with a 12" diameter bladder and rated for the capacity of four men. The flotation tube was separated by vertical bulkheads at the bow and stern, colored what they referred to as chrome yellow with a light blue bottom. This new raft was designated as the Type A-1, Specification No. 94-40068, and adopted as standard, 20 November 1928. By description it closely resembled the raft used by Byrd and NavAer Specification M-3. The USAAC and BuAer were consulting with the same companies, and in turn, as documented, the USAAC did consult the BuAer on design principles.

The Materiel Division felt that the US Navy had more experience in the matter. Like the BuAer M-3, the Type A-1 came equipped with a simple compliment of two oars and a hand pump. Records for the procurement of the Type A-1 indicate that Airships Inc., Air Cruisers Inc., and the Rubbercraft Corp. received contracts. Up to its cancellation in December 1935, 410 Type A-1 life rafts were manufactured, 324 of those by Air Cruisers Inc. Airships Inc. were the first to receive a contract and only produced 21 rafts. The original specification was Airaft, Type A-1, which led to some confusion. It was reported that "Airaft" was a copyright held by the J. Lansing Co. and did not apply to Air Corps equipment or Airships Inc. No rafts were ever procured from Lansing. The title was changed in Amendment-C, 11 July 1931, to Raft, Type A-1.

The Type A-1 life raft was made available at air fields where aircraft may be required to fly over water, supplied as needed. As air operations expanded overseas, first in Hawaii and later in Panama, those commands requested that all pursuit planes, and two-place attack bombers and reconnaissance airplanes,be equipped. The Type A-1 was found too bulky to incorporate, prompting the Materiel Division to develop a raft suitable for use in aircraft with three persons or less. The Materiel Division streamlined the Type A-1 to 6' x 3', assigning the designation Type B-1, detailed under Specification No. 94-40099 and adopted as standard 20 October 1931. Each would be provided with the basic compliment of two sectional oars, a hand pump and one seat with a rated capacity of two men, being principally designed for use in pursuit aircraft.

An improved version of the Type B-1 was adopted as standard 24 October 1933, designated as the Type B-2 under Specification No. 94-40203. One feature from the Type A-1 not carried over to the Type B-1, was the use of bulkheads. These were integrated into the Type B-2, completely revising the bladder. The USAAC did not technically use a person placement designated system but a pound capacity. Type-A was rated for 1,000 pounds and Type-B rated for 400 pounds however there was a suggested person capacity with the Type-A rated for four-five men and the Type-B for one-two men.

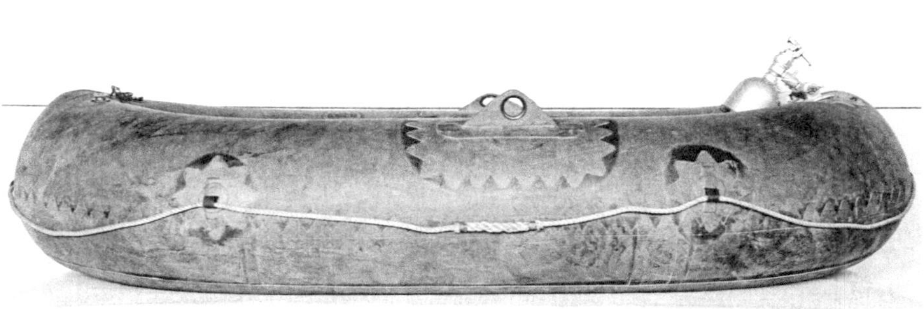

*(Left and below left) Amelia Earhart checks her safety equipment, which happens to be a USAAC Type B-1 life raft, sometime in 1937 before her ill-fated flight around the world with Fred Noonan. To top off the raft she is using the BuAer hand pump M-197. The original detail calls the shape of the boat to be oblong, as with the Type A-1, but the USAAC often utilized the term "canoe shaped" which was more like its descriptive appearance. Though the raft was rated for two men, its primary purpose was for one man, which is noted in the nomenclature stenciled on the bladder.*

*(Right) The Types A-1, B-1 and B-2 were stored in a zippered carrying bag, as shown. To inflate the raft, it would be completely removed and the CO2 released manually by turning a wheel valve.*

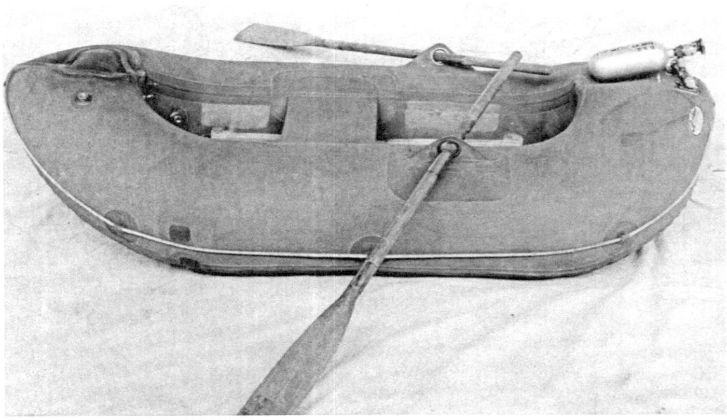

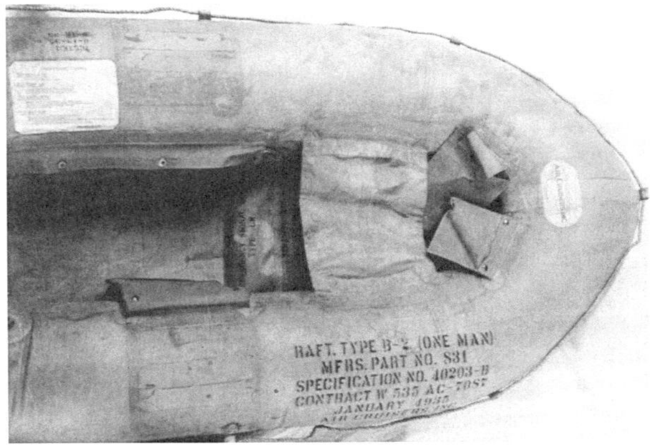


(Above) These two images illustrate the Type B-2 raft. Though intended for one man, it was rated for use by two persons, if necessary, for attack and reconnaissance aircraft. Left is a pre-production prototype illustrating the overall basic constructions features. (National Archives)

(Left) Two volunteers demonstrating the Type B-2 life raft, c.-1934. The sectional spruce or fir oars can be used in the oar-locks or connected together to be used like a canoe or kayak type paddle. (National Archives)

In 1935, new specifications were drafted to address the deficiencies of the Types A-1, B-1 and B-2 life rafts. A preliminary draft was released in November of that year, titled Rafts, Pneumatic Bladder Type, General Specification, under Specification No. 94-40234. Under this specification, three new designations were detailed: Types A-2, B-3 and C-1 life rafts. Type-C was rated for the capacity of 250 pounds. The new bladder type construction consisted of a heavy rubberized duck casing containing a separate latex rubber bladder. It was the initial belief that this new construction would be more durable and provide a longer life span. The Types B-1 and B-2 were found to have stability issues, and were revised again under this specification to increase the size to 78" x 40", increasing its capacity to a 500-pound rating and redesignated as Type B-3 under Specification No. 94-40235. With the added length, an additional seat was included, with one at the stern and one at the bow. With the added designation of Type C-1 for one man, the B-3 would be upgraded to the rating for two men. The Types A-2, B-3 and C-1 were field tested at Langley Field for 6 months in 1936, ending with favorable results. The Type B-3 was adopted as standard, superseding the Types B-1 and B-2, 13 April 1937. As part of the amendment, more accessories were added, comprising two aluminum oars, one hand pump, one repair kit, one distress signal kit and 40' of line for lashing. Three pockets were cemented to the bladder, one for the repair kit, an accessory pocket and a canteen pocket for the stowage of one M-1910 canteen.

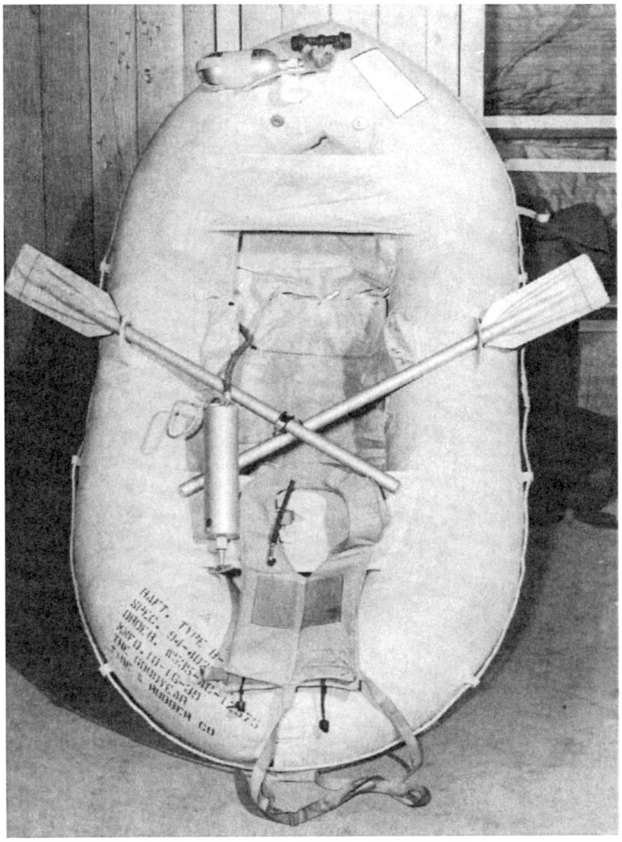

*(Left) Type B-3 life raft being displayed with its standard compliment of accessories and a Type B-3 life preserver. Cemented to the floor of the raft is the distress signal kit Dwg. No. 37G2993, pockets for stowing the two aluminum oars and hand pump.*

*(Below) Type B-3 unpacked and used for demonstration to USAAF personnel operating in Hawaii, April 1942. The oars and pump would be secured to the floor by pockets with strap fasteners. Also illustrated is the 40' of line being utilized for lashing most of the equipment to prevent loss. The detailed pump to be used was dual action and could be used as a bilge pump if needed. Of particular interest is the use of the early Air Corps Emergency Ration. A quantity of 2,420 Type B-3 life rafts were procured for the war effort in 1942, prior to being superseded by the one-man parachute type.*

By the late 1920s, pneumatic bladders were installed on US Navy carrier aircraft, essentially transforming the plane into a raft so that it could remain afloat for an indefinite length of time. The same is true for aircraft of the USAAC in foreign service in Panama and operating in Hawaii.

USAAC pursuit or basic combat aircraft were equipped with flotation bladders, but by the mid 1930s it was announced that, due to increase in weight and constant maintenance and inspection, they would not to be installed on later types. The preferred method of flotation was the integration of watertight compartments in the wings, tail section or fuselage. This method overcame the need for maintenance of the gas inflated bladder types as well as the increase in weight. Bladders were phased out of service after Air Corps Policy No.159 of April 1937, making it mandatory for aircraft to have compartments for life rafts, except pursuit types. Planes in development from that point included the watertight compartments.

*(Above) Pneumatic bladders being tested by US Navy mechanics, March 1934. Bladders would continue to remain in service into WWII, but installation was discontinued on most new aircraft by 1941.*

*(Below, left and right) Inflatable bladders would be manufactured in varying sizes relative to the aircraft they were used with. Some aircraft retained these bladders after 1937, but due to their inherent failures under the increased number of emergencies, they were eventually discontinued for installation by the USAAF in favor of built-in watertight compartments. (National Archives)*

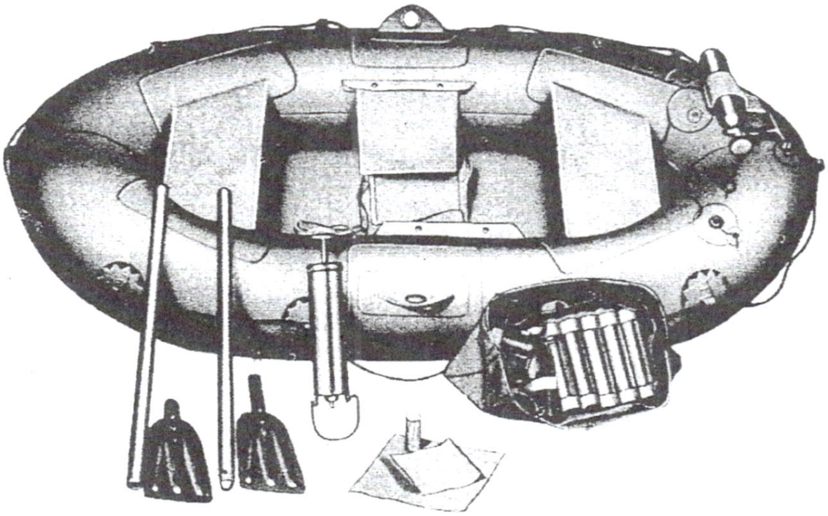

*(Right) Drawing showing the accessories of a Type B-3 and A-2 life raft as per specification from 1936 to 1942. Two 4' 6" aluminum oars are included. The pump was revised in 1938 to include a foot stirrup. The distress signal kit can be seen secured to the floor in the raft center mass and displayed on the outside. The accessory and canteen pockets are above the signal kit. The repair kit consisted of rubberized fabric 6" x 12", 1/2 oz. tube of rubber cement, 6" square piece of sand paper and a 5" pair of cadmium straight nosed pliers.*

*(Left) The last of the B series to be procured was designated as the Type B-4, adopted as standard 25 February 1942, under Specification No. 94-40419. This raft and the Type A-2 would be revised to conserve the use of rubber by adopting a rubberized fabric for the bladders as detailed in NavAer Specification M-3. In the case of the Type B-4 it reduced the use by 5 pounds of rubber per raft. Procurement would be limited, as by the end of the year all rafts under a 1,000-pound capacity would be cancelled. All details from the Type B-3 were carried over, differing only in the bladder manufacture. For wartime conservation, the USAAF went back to the use of wooden oars as seen in this wartime color image. A quantity of 7,795 Type B-4 life rafts were procured in 1942 before being superseded by the AN-R-2 pararaft. (National Archives)*

*(Right) Adopted to supersede the Type A-1 was the Type A-2, on 9 April 1937, under Specification No. 94-40240. This new bladder raft was determined stable for use by five men at 9' x 5'. All the same features and accessories were applied to this assembly from the Type B-3 and General Specification 94-40234.*

The Types A-2 and B-3 were the standard life rafts in service for the USAAF upon entering WWII. As directed in November 1942, it was the responsibility of Air Service Command to supply complete life rafts for all airplanes in which overseas service was anticipated. A specified stock would have been maintained at points of embarkation. Life rafts were no longer shipped to aircraft manufacturers, instead, contractors were directed to ship assemblies to designated Air Service Command Depots. From the issuance of Policy No.159, work had been conducted to incorporate life raft compartments in airplanes which became significant upon entering the war. The guidelines for issuance in 1942, for all intents and purposes, eliminated the Type-B life raft but it was utilized until sufficient quantities of the Types A-2 and A-3 became available and stocks were depleted. They were used to supplement the five-man rafts stored in the airplane. Military Requirement Policy No.33, of 24 November 1942, directed that Heavy and Medium bombardment aircraft were to be supplied a standard compliment of one Type A-2 or A-3 life raft, and that additional rafts were to be provided so that there was sufficient capacity for all crew members. Transport aircraft were to be supplied with sufficient quantities for the total number of crew and passengers. The Type A-2 or A-3 was now the standard rated for five men. This requirement would be amended with the introduction of the Type-E 2,500-pound capacity, or seven-man, raft.

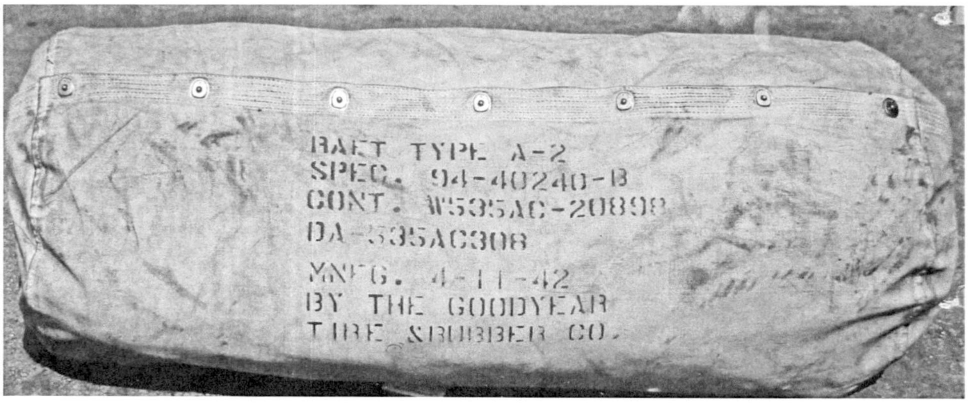


*(Left) The carrying case provided for both the Types A-2 and B-3 was also revised to a cylindrical shape and fitted with lift-the-dot fasteners for ease of removal. (National Archives)*

*(Below) Type A-2 life raft as manufactured under Supplement-B, 4 October 1940. Added was a bailing bucket stowed in the accessory pocket. Construction details were slightly revised, making no real difference in physical appearance. On each side of the bladder, under the accessory and canteen pockets, were slots provided for the removal and repair of the inner latex bladder. The construction of an outer casing and separate bladder proved to be problematic, as air would often get trapped between the two, making complete deflation difficult. It was also found near impossible to utilize the leak plugs. The accessory pocket, canteen pocket and repair pocket can be seen. Also included was the standard compliment of two 4' 6" aluminum oars. 20,000 Type A-2 life rafts were procured in 1942, with production spanning into 1943. (National Archives)*

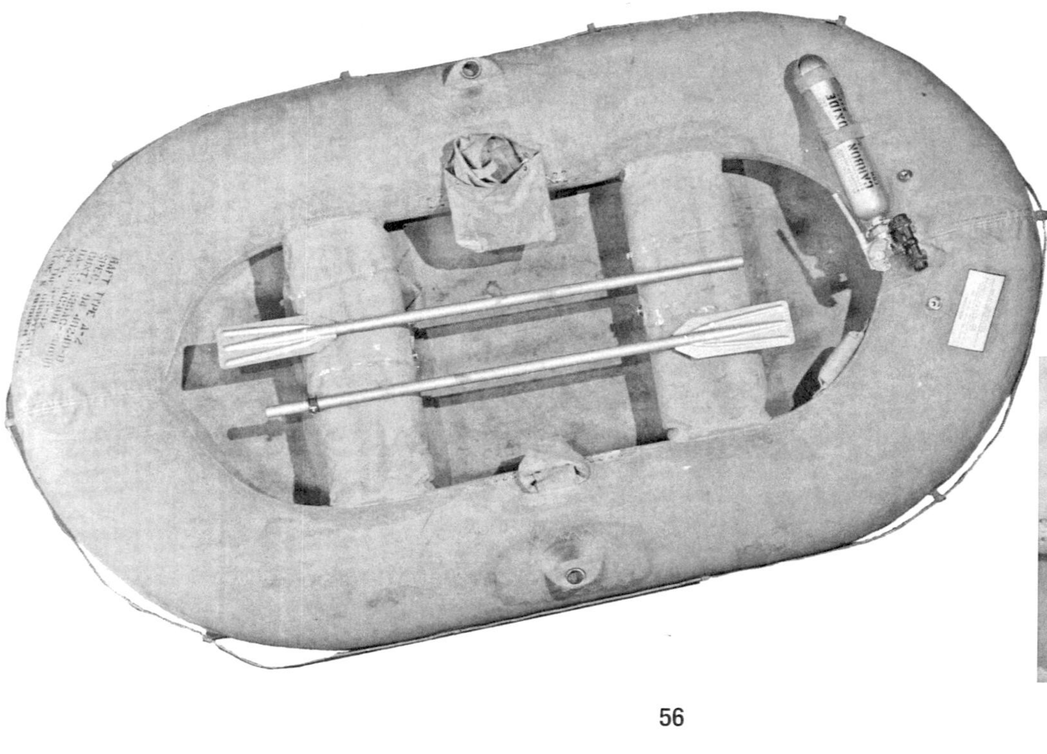

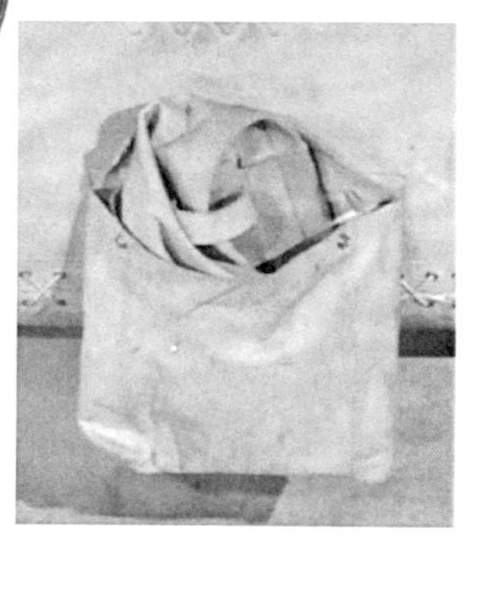

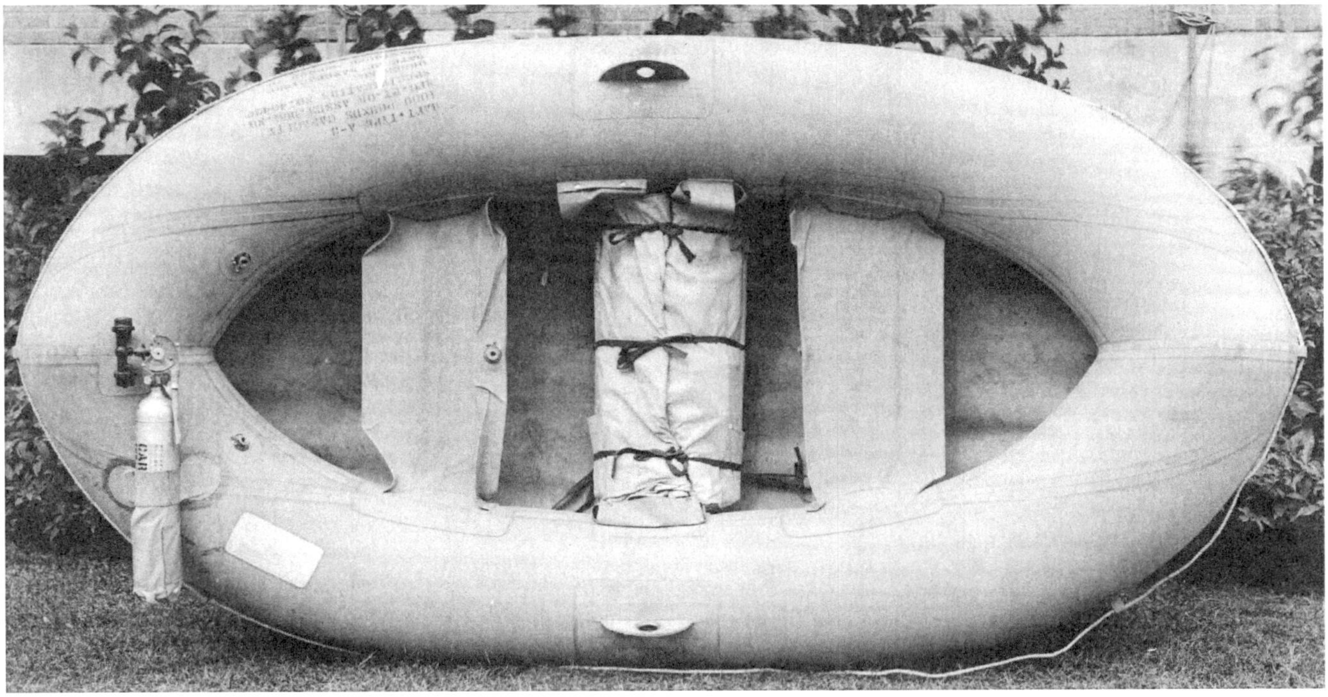

*(Above) General Specification No. 94-40418 covered the details of conversion, transitioning from a bladder to rubberized fabric for the Types B-3 and A-2 life rafts. This revision covered two new designations as Types B-4 and A-3, the Type B-4 as detailed previously. The Type A-3 retained the same specification No. 94-40420, and was adopted as standard, 25 February 1942. The other principle changes were the inclusion of two accessory pockets, one canteen pocket and a central accessory container. By this point, USAAF life raft development had mirrored that of the BuAer, and the USAAF began the integration of accessories such as water, rations, first aid kit and sails, to name a few. All these items were to be stowed in the accessory container secured to the floor of the raft. (National Archives)*

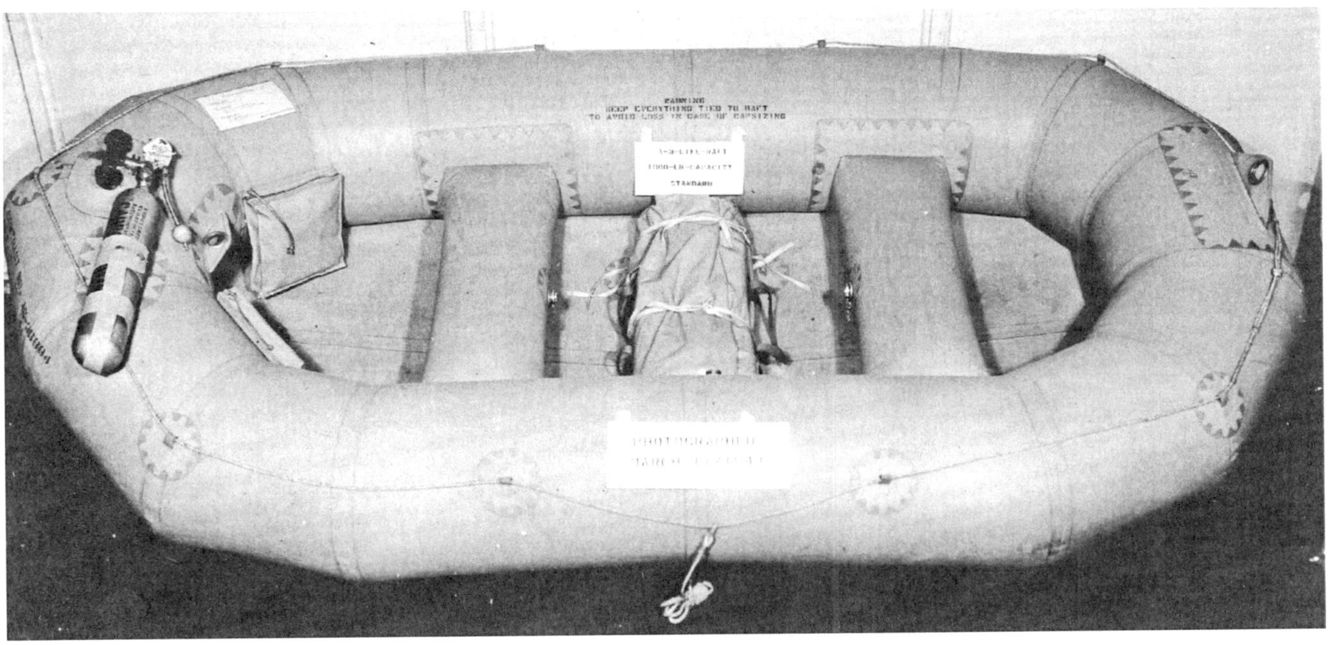

*(Above) The Types A-2 and A-3 were found to be inherently unstable in rough seas, and influenced by service recommendations and BuAer developments, Materiel Command made further refinements to the Type A-3. Effective 5 December 1942, under Amendment-1, the raft was revised to an oval shape, retaining the same overall outside dimensions. The other changes took characteristics of the BuAer M-3Q life raft by adding a mast socket and placing the two accessory pockets at the bow. It also added an oarlock at the stern and removed the locks on the side. Three oars were now part of its compliment, two at 6' and one at 4' in length. The two longer oars would form the sailing mast, with the shorter used for a rudder. This revision also included the use of horizontal bulkheads. Also included were two righting lines attached to the encompassing lifeline along the bladder, in case the raft were to capsize,. (National Archives)*

As mentioned previously, a series of field tests took place in the spring and summer of 1943 in which the USAAF had representatives present, or conducted their own surveys. They reached the same conclusions as to the present status of life rafts. Seats would interfere with sleep, rest and the treatment of the injured. It was also felt more could be done to make life rafts more seaworthy. With suggestions and recommendations by coordinated committees and the 8th AF Service Command, authority for further revisions was authorized by Materiel Command. These revisions were first executed by Technical Maintenance Instruction 04-15-1, on 5 January 1944, to modify Type A-3 rafts in service in the ETO. The new revised Type A-3, redesignated A-3A, would be adopted as standard on 15 July 1944.

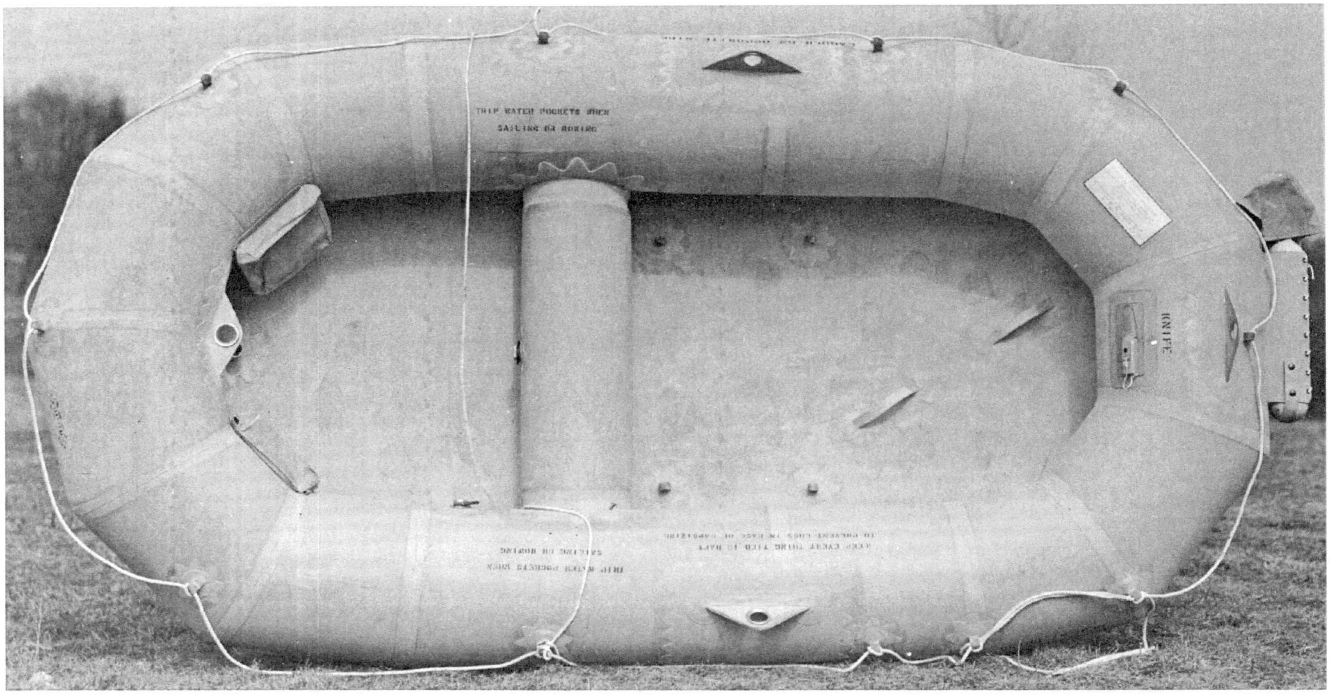


*(Above) The Type A-3A took the characteristics of the BuAer M-3R. It retained the oval shape for stability and eliminated the seats but had a support cross member. To the right of the support member are four tabs for securing the central accessory container, and to the right of those tabs are two boarding handles. A boarding rope ladder was incorporated into the lifeline seen at the lower right. A righting line can be seen spanning from port to starboard near the cross member. Also added as a new standard was the inclusion of the floating life raft knife seen at the stern. These features were detailed to be added in TMI 04-15-1. Production of the Types A-3 and A-3A would reach a total quantity of 150,000. (National Archives)*

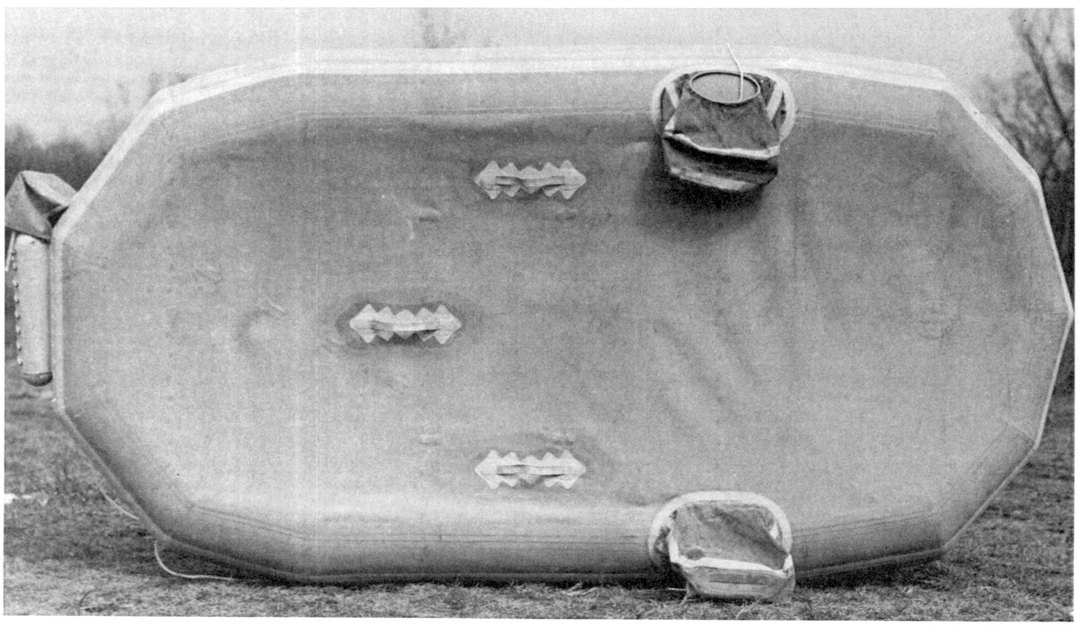

*(Left) Two water pockets were added to help eliminate the risk of capsizing. If capsized, righting handles were additionally fitted to the bottom. The Type A-3A would remain the standard through VJ-Day. (National Archives)*

In 1938 a sample ten man life raft was tested. Fabricated the same way as the Type A-2, but larger, it measured 16' x 6'. Based on a satisfactory evaluation, it was recommended for a service testing to replace the two Type A-2 rafts per large aircraft. This raft was designated as the Type D-1, 2,000-pound capacity, under Specification No. 94-40282. Samples were received at Mitchell and Langley Fields for a six-month long service test beginning in October 1939. In March 1940 the final report was released finding the Type D-1 unsatisfactory. It was considered too large for inflation and launching from airplanes, and not recommended for adoption.

*(Left and below left) The Type D-1 being demonstrated, with ten men aboard, somewhere near Wright Field in 1939, during which time it was accepted for field testing. The raft was tested in bombardment aircraft at Mitchell and Langley fields.*

The wartime environment prompted Materiel Command to revisit the possibilities of a larger capacity raft. From service experience it was found desirable to a provide a seven man raft in heavy bombardment aircraft, so that if one raft was lost, the remaining raft would accommodate the entire crew.

In October 1942, Material Command evaluated several samples including one created by them and another submitted by the commercial airline Trans World Airlines (TWA) which they used as standard on their aircraft. The Type A-3 at this time was being revised to address specific issues, with larger rafts considered more seaworthy and providing more room for comfort. As new features were being integrated into to the Type A-3, specifications were drafted for the fabrication of what would be designated the Type-E, 2,500-pound capacity life raft. The Type-E was to be a larger, more seaworthy, raft for the same number of occupants as the Type A-3 respectively. It would be rated for seven men, more comfortable for six, but could support up to twelve occupants if necessary. The basis of distribution would be aircraft with more than a seven man crew. being principally reserved for heavy bombardment aircraft.

In January 1943, Type-E life rafts became available for field testing, the first of which was designated as the Type E-1. This raft was designed to be an air droppable assembly for use with air sea rescue organizations. A 16' static line was provided which actuated the valve release cord and inflated the raft, allowing the raft to clear the airplane before unfolding and inflating.

The Type E-1 was adopted as standard on 1 July 1943, under Specification No. 94-40490. Only one procurement contract was awarded, for a quantity of 1,000, as it fell from favor with the decision to adopt the BuAer Type-D series for use in air-sea rescue operations.

*(Left) USAAF personnel prepare for field testing the Type E-1 and its complete compliment of accessories on the Potomac River, January 1943. The E-series rafts would contain the same components as the Type A-3 as it was intended to replace it on heavy bombardment aircraft. The theory was that if a ten man crew were able to retain both rafts, five men in each would allow sufficient comfort at sea. In the event only one raft would be retained it could support the entire crew. (National Archives)*

*(Left) Seven airmen field testing the Type E-1 life raft. The Types E-1 and E-2 were the same, differing only in the placement of the CO2 cylinder. On the E-1, the CO2 is on the inside, in the middle of the bladder. (National Archives)*

*(Left) Volunteers utilizing a Type E-1 for anti-exposure suit field testing in Canada, January 1944. The placement of the CO2 cylinder can be seen. (National Archives)*

*(Below) Other images from the same series allow the nomenclature to be seen. (National Archives)*

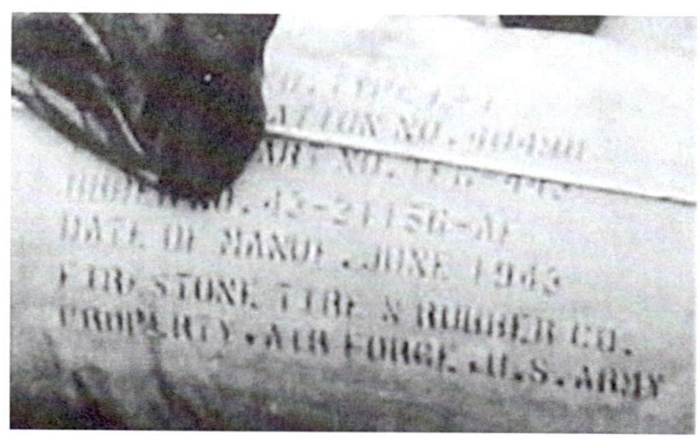

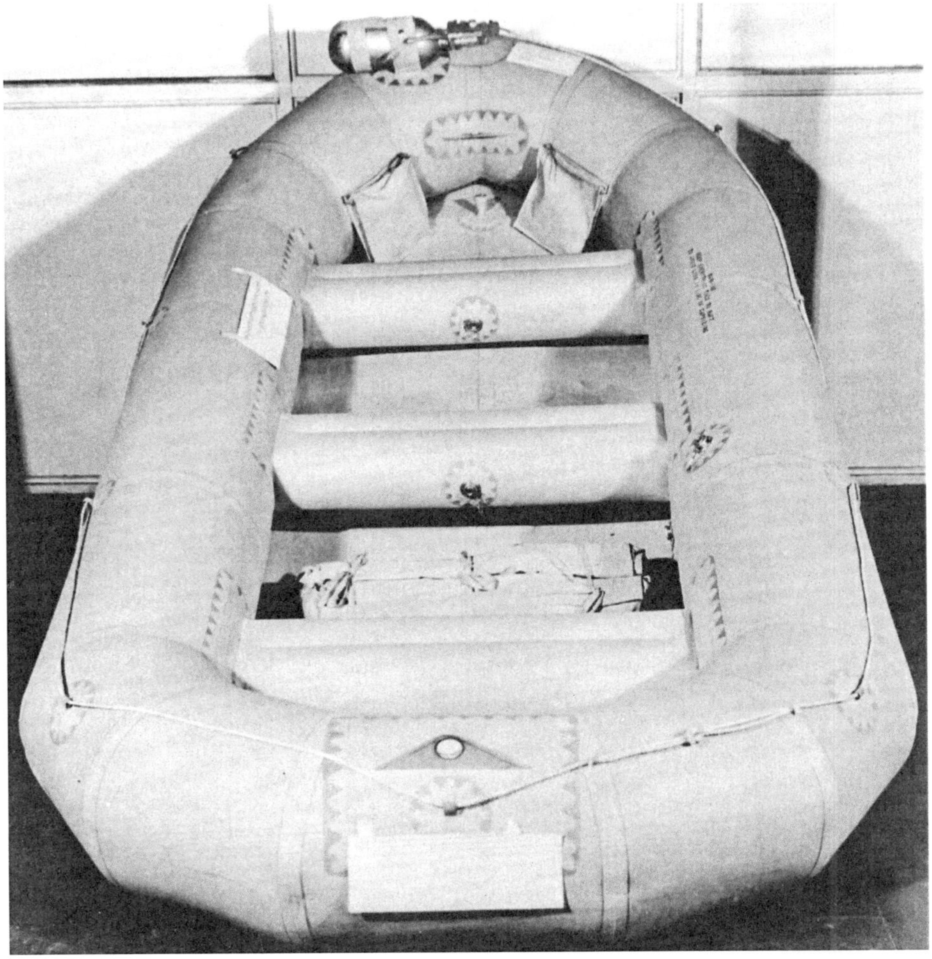

*(Left and below) The Type E-2 was adopted as standard, 8 May 1943, under Specification No. 94-40618. This raft shared much of the same characteristics as the revised Type A-3, but in a larger size of 12' x 5' 8" wide. Three seats were provided and it retained the compliment of only three oars. The Types A-3 and E-2 would share the same number and types of accessories. The Type E-2 would not enter service until the close of 1943 and only in limited quantities. (National Archives)*

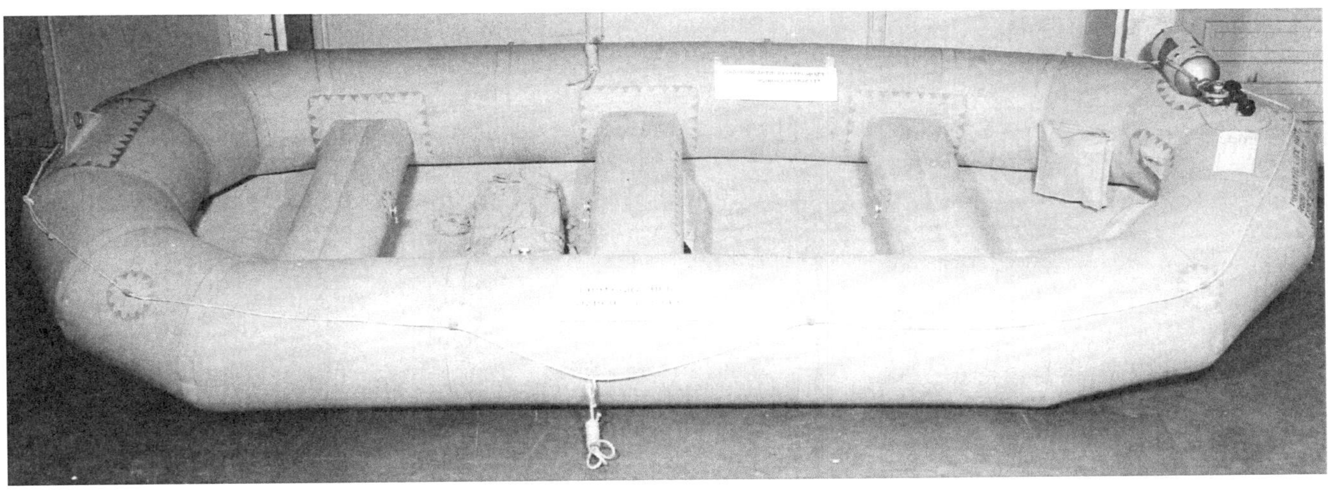

To assure the safety of the crew, several redundant measures were incorporated into aircraft. Eventually the Type E-2 or E-2A would supplant the Type A-3/A-3A for use in very heavy, heavy and medium bombardment airplanes upon availability. With the introduction and availability of the pararaft, multi-person rafts became a secondary means of emergency provision.

By 1945, emergency equipment carried in airplanes was directed as per revised Table of Organization and Equipment for the respective classification: Very Heavy Bombardment were to have a compliment of two E-2/E-2A life rafts stored in the life raft compartment with one A-3/A-3A stored in the aircraft. Each crew member was to be equipped with a pararaft. Heavy Bombardment aircraft were outlined for the same complement, except no Type A-3 was included. Medium Bombardment was to include one E-2/E-2A raft with each crew member equipped with a pararaft. Transport aircraft were directed to simply supply sufficient E-2 or A-3 life rafts to accommodate all passengers.

Such measures encompassed a wide variety of scenarios. If the crew were forced to bail out they would have their pararafts, but in the event of ditching, they would benefit from the deployment of a multitude of life rafts at their disposal.

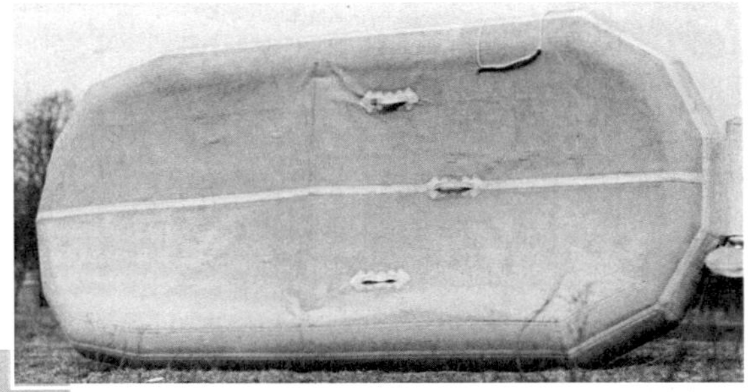

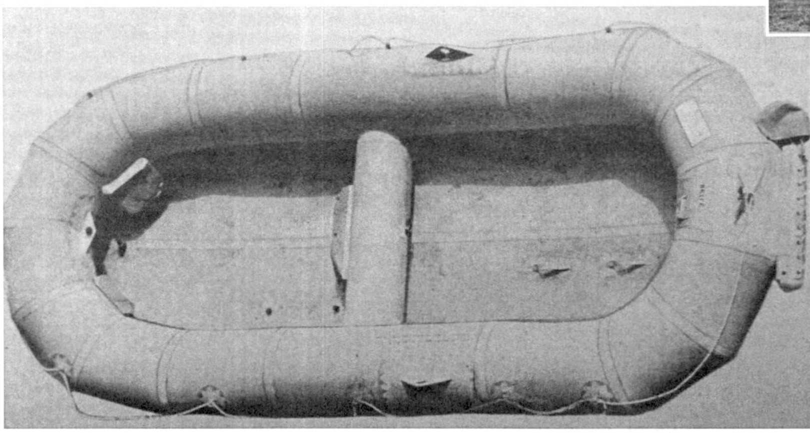

*(Left and above) The Type E-2 was revised with the same new features as detailed for the Type A-3A, adopted as standard 13 July 1944 and designated Type E-2A. The pump was removed from the central accessory container and placed in its own bag attached to the cross member. The two rafts were almost identical in design, differing only in size. Due to the excess width, water ballast pockets for stability were not a necessary application.*

Increasing flights over water had highlighted the importance of life rafts and their accessories. The actual use of life rafts in saving crews of airplanes forced to land on water had made it necessary to add certain items of equipment. As illustrated previously, life rafts were only provided with a basic complement, beginning with oars and a hand pump. With adoption of the Types A-2 and B-3, those components were updated to include a repair kit, lashing line and a distress signal kit. The only other item added before declaration of war was a bailing bucket. These were the basic provisions for aircrew in the wartime environment through 1942. In a period interview with Rickenbacker about his experience, he noted that, combined with their three rubber rafts, they had a small first aid kit, 18 flares with one Very pistol, two hand pumps, two sheath knives, a pair of pliers, a small pocket compass, two bailing buckets, three sets of patching gear, several pencils and a map of the Pacific. These were the articles at their disposal for surviving 21 days at sea.

Materiel Command progressively added items deemed necessary for survival situations. The first official addition of items was detailed in Technical Order 03-55A-1, 6 July 1942. In this order the Types A-2 and A-3 were to include five cans of water AN-W-5, five cans sea marker M-528, five US Army Field Rations D and one Type A-7 floating lamp. Type B-3 and B-4 life rafts were to include three cans water, three cans sea marker, three Field Rations D and one floating lamp. These were to be stowed in the accessory pocket or central accessory container. As per updated Technical Order 04-15-1 (formally 03-55A-1) of September 1942, on the inspection and maintenance of multi-person life rafts, the water was increased to seven cans per A-2/A-3, decreased to three cans of sea marker, and an increase to seven D-rations. The B-3/B-4 was in turn increased in the count of water to five cans, and five D rations, with the sea marker remaining the same quantity. Early deliveries of rafts with rations were reported to contain the Bail Out ration in lieu of the D ration.

*(Left) The display of components illustrated here is for the Type A-2 life raft representative of a January 1943 packing. This could be best described as transitional, as it included items conforming to those detailed in revised specification of December 1942. The fishing kit is an improvised type and includes a TL-122A flashlight. The fishing kit was authorized prior to its actual procurement, hence the type shown here. D-rations were also included, but on a technical standpoint the K-ration was to supersede it. Components also included the paulins, Size-77 and Size-18. (National Archives)*

*(Above) Authorized additions were not always available, leaving contractors and service personnel to improvise. This Type A-3 includes a conglomeration of items that loosely conform to those specified in the December 1942 amendment. Some of the more obvious items include the flat AN-W-5 water cans, round M-528 sea markers, Field Rations D, distress signal kit, aeronautical first aid kit, two-cell flashlight, scout knife, whistle, paulins and pump, among other items. This particular raft was being used for field tests as a sailing rig, having rigid cross members. (National Archives)*

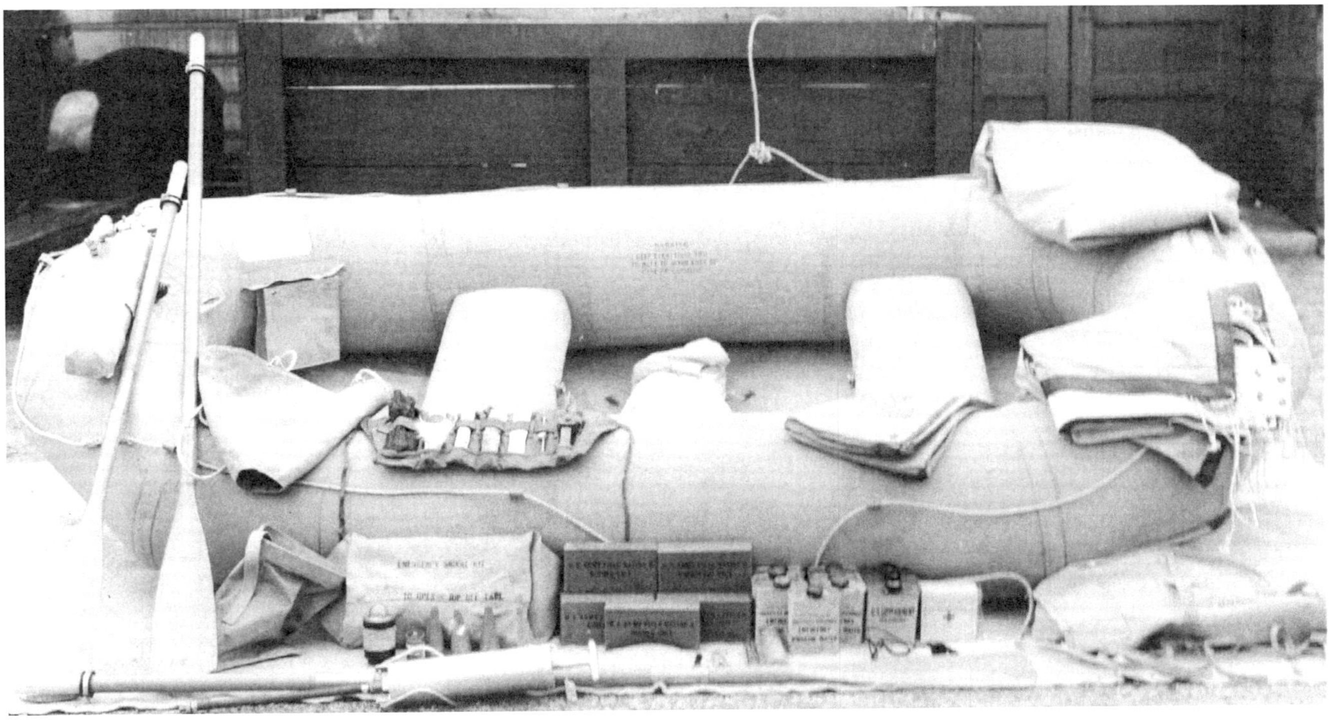

*(Above) A complete assembly of the Type A-3 as per amendment-1 of December 1942. All items pictured conform to those as detailed for the accessories. Of particular note is the use of the five unit US Navy first aid kit for pneumatic life rafts S2-1055. USAAF multi-person life rafts utilized both the aeronautic and US Navy first aid kits. The initial first aid units included would have been the USMD aeronautic type, as the US Navy type was not permitted until 1943. (National Archives)*

*Components list as detailed in the amendment of December 1942 for the Type A-3 life raft. At the time of draft, new items were being prepared for standardization and included. This list would carry over to the Types E-1 and E-2 life rafts. Stock number for the complete E-2 would be 6600-663560.*

| ITEM | STOCK NO. | SPECIFICATION NO. |
|---|---|---|
| Raft, Pneumatic, Type A-3 Complete | 6600-661100 | 94-40420 |
| Oars, Wood** | (2) 6'-4500-618000<br>(1) 4'-4500-618100 | None |
| Sea Anchor | 4500-036450 | None |
| Hand Pump** | 4500-705000 | None |
| Repair Kit* Kit contains;<br>-Patching material 6"x 12"<br>-4" square sand paper<br>-2 oz. tube rubber cement<br>-1-pair 5" adjustable pliers<br>-blunt nose scissors<br>–4 sets wood repair plugs | 4500-526600 | None |
| Bailing Bucket* | 4500-113100 | None |
| 40' Line 75# test* | 7100-26200 | T-T-88 |
| Pyrotechnic Pistol M2*** | Ord. Stk. No. M-2 | |
| Distress Signal M11 (qty. 6)*** | Ord. Stk. No. S005-90-24050 | Ord Dwg. 78-0-10 |
| Water, Emergency Drinking (qty. 7)** | 6700-171480 | AN-W-5 |
| Sea Marker (qty. 3)** | 7300-379000 | M-528/AN-S-10 |
| Rations Type-K (9 pkg. 5-dinner 4-breakfast)** | 8300-669200 | C.Q.D. 28F |
| Flashlight Type A-7** | 7700-518591-5 | 32410 |
| Scout Knife, 3 Blade* | 7900-473900 | None |
| Whistle, Police* | 6700-997970 | None |
| First Aid Kit, Aeronautic** | USMD Stk. 97765 | |
| Fishing Kit Type-5** | 8300-559020 | M-552A/AN-K-2 |
| Paulin, Life Raft Size-77** | 4500-635850 | 26756 |
| Paulin, Sail Size-18** | 4500-635825 | 26756 |

**Stored in accessory pockets*
***Stored in central accessory container*
****Stowed in Emergency Signal Kit container 37G2993; Contractors had been directed to change to the use of the pyrotechnic pistol AN-M8 in October. Complete signal kit stored in central accessory container.*
*Upon availability of the Size-35 Paulin, sail, it would replace the Size-18 for use in updated A-3 rafts with mast socket provisions. Size-18 remained in use with A-2 rafts.*

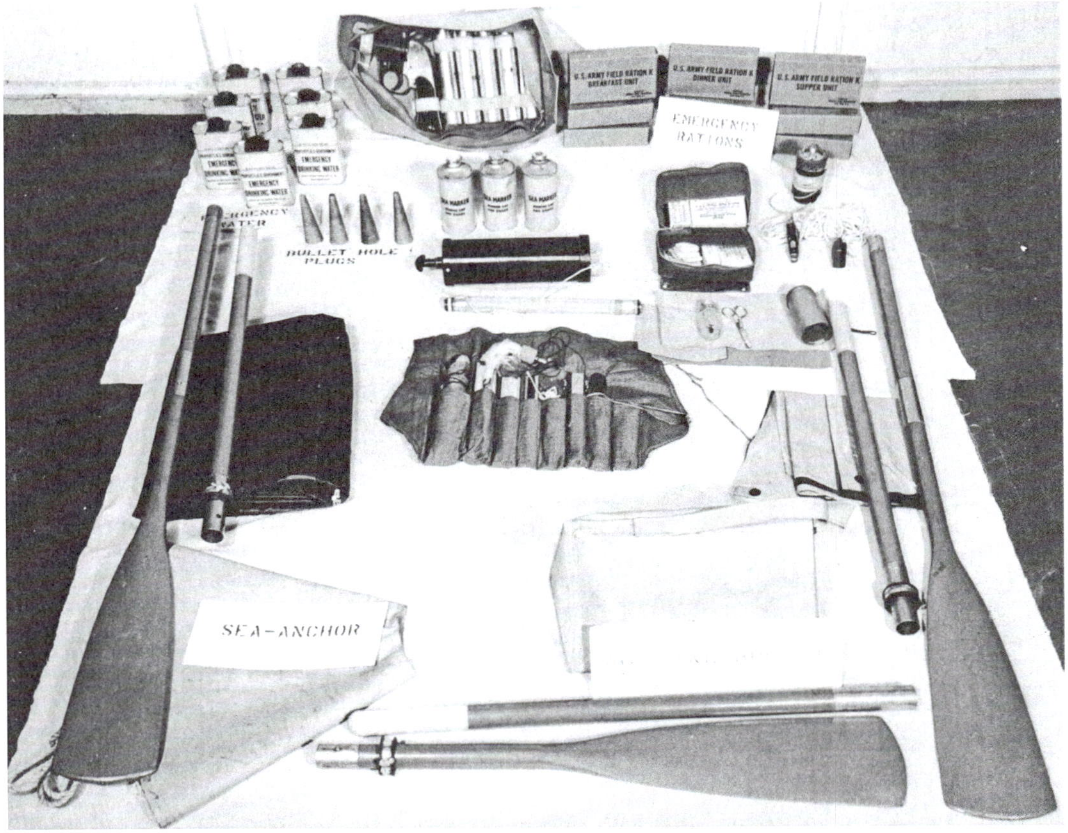


*(Left) Display representing those items contained on the component list. This complement includes the AN-M8 pyrotechnic pistol and round M-528 sea markers. By February 1943 contractors were directed to switch to the use of the BuAer M-550 hand pump which is illustrated here. (National Archives)*

*(Below) Another look at typical packaging for 1943 and 1944. Note the inclusion of a pocket compass next to the first aid kit. (National Archives)*

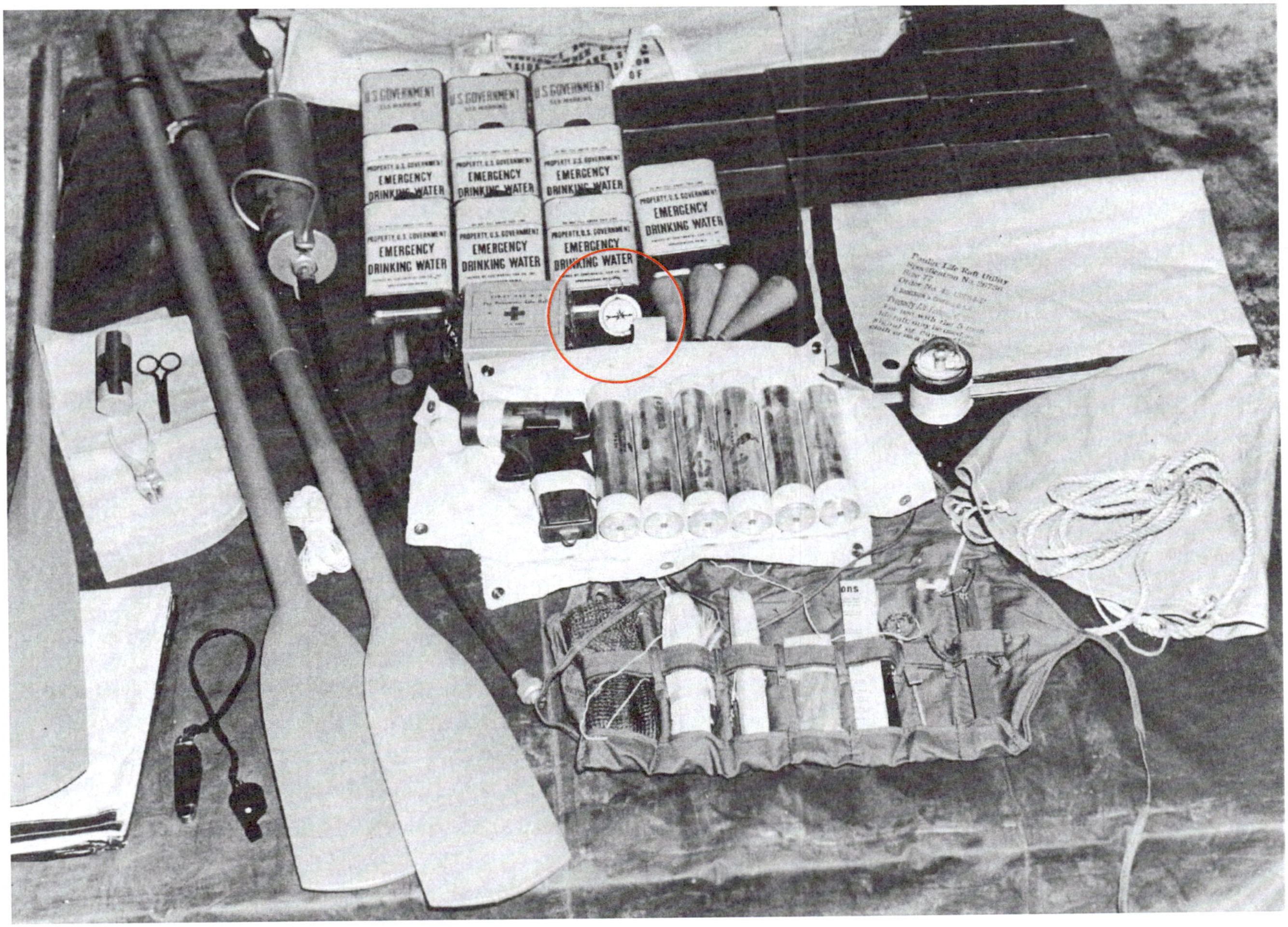

*(Right) Revision of May 1943 would include the following new items for multi-person life rafts as standard.*

| ITEM | STOCK NO. | SPECIFICATION NO. |
|---|---|---|
| Sunburn Ointment (qty. 4 tubes) | 8300-605500 | Mfg. Pt. No. P-40469 |
| Compass, Match Safe | C.F.E. | AN-C-101 |

*(Right) Revision of September 1943 would update life rafts to include the M-9 hand projector, life raft rations Type-A and engineer wrist compass as the standard compliment. The mirror and booklets were new items.*

| ITEM | STOCK NO. | SPECIFICATION NO. |
|---|---|---|
| Pyrotechnic Pistol M9 | Ord. Stk. No. M-9 | Ord. Dwg. 78-6-1 |
| Rations, Life Raft, Type-A (qty. 7) | 8300-680300 | C.Q.D. 164-A |
| Mirror Emergency Signaling | 4500-580650 | 40653 |
| Compass, Liquid Filled, Wrist Type | 6200-061700 | EBP-292-A |
| Religious Booklets (1-Set) | None | None |

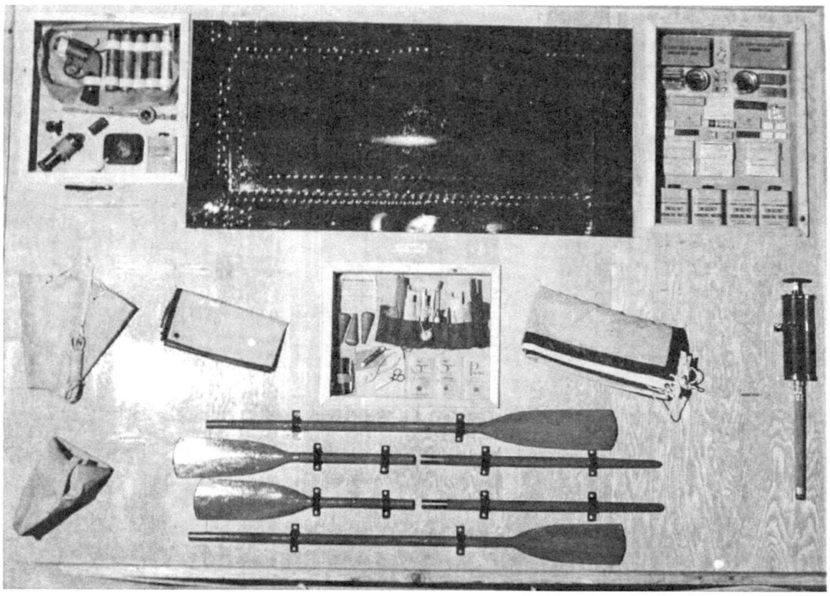

*(Left) Components were more of an acceptation than a rule. This display board includes some of the new items as detailed in the amendment of September, but retains some older items. The signal mirror, religious booklets and wrist compass are the updated items. (National Archives)*

*(Below) This display of components represents the selection of accessories as per the September 1943 amendment, the only item not included is the set of religious booklets. (National Archives)*

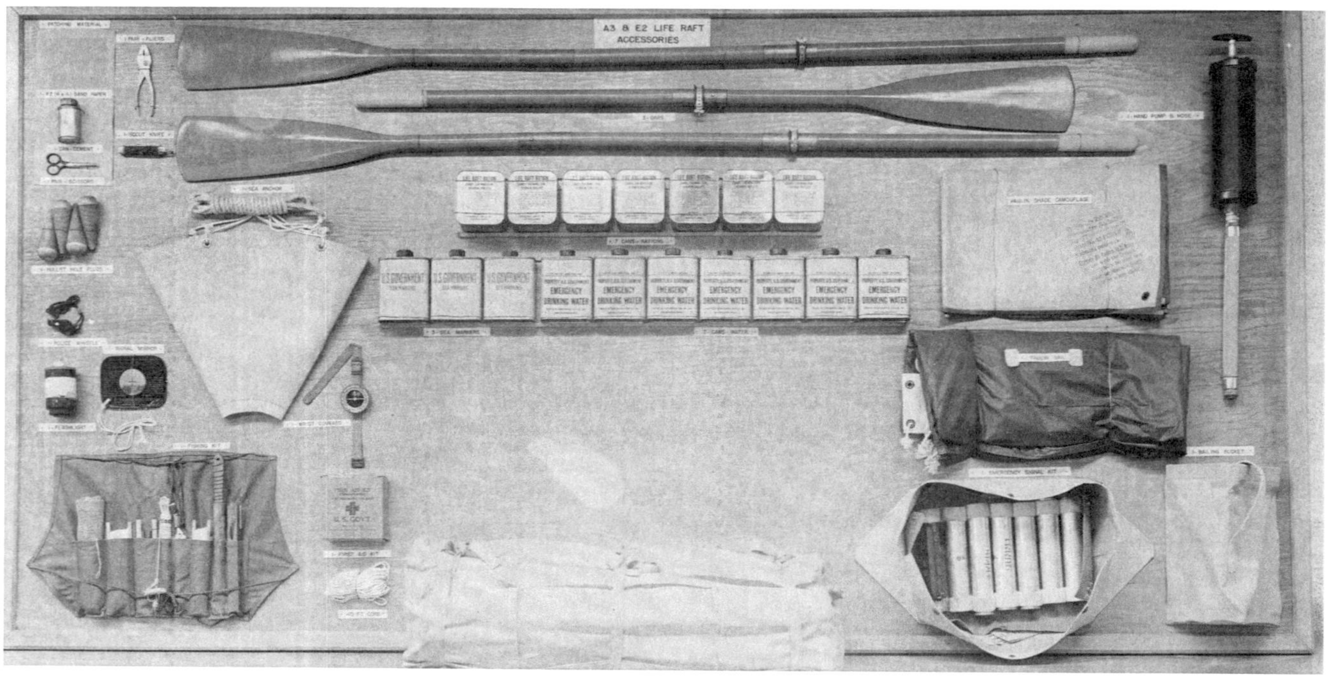

*Updated component list authorized as standard for Types A-3A and E-2A life rafts upon their standardization. As was typical, some items were not immediately available for issue. The stock number for the complete rafts remained the same.*

**Stored in accessory pockets.* ***Stored in central accessory container.*
****Stowed in Emergency Signal Kit container 37G2993.* *****Stored in pump pocket.*

| ITEM | STOCK NO. | SPECIFICATION NO. |
|---|---|---|
| Raft, Pneumatic, Type A-3A Complete | 6600-661100 | 94-40420-A |
| Oars, Aluminum** | (2) 6'-4500-617200<br>(1) 4'-4500-6617250 | Dwg. 44G18937 |
| Hand Pump**** | 4500-705000 | AN-P-48 |
| Repair Kit* Kit contains;<br>-Patching material 6"x 12"<br>-4" square sand paper<br>-2 oz. rubber cement<br>-1-pair 5" adjustable pliers<br>- 1-pair blunt nose scissors | 4500-526600 | None |
| Repair Plugs, Metal (qty. 4) | 4500-683775 | Dwg. 45B18979-1 |
| Bailing Bucket* | 4500-113100 | Dwg. 44D9375 |
| Sponge, Bailing* | 6700-868980 | L-S-626 |
| 40' Line 75# test* | 7100-26200 | T-T-88 |
| Pyrotechnic Pistol M9*** | Ord. Stk. No. M-9 | Ord. Dwg. 78-6-1 |
| Distress Signal M11 (qty. 6)*** | Ord. Stk. No. S005-90-24050 | Ord Dwg. 78-0-10 |
| Water, Emergency Drinking (qty. 2)** | 6700-171480 | AN-W-5B |
| Sea Marker (qty. 3)** | 7300-379000 | AN-S-10 |
| Rations, Life Raft, Type-A (qty. 7)** | 8300-669200 | C.Q.D. 164-A |
| Kit, Water, Desalting JJ-1 (qty. 10)** | 4500-526700 | 40725 |
| Water Stowage Containers (qty. 4)** | 4500-240000 | Mfg. Pt. No. 283-A |
| Flashlight Type A-9** | 7700-518591-65 | 32454 |
| Knife, Floating | None | Dwg. 45B18786 |
| Whistle, Police* | 6700-997970 | None |
| Sunburn Ointment (4-tubes)** | 8300-605500 | Mfg. Pt. No. P-40469 |
| Mirror Emergency Signaling** | 4500-580650 | 40653 |
| Compass, Liquid Filled, Wrist Type** | 6200-061700 | EBP-292-A |
| Navigation Charts (5-oceans)** | None | None |
| Religious Booklets (1-set)** | None | None |
| First Aid Kit, Pneumatic Life Raft** | USMD Stk. 9776900 | JAN-K-87 |
| Fishing Kit ** | 8300-559020 | AN-K-2 |
| Paulin, Life Raft Size-77** | 4500-635850 | 26756 |
| Paulin, Sail Size-35** | 4500-635825 | 26756 |

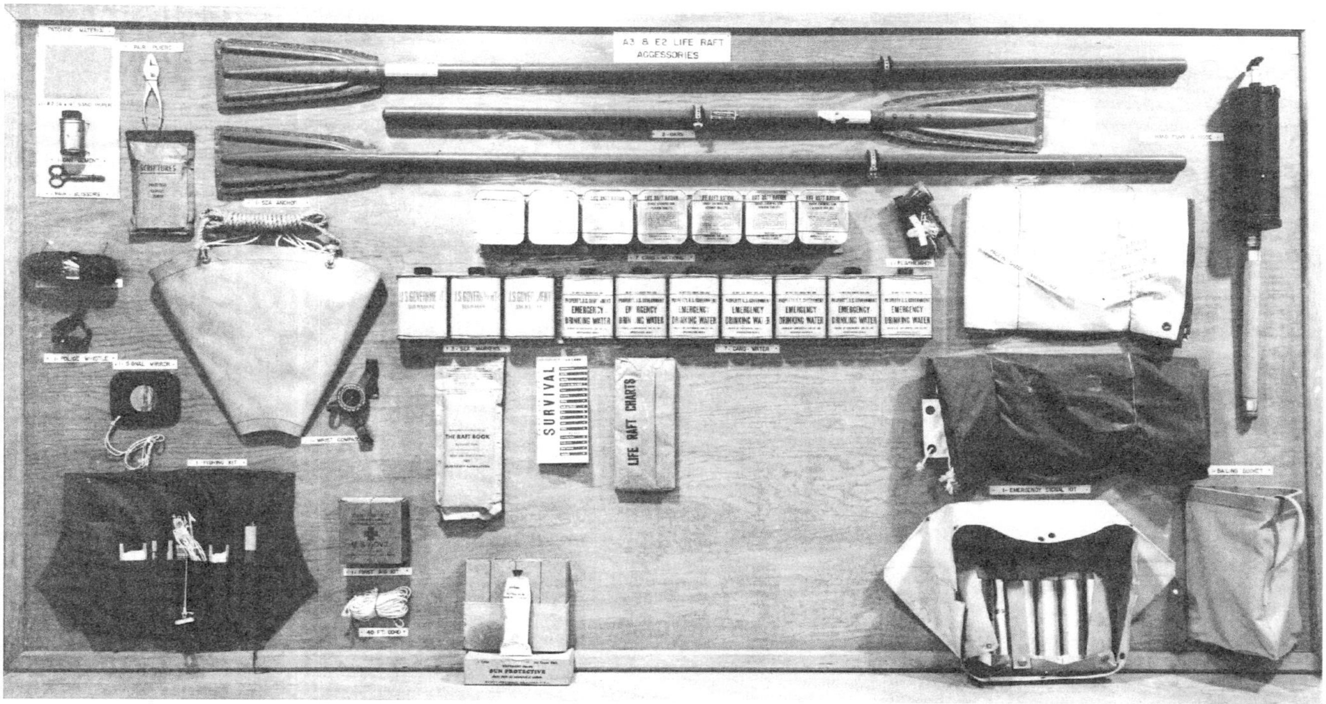


*(Above) As was typical with new items, there was a grace period for when they entered service. This equipment display is representative of late 1944 through 1945 packing, which includes some items as detailed in Amendments-A for the Types A-3 and E-2 life rafts. It includes the new aluminum oars 44G18937, flashlight Type A-9, metal repair plugs and life raft charts. As per directives, the manual 21W was to be included in all emergency equipment and is represented here. Also included is the condensed version of The Raft Book by Harold Gatty. (National Archives)*

*(Right) At this point there was an established list of standard components, but those in service that had undergone inspections loosely followed those guidelines, leaving some diversity in what was included. This display, c.-1945, includes new and older equipment, including the new AN-K-2 fishing kit in metal can, sponge, drinking water kits and the water storage bags. Two items of particular interest are the inclusion of M28 through M42 signal cartridges in the signal kit which would typically include M11s. Also included are two experimental Gallowhur sun stills.*

*(Below) As per updated Technical Order 04-15-1 of February 1945, the MX-138/A radar reflector was to be installed when supply permitted. Solar stills were also directed to be included, changing some quantities. Only one can of drinking water AN-W-5 was to be included, six cans desalting kits JJ-1 and 2 distillation kits LL-1. When available, distillation kits LL-2 were to supersede the LL-1. These would be the last principle changes before VJ-Day. Below is a list of pertinent technical orders pertaining to USAAF multi-person life rafts.*

| | | |
|---|---|---|
| 01-1F-1 | General-Pneumatic Rafts, Types A-1, B-1 & B-2– Use, Maintenance & Inspection. | 30 August 1934<br>14 September 1937 |
| 04-15-1 | Distribution , Inspection, Installation & Maintenance of Multi-Place Life Rafts.<br>(Primary Technical Order utilized. Title would include type designation of rafts and would be adjusted accordingly through its service life)<br>* Originally designated as 03-55A-1 | 6 July 1942*<br>18 September 1942<br>20 December 1942<br>26 May 1943<br>15 August 1944<br>22 February 1945 |
| 04-15-4 | Life Rafts-Use, Maintenance & Installation of Accessories | 21 October 1943 |
| 04-15-6 | Inspection & Modification– Type A-3 | 20 December 1943 |
| 04-15-8 | Inspection of Hand Pump– Types A-2, -3, E-2 Life Rafts | 28 March 1944 |
| 04-15-9 | Replacement of Type A-2 Life Rafts & CO2 Cylinder Assemblies | 25 September 1944 |

*(Above) Type A-3B or E-2B life raft, c.-1950. Even though taken five years later, the image illustrates what a complete multi-person raft would be comprised of as detailed in updated technical orders by VJ-Day. In the immediate post war years, further amendments occurred to the designations of Types A-3B & E-2B and were used throughout the Korean war. Many WWII components can be seen, which by this time would have been superseded by other items. (National Archives)*

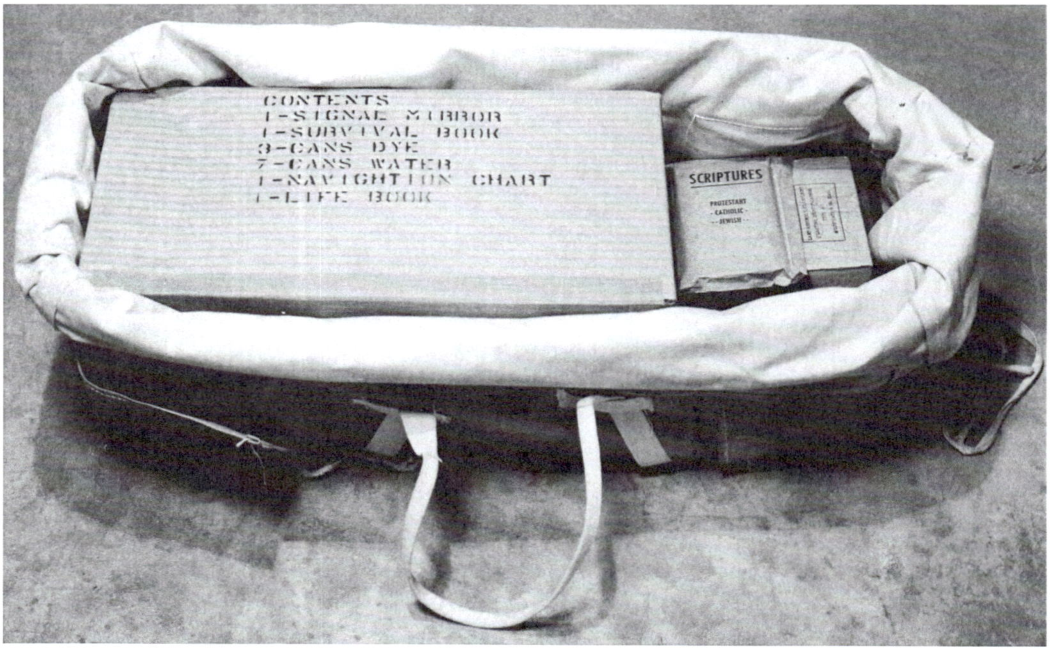


*(Above) As part of the amendment of February 1942, an accessory container was added to the Type A-3 for the bulk of components, to help reduce the possibilities of loss of accessories and to allow for greater quantity. As of the amendment of December 1942, it was referenced as Central Accessory Container Dwg. No. 43G17646, and was available under stock No. 4500-228000. All accessories except those stored in the pockets provided were stowed in this bag.*

*(Left) The two images illustrate the packing inside the container. The oars would be stowed in pockets provided on each side. 9" tie tapes were provided at the four corners for securing to the raft floor. Three sets of 24" tapes were provided, one at each end and one in the middle, for securing the bundle closed.*

*(Right and below) Pursuant to Air Corps Policy No.159, requiring all airplanes except pursuit and training to be equipped with pneumatic life rafts, considerable design changes had to be made to aircraft to incorporate life raft compartments. Heavy and eventually Very Heavy bombers had two compartments, typically along the spine of the fuselage. These two images illustrate the two compartments on a B-17 Flying Fortress. (National Archives)*

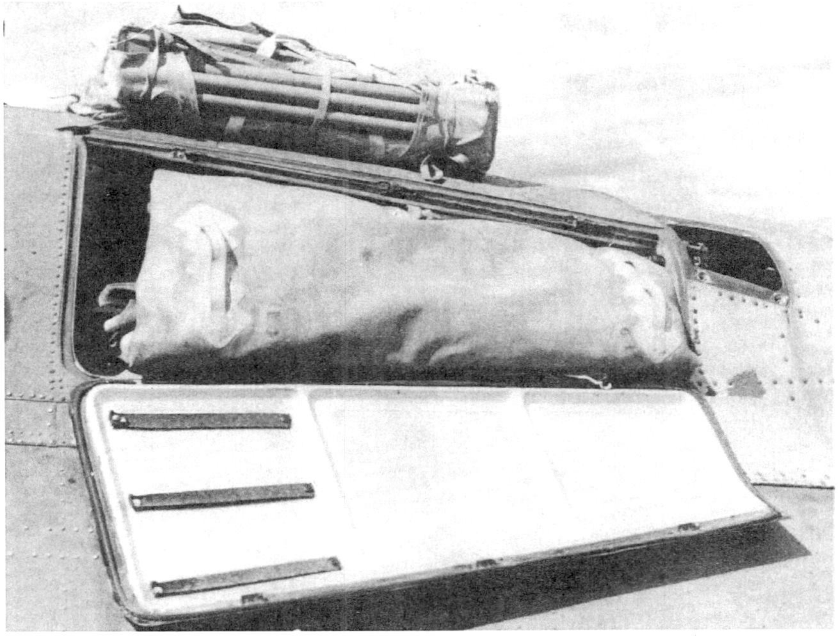

Rafts were stored in their compartments without a case, being folded and rolled for automatic ejection. Automatic ejection systems remained in use with the USAAF after being discontinued for use in BuAer aircraft. Manual release systems were also provided, inside and outside the airplane, in case the automatic failed to engage – which they often did.

These compartments could not accommodate rafts with the accessory container and additional components. The accessory container was to be stored in a convenient location inside the fuselage.

Stowage of life rafts was relative to the type of aircraft, with some models having them installed in the doors and hatches. When the rafts were ejected they were tethered to the fuselage by 16' of rope, so that upon exiting the aircraft, occupants could, in theory, retain the raft and load the appropriate supplies removed from the airplane.

By late 1942, drawings for the modification of compartments large enough for both items were distributed to aircraft contractors to address the issue.

*(Right) Life raft compartment in a B-29 Superfortress (Very Heavy Bomber).*

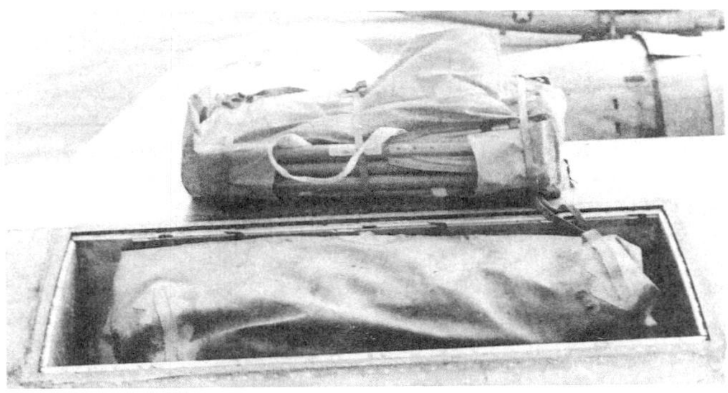

*(Below and below right) Two images of the life raft compartment in a B-25 Mitchell (Medium Bomber) showing the compartment open and closed. On the closed compartment, the exterior manual release handle can be seen. (National Archives)*

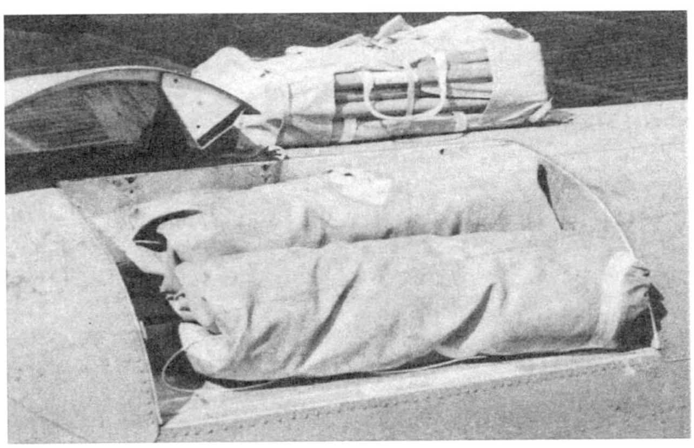

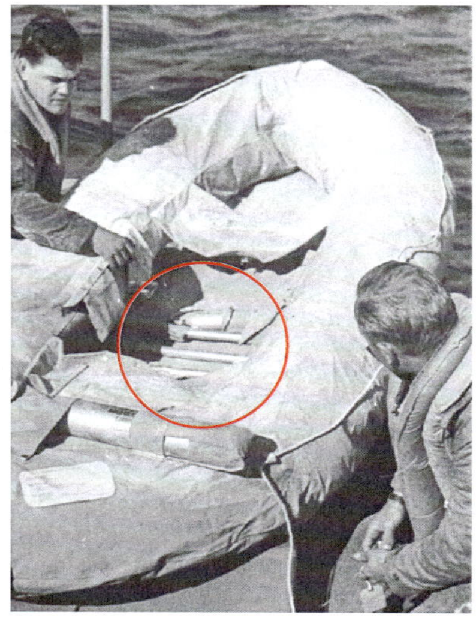

(Above) Prior to the adoption of the central accessory container, the oars and pump would be secured to the floor as shown. (National Archives)

(Left) Color image of a Type A-2 life raft with a compliment of two aluminum oars, showing what would have been provided in the pre-war years. Upon adoption, the Types A-1, B-1 & B-2 life rafts were equipped with two spruce or fir oars. This was revised to aluminum oars at a length of 4' 6" with the adoption of the Types A-2 and B-3. This would remain standard until entering WWII when they went back to wooden types, aluminum being a critical war material. (National Archives)

(Above) For the initial wartime production of wooden oars they would have been a natural color with several coats of clear varnish. The switch to wood occurred with amendment of February 1942, listing the approved materials of linden, poplar, maple, cherry or spruce. In the amendment of December 1942, oars were directed to be painted a light blue or sea green. This set were originally painted a light blue but have been over-painted with a dark OD sometime in their lifespan.

*(Right) The two sectional oars illustrated conform to the 4' 6" type provided with the Type A-2, early production A-3, B-3 and B-4 life rafts.*

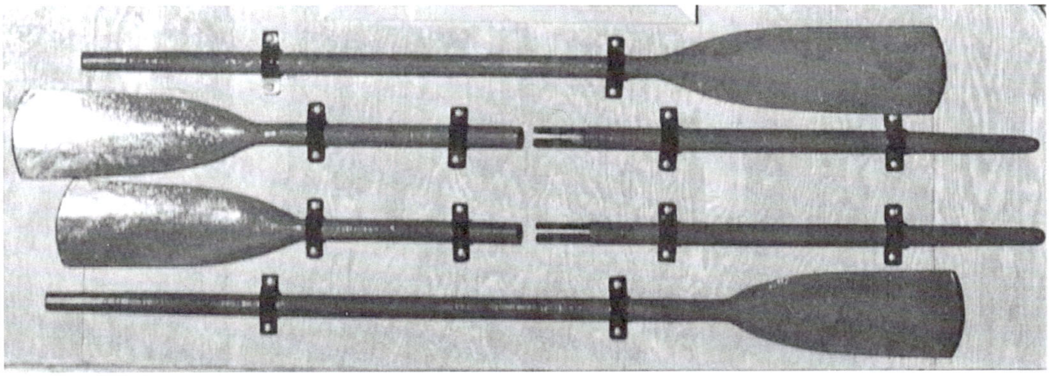

*(Below) Color wartime image showing the oars painted. (National Archives)*

*(Right) Amendment of December 1942 changed the oars to lengths of 6' and 4', provided at a ratio of 2-6' and 1-4' per A-3 life raft and eventually the Type E-2. Shown are examples of wooden 6' and 4' oars painted in what is referenced as Sea Green. One paddle has had an acceptance/inspection stamp applied (far right).*

*(Below) Clamps were provided with oars to prevent loss when in the oarlocks. Several variations have been noted to exist with two examples illustrated here.*

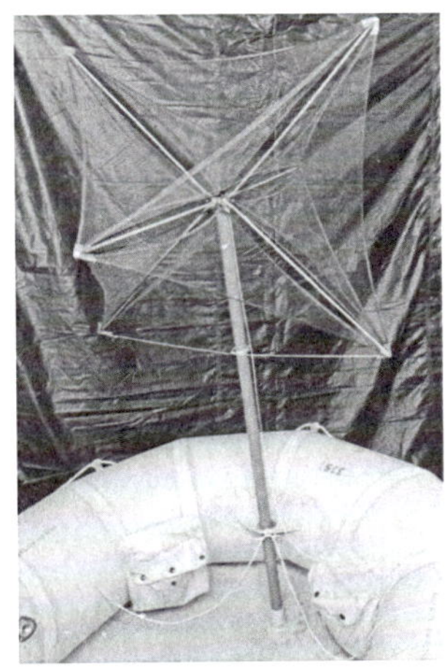

*(Below) For the USAAF, oars were detailed as part of specification for the life raft, not a separate specification like that of the BuAer. By 1944, restrictions on the use of aluminum had been eased and, aluminum oars being preferred, the USAAF drafted formal drawings closely following those details outlined in NavAer Specification M-162. New oars were provided and detailed in revised specifications for the Types A-3A and E-2A life rafts under Dwg. No.44G18937. The lengths of 6' and 4' remained standard, but they did include the application of a Scotchlite material for night rescue. The paddle sections have an applied decal with basic instructions. The examples below illustrate the standard compliment provided for the Types A-3A and E-2A life rafts, they were made available through standard supply channels to update all rafts in service. For the erection of the radar reflector two sections of the oar would be utilized (right).*

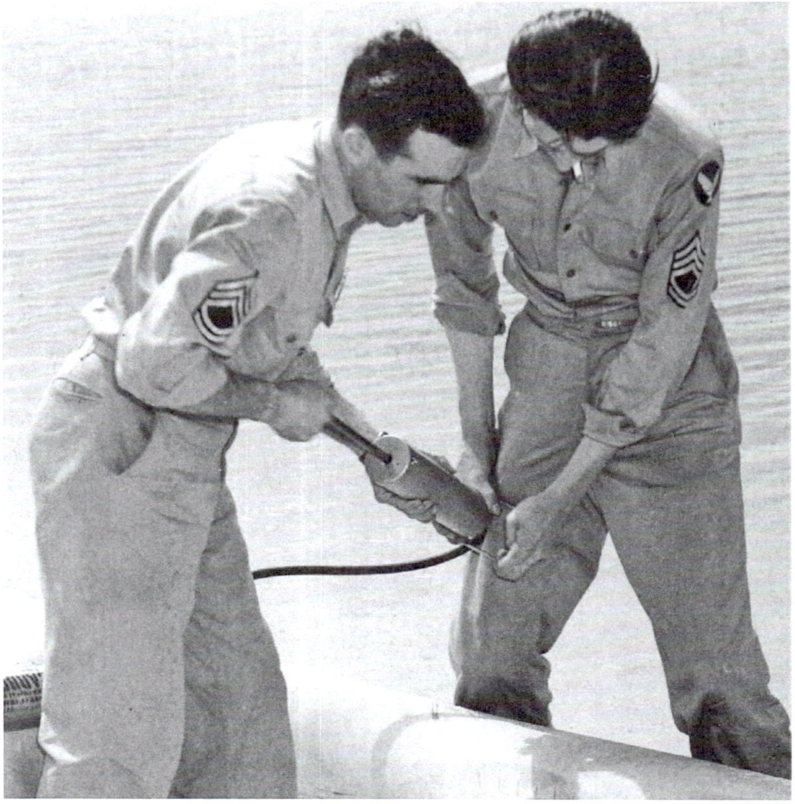

*(Left) USAAF personnel manually inflating a Type A-2 life raft, using the foot stirrup as a brace r. The original pump was a dual action air pump with bilge capabilities. The a foot stirrup was added in 1938. Pumps were supplied with a 24" length of rubber hose. (National Archives)*

*(Above) Hand pump detailed in USAAF specifications (the original 24" hose is not present). By the end of 1942 the M-550 hand pump was integrated into USAAF life rafts, with eventual joint Army-Navy standardization as the AN-P-48 remaining the standard well through the 1950s. For Amendment-A, the pump was removed from the accessory container and placed in a separate pump bag that would be attached to the raft cross member.*

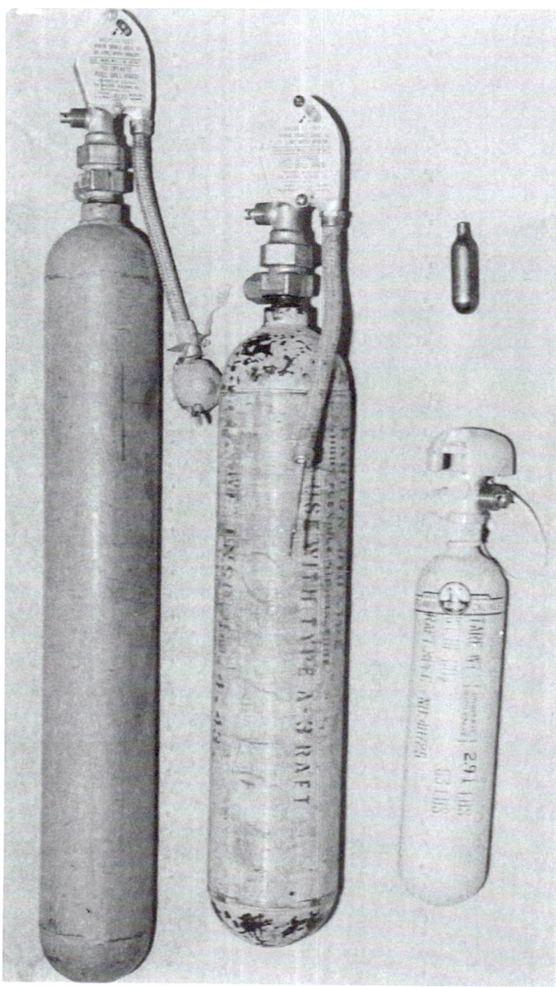

*(Left) USAAF life rafts are best described using BuAer designations as the Type-S and -A. Shown are three CO2 cylinders, The one on the right is for the pararaft and life preserver, the other two are for the Types A-3 and E-2 multi-place rafts.*

(Above) Bailing buckets for USAAF life rafts. On the left is a pattern detailed at the adoption of the Types A-2 and B-3 life rafts, which remained in continuous use throughout the war. The throat and bottom were triangular in shape, reinforced along the opening, which made it difficult to collapse for compact stowage. With the adoption of the Types A-3A and E-2A life rafts, a new bailing bucket was adopted as detailed in Dwg. No. 44D9376. This new design addressed the folding issue. Two examples are illustrated above.

(Right) A cone shaped sea anchor was provided for each life raft. It was made from waterproof fabric. 13-1/2" at the top, tapering to 3-1/2" at the bottom. It is secured to the raft life line by 10' of sisal rope.

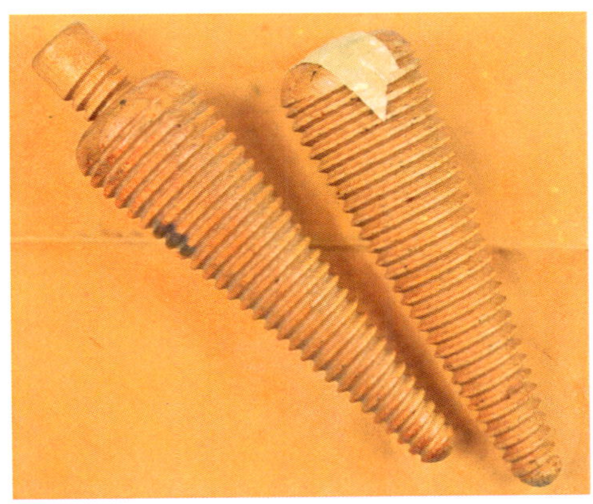

(Right and below) Typical repair kit for USAAF life rafts.

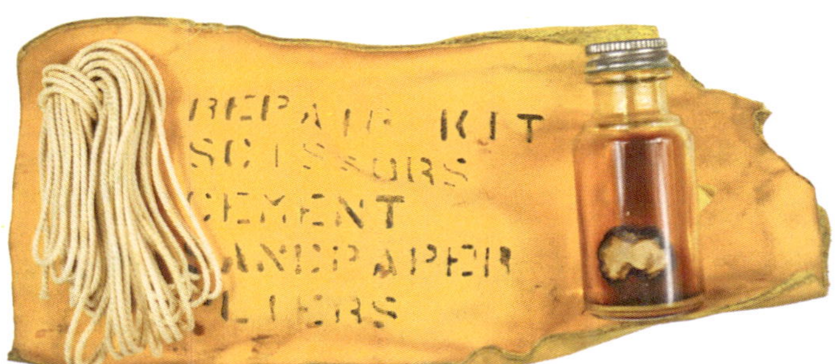

In June 1943, H.Q. 8th AF released a series of recommended modifications based from experience in that theater. The list included changes in both aircraft and equipment considered necessary for the proper protection of aircrews. In this recommendation, additional emergency equipment was to be provided with the Royal Air Force (RAF) Emergency Ration Pack Type-IV distributed two per airplane for the B-17 and B-24. These ration packs were directed to be modified with provisions supplied in the Types A-2 or A-3 life rafts. Approval for these modifications were granted, with drawings prepared and distributed with TMI 04-15-1, January 1944.

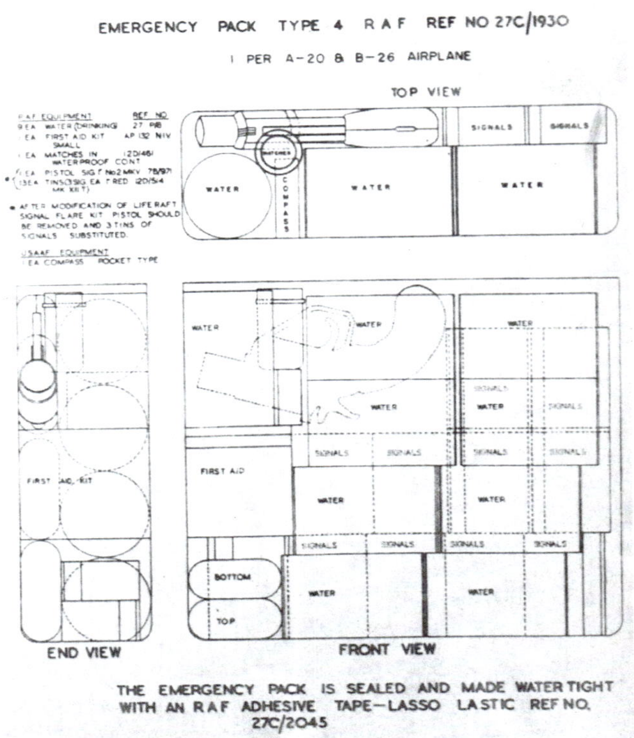


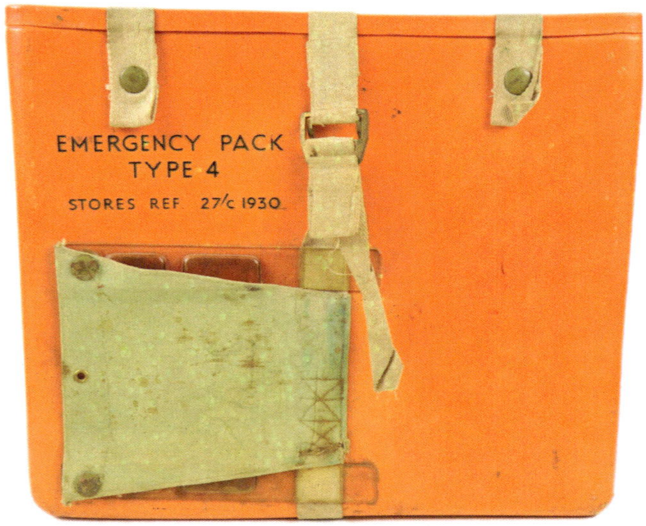


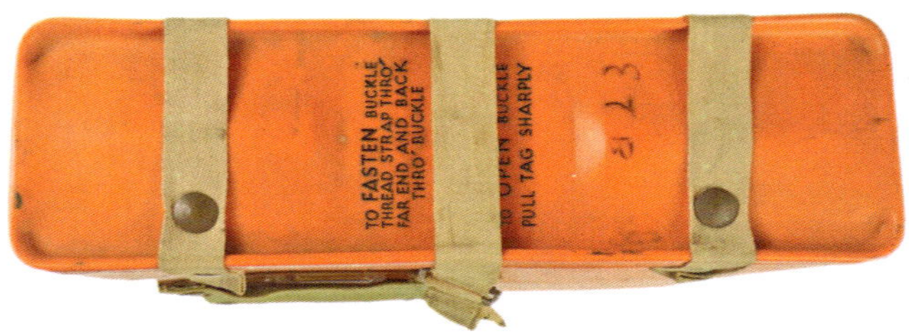


*(Above) The RAF ration pack was available in two sizes, designated as the MK-VII and MK-IV, the MK-IV was chosen for modification. The pack measures 14-1/2" x 12-1/2" x 4-1/2", weighing 21 pounds when full. The typical RAF complement included 12 cans water, 4 tins flying ration, 4 skull caps, 3 sea markers, 1 telescopic mast and flag, 1 tin waterproof matches and 6 tins extra flares. Modification for USAAF use included the addition of 9 cans British or US drinking water, 1 US match safe with waterproof matches, 13 tins 1" cartridges, 1 flare pistol (1"), 1 RAF small first aid kit and 1 US pocket compass. It was directed that two kits be supplied for heavy bombers and 1 for medium and Light bombers, to be stored in the radio compartments.*

*(Right) The modified ration pack is illustrated with this USAAF Type A-3 life raft of the 303rd BG, c.-1944. As part of the modifications, a quantity of 5 Emergency Flying Rations MK-II were to be included in the central accessory container, replacing QMC types, and one RAF heliograph mirror was to be included. Also seen is the modified signal kit including the 1" British flare pistol and cartridges. (National Archives)*

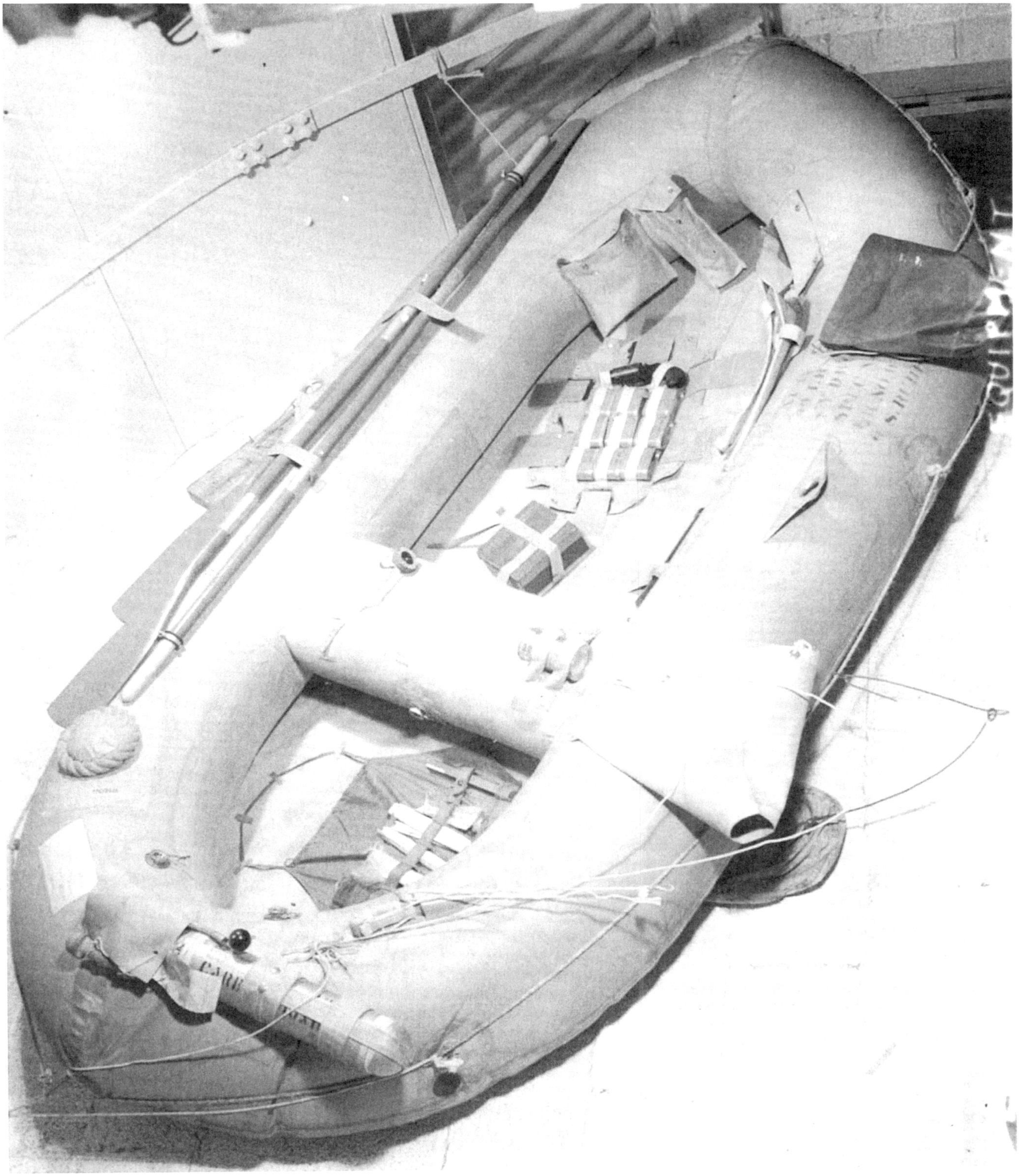

*(Above) As part of TMI 04-15-1, A-2 and early generation Type A-3 life rafts were significantly modified to resemble that of the Type A-3A. Added was an oarlock at the stern, boarding handles and ladder, righting handles and ballast pockets underneath, with one seat being eliminated. Seen at the stern is a pocket for the pump. This Type A-3 life raft, hanging in the equipment room of the 490th BG c.-1944, has had many modifications applied. A RAF dinghy knife has been added to the bladder at the bow. Opposite is a heaving line (resembling a small pie in this image). Heaving lines were standard equipment for RAF life rafts. Colored for visibility, they were buoyant and could be tossed further than other lines on the raft. Attaching it to the raft made it easier to reach and toss. Also included is a RAF Large Dinghy First Aid Outfit. (Courtesy Gregory Pons)*

First-generation Type A-3 life raft. It retained the same overall design features of the Type A-2. Two oars was the standard complement, the raft being provided with only two oarlocks. Attached to one side of the bladder were two pockets, one for the repair kit and the other for a canteen. On the opposite bladder was a pocket for the bailing bucket. An added feature was the accessory container under the bow seat, secured to the floor. At the date of standardization, when this example was manufactured, a limited number of accessories were authorized. As components were added, the accessory container would be moved towards the center, in between the seats.

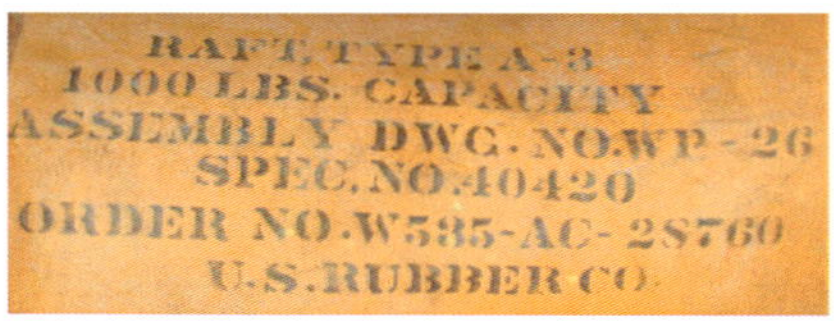

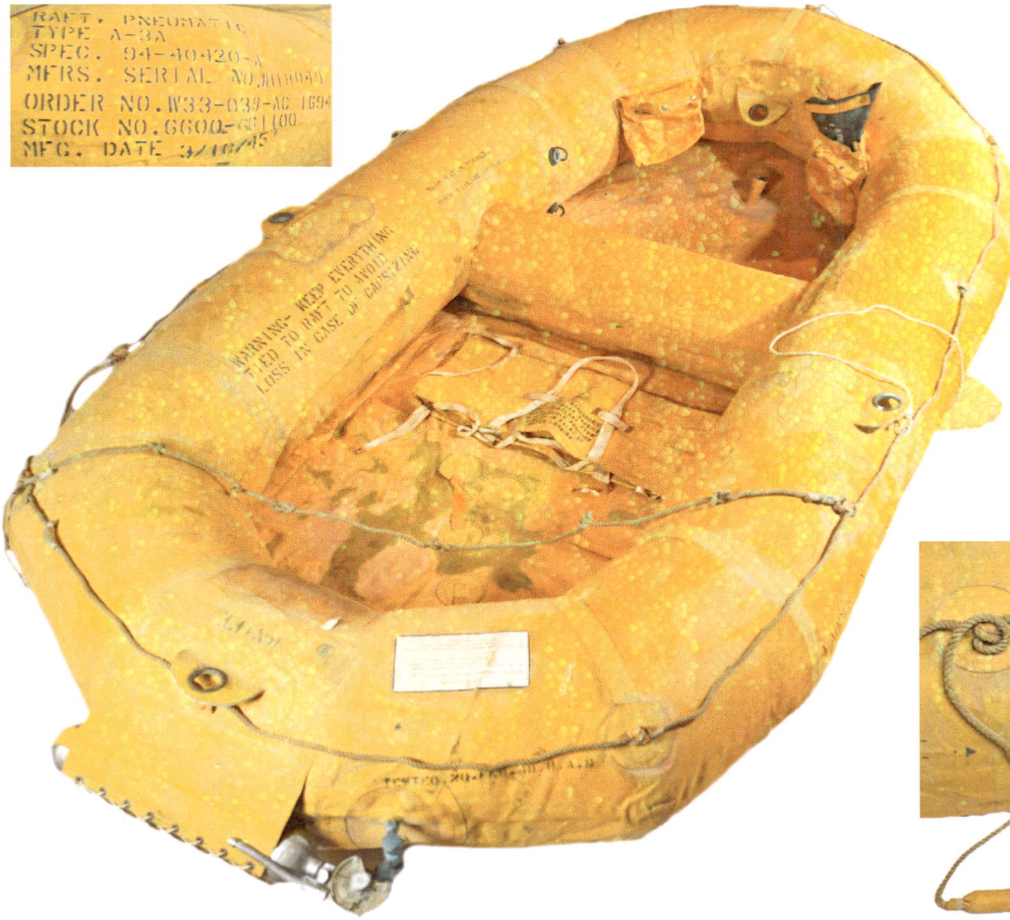


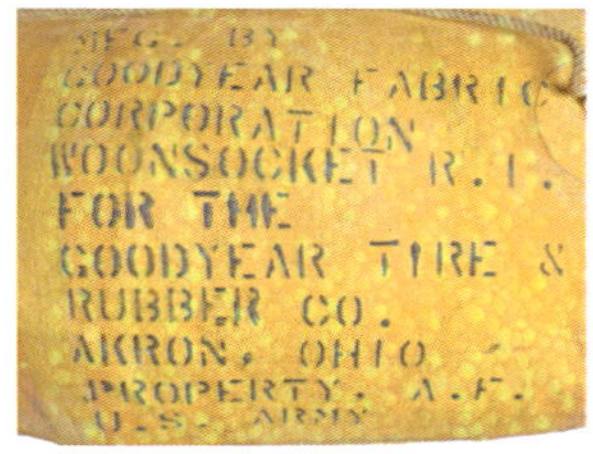

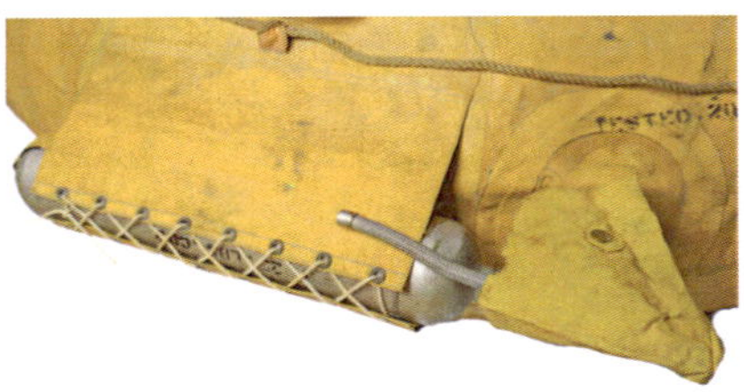

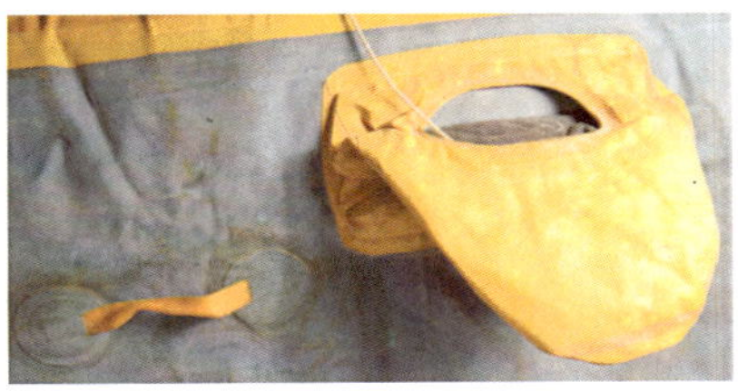

*(Above) A later Type A-3A, significantly revised. Other than the overall change in the shape of the raft and the deletion of one seat, two water ballasts were added to help prevent capsizing. A rope secured to the pocket and accessed from inside the raft secured the ballast in the up position for sailing and lowered when at drift. If the raft capsized, handles on the bottom were used with a knotted righting rope across the top of the raft for leverage. For boarding, a ladder was integrated into the life line. The pump had its own pocket attached to the seat/support beam. At the bow were pockets for the repair and accessory kit and bailing bucket. In between these pockets was the support system for the sailing mast. Covers were applied to the CO2 valve before packing, to prevent chaffing. For the Amendment-A rafts, the previous standard of a jackknife was superseded by the adoption of curved life raft knife, secured to the stern.*

In November 1943 Wing Commander R. Bicknell of the Deputy Directorate, Air Sea Rescue, Air Ministry, brought the drawings and a physical example of the Mark-I Airborne Lifeboat to the United States.

Development was immediately initiated to build a suitable model for USAAF Air Sea rescue organizations. It was felt that the two main benefits served by a lifeboat would be rescue from the sea and escape from the enemy. Test drops were conducted of the British model, along with other models presented by several companies at Eglin Field, from late 1943 through early 1944.

The primary partner for coordinating development was Higgins Industries Inc. From trial and error, and results from field tests, a 27' mahogany plywood torpedo shaped boat was adopted as standard, 22 May 1944, as the Airborne Lifeboat, Model A-1 USAAF Specification No. 94-40743. An award of contract for the manufacture of 300 were approved to Higgins the following month.

The first operational use of the A-1 lifeboat was conducted by the 5th Emergency Rescue

Squadron in April 1945, rescuing six men off the coast of Denmark. It was recorded that in 90 seconds the men were aboard and 12 minutes later they were under way heading to England. While en route, a gale force storm swept in with winds over 60 mph and 25' waves. The boat withstood the weather with the six crew members being picked up on the third day.

*(Above) The boat was specifically designed to be dropped from a modified B-17 Flying Fortress. The bomb bay doors would be removed and the boat suspended by four cables running up into the standard bomb shackles. For air dropping, the bombardier would release the boat. The B-17 for air sea rescue activities was nicknamed the Flying Dutchman. Initially it was felt the boat was not adaptable for the B-29, since when attached, there was only a 4" clearance between the boat and ground. By war's end, a modified aluminum boat, Type A-3, was developed for the B-29 which was assigned the new designation as SB-29 "Super Dumbo." (National Archives)*

*(Right) When released, a static line would deploy the 48' cargo parachutes. Upon striking the water, rockets would fire, trailing a 200-yard lifeline on each side, allowing the survivors to swim to the lines and pull themselves towards the boat. (National Archives)*

*(Above) The Type A-1 lifeboat measured 27' x 7' and weighed 3,250 pounds when fully equipped. The boat featured self-righting characteristics and included 20 watertight compartments for gear. It was painted blue for camouflage. It was fitted with two air-cooled four-cycle engines that could run independently of each other, and under power, had a 500-mile range. A mast and sails were provided in the event the power plant should fail, and oars were also included. Accessories included a long list of items to sustain twelve survivors for five days, and included many of the same accessories found in pneumatic life rafts, with the addition of a clock, fire extinguisher, blankets, soap, socks, rubberized suits and boots, underwear, hats, under-shirts, rations, water, air mattresses, blood plasma and first aid kit.*

*(Below) Details of the inside of the life boat. (National Archives)*

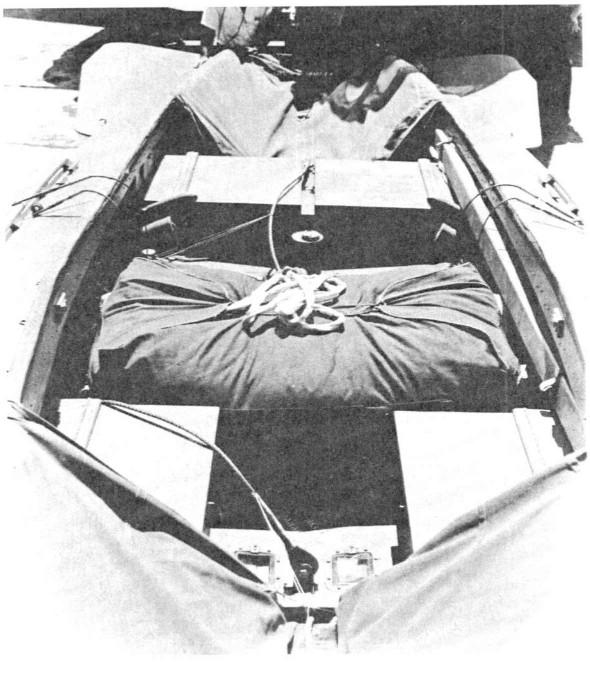

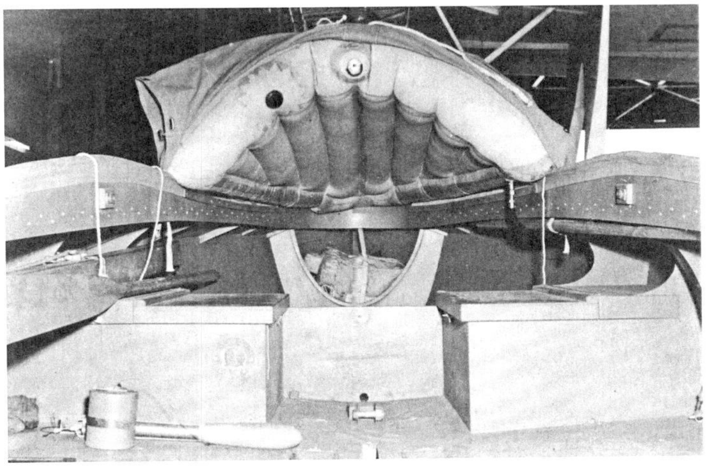

(Above) The USAAF incorporated air droppable assemblies for air sea rescue activities as well. Major James Tapp, stationed on Iwo Jima, devised an assembly utilizing a fuel tank designed for P-51 Mustangs. Air sea rescue was a large component of units stationed in the Bonin Islands, supporting long range bombing over the Japanese mainland. Fighters were adapted and dedicated for locating survivors. This unit allows these reconnaissance aircraft the ability to drop provisions. Though the unit was not officially adopted, it was designated as Gear, Fighter Rescue, and Materiel Command distributed a technical order for service activities detailing the improvisation under Technical Order 01-1-222. Items included in the drop tank would be as outlined by the local command. (National Archives)

(Right) Air droppable units were improvised in all theaters of war, as they all experienced the same necessity for the rescue of downed airmen. A life raft bundle is being loaded under the wing of a 7th AF P-47 Thunderbolt delegated for life guard duty prior to strikes on Japanese bases in the Marianas, c.-1944.

(Bottom right) Life raft in the bomb bay of a B-25 Mitchell operating in North Africa, March 1943 (National Archives)

(Below) By the end of the war, the USAAF developed an air droppable assembly designated as Gear, Sea Rescue, Type A-1, under interim Specification Y-40889. The complete unit consisted of five cases weighing a total of 200 pounds. One case contains a modified Type E-3A life raft with the other four containing supplies of the standard accessories for the raft and the addition of clothing and other misc. articles enough to last four men from ten days to two weeks. It was assigned stock number 8300-286100. Only 492 units were procured pending further development.

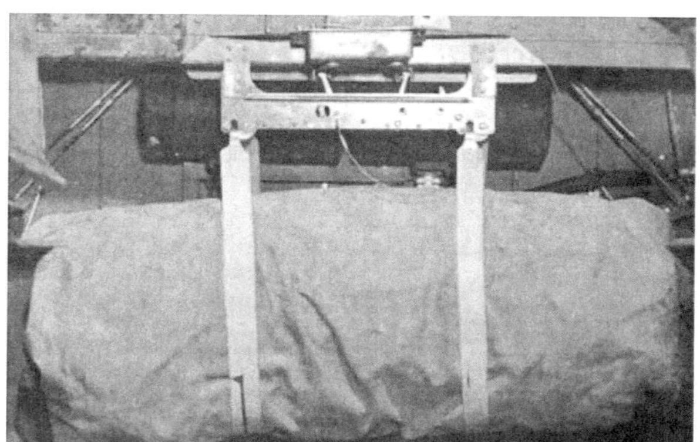

*(Above and left) A comparison study between USAAF and BuAer life rafts. Each service in turn influenced the other in terms of design and modifications. Navy characteristics of the M-3 were integrated into the USAAF Type A-3, then it was revised to an oval shape. This oval design ultimately influenced Navy development and was integrated into their M-3R. In turn USAAF life rafts adopted Navy features with their Amendment-A. Above is shown the seven-man USAAF Type E-2 with the BuAer MK-VII M-3Q. Left is the four-man US-AAF Type A-3 alongside the BuAer MK-IV M-3Q.*

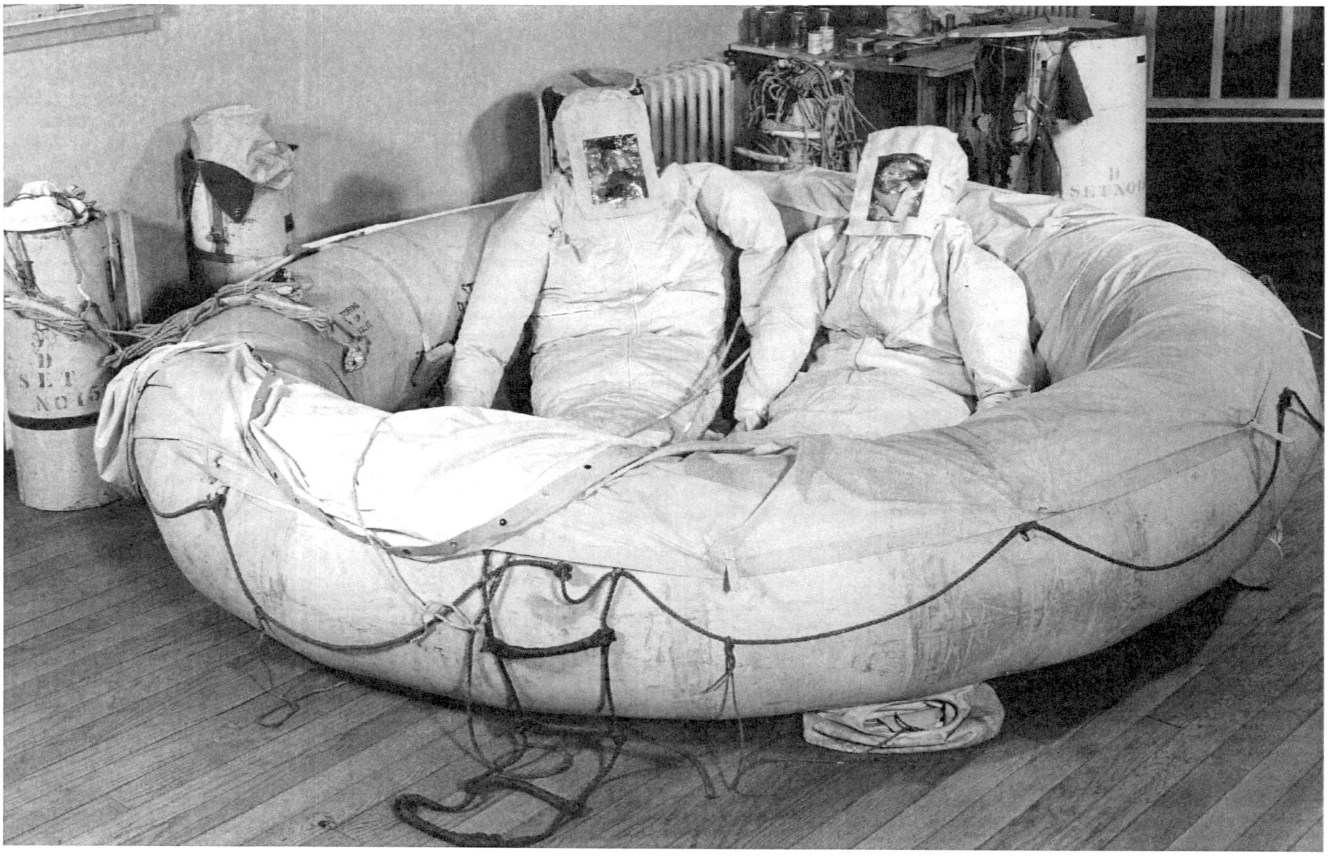

Above and below) Shown for comparison purposes, the standard multi-person life raft for the RAF was the Lindholme Type-H Dinghy. It had a circular bladder for stability. Both the USAAF and BuAer evaluated and field tested this model in 1940 and 1941 for possible adoption, but both services opted to retain their designs. They did find use with USAAF units operating in the ETO. Their rafts came fully equipped with two accessory containers, protective suits and spray shields. With this style the occupants were left to the mercy of currents and drift, but suited for the environment in the English Channel and established air sea rescue organizations operating from England. The image below illustrates many of the principle components included in the equipment containers. (National Archives)

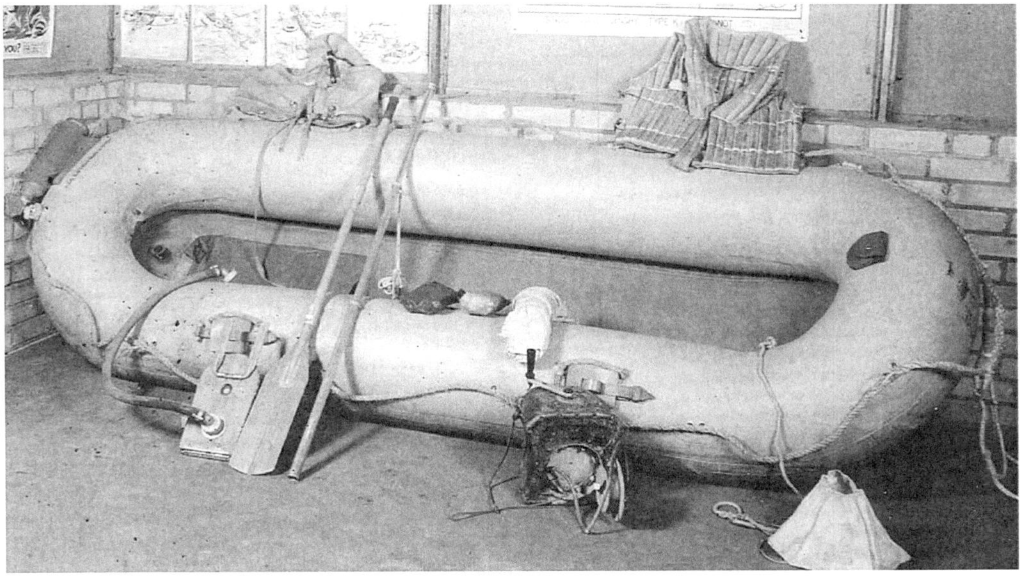

*(Left and below) For additional comparison purposes, Luftwaffe multi-person life rafts are shown. Left is a prewar model having a unique shape with what appears to be a very basic complement of supplies. Below is a later, updated model, that for a lack of a better description, has all the bells and whistles, much of which were incorporated in BuAer and USAAF life rafts. (National Archives)*

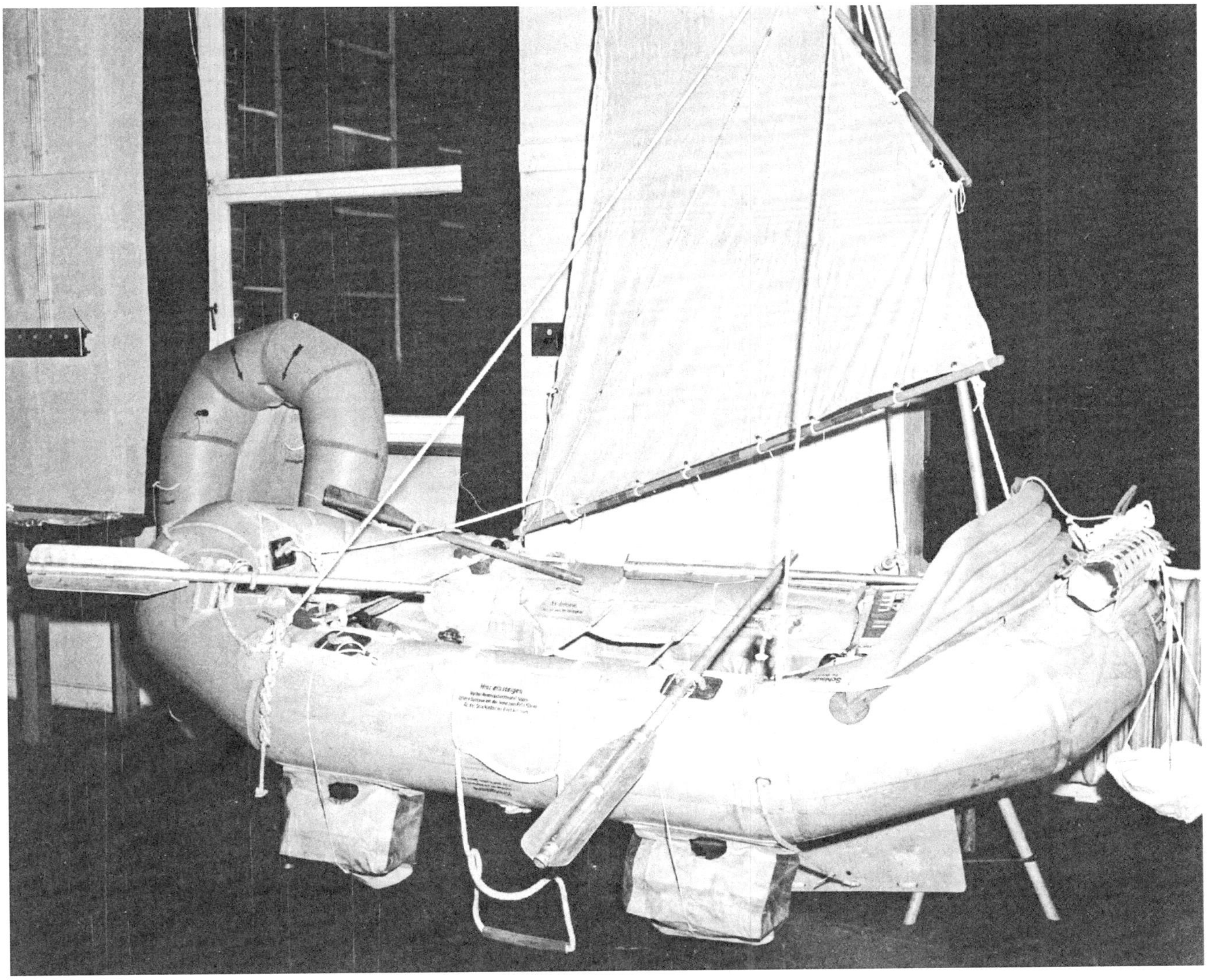

# LIFE RAFTS

## Part II.
### One-Man Types

The development of the one-man pneumatic life raft paralleled that of the multi-person types in both aeronautical services. Their development, like most emergency equipment, was principally based on service requirements. It was not until the late 1920s that either service began the initial steps of a standardized type of life raft for aircraft. The BuAer was the first to incorporate a designated capacity for one-man as detailed in NavAer Specification M-3, dated 20 April 1929.

The one-man raft consisted of a rubberized fabric bladder, separated into two internal bulkheads, of an oblong shape and colored orange-yellow. The one-man raft was 6' x 3' wide with a 12" diameter bladder, having a rated capacity of 700 pounds.

When deflated, the raft was stored in a khaki duck waterproof carrying case, consuming a storage space of two cubic feet (20" x 18") and weighing 24 pounds. Accessories through its life span would parallel its larger capacity counterparts.

*(Left) Chief Aviation Pilot Verne Harshman of VF-2 would be the first Naval aviator recognized to have given the standard pneumatic life raft a practical service test. In March 1931, he led a flight section from the USS Langley on an exercise off the coast of Honduras and was forced to ditch his Boeing F2B-1 fighter, spending 5 days adrift until being rescued. Harshman poses for a PR photograph in Panama with an early generation M-3 one-man life raft with the distinctive feature of having metal ring oarlocks. He additionally wears the M-5 life preserver.*

*(Below) An aviator demonstrates another early generation one-man raft. He wears a Navy type kapok life preserver for aviation.*

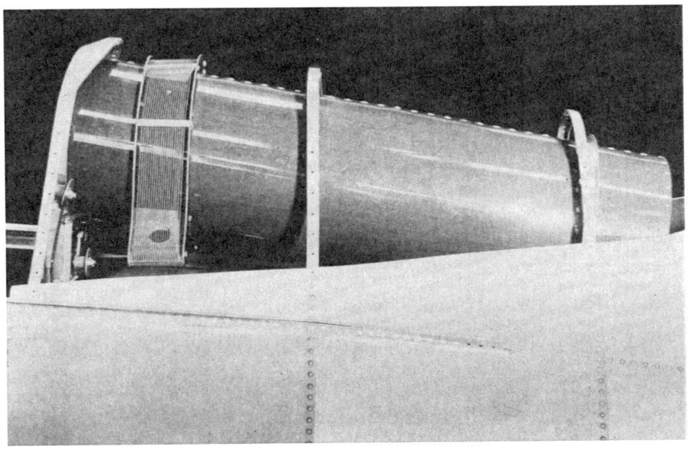

*(Left) Stowage of life rafts was relevant to the model of aircraft, but for most of the era it was referenced as the pilot's luggage compartment, which conveniently doubled as life raft storage. With the integration of life rafts as standard for all aircraft, new designs would incorporate specially designed compartments, as can be seen in this Grumman F3F-2 fighter, c.1934. The life raft was stowed in a cylindrical tube behind the pilot's head rest. (National Archives)*

*Above) Naval aviator demonstrating the M-3 one-man life raft in the revised pattern of a boat shape, April 1934. (National Archives)*

*(Above) A factory worker demonstrates the raft inflated. In her left hand is the raft packed in the orange-yellow carrying case marked 1 man raft. Provided on one end is a grab handle, c.-1934. (National Archives)*

*(Left and below left) Demonstrating the correct way to get into the raft. Note stenciling; as part of Amendment-M of December 1935, M-3 rafts were officially designated under a Mark nomenclature. (National Archives)*

*(Above) Detail of stenciling. Contracts awarded after December 1935 would reflect the new property Mark designation. One-man rafts could be requisitioned under ASO stock number R83-R-250. (National Archives)*

Life raft compartments for US Navy fighters were integrated into the engineering of the fuselage. Aircraft were also provided with flotation bladders that would automatically inflate upon contact with water.

The F2B-1, flown by Harshman, was so equipped with fabric bladders which kept the plane afloat allowing him to inflate the raft and make a safe exit.

Automatic inflation bladders were installed on the new US Navy carrier fighter the Grumman F4F-3 Wildcat which would be the standard carrier fighter until 1943 when it was replaced by the F6F Hellcat.

(Above) Ens. H.E. Tennes of VF-6 had to ditch his F4F-3 Wildcat in San Diego Bay due to damaged landing gear, which prevented him touching down on a land-based airstrip, May 1941. Two automatic bladders were provided in the wings. The partial concept was that the aircraft could potentially be salvaged after a ditching, remaining afloat for an extended period of time.

(Above) The MK-I life raft was stored in a compartment along the spine of the Wildcat, behind the cockpit. The design emphasis was that, during a ditching, contact with water would actuate the inflation system for the bladders in the wings, keeping the plane afloat. Additionally, a water-sensitive switch on the underside of the fuselage automatically opened the release valves on two CO2 cylinders, actuating the Type-A (Automatic) life raft. The gas from the first of these cylinders opened the hatch of the life raft compartment and gas from the second cylinder inflated the raft. In the event the automatic switch failed, a manual lever was provided in the cockpit behind the pilot. A lanyard would keep the raft secured to the aircraft, allowing the pilot to exit onto the wing, retrieve and board the life raft. Life raft compartments were deleted in later production VF class aircraft, such as the F6F Hellcat and F4U Corsair, since the MK-I was superseded by the pararaft. (Walter Kidde & Co. press release)

*(Left) This F4F Wildcat was forced to ditch after a failed catapult launch, and upon contact with water, the MK-I life raft automatically inflated. The Wildcat nose-dived and began to sink rapidly, leaving the pilot only seconds to evacuate and place himself at a safe distance before it plunged to the bottom. This scenario proved problematic during wartime conditions and resulted in many fatalities, partly because it did not allow sufficient time for the pilot to gather emergency equipment.*

*(Right) A MK-I life raft resting against a palm tree near Henderson Field, Guadalcanal, c.1943, with two squadron members of VMF-124, which had the distinction of being the first Corsair squadron arriving at the Canal in February 1943. The Corsair did not have provisions for stowage of the MK-I life raft, because emphasis had shifted towards the integration of the parachute life raft. The raft pictured could had been used for training or recreational use, which was not uncommon in the tropical Pacific. (USMC)*

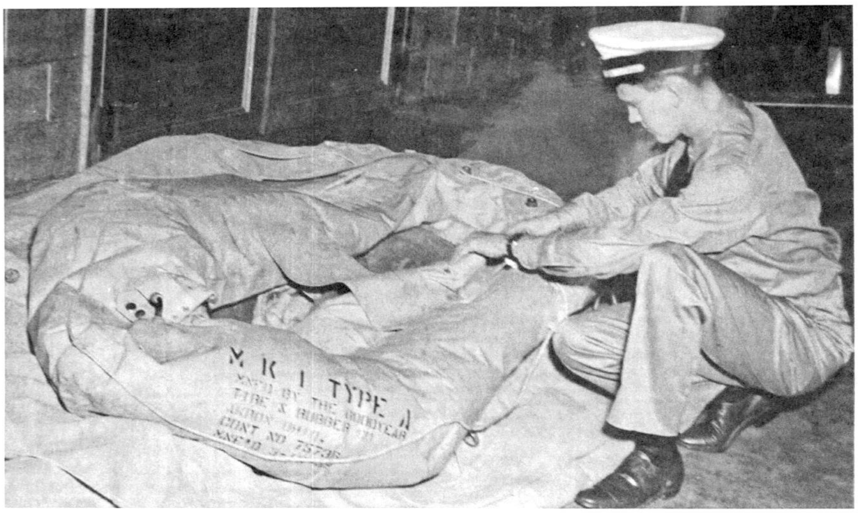


*(Left) The MK-I being inspected is designated as a Type-A. It became common practice to replace the CO2 systems eliminating the problematic automatic inflation.*

*(Facing page, top) An aviator in a MK-I raft being assisted by the crew of a J2F Duck. When the MK-I was phased from service, many of the surplus rafts on hand were transferred to the Civil Air Patrol. (National Archives)*

*(Right) In May 1941, installation of automatic inflation bladders in the wing of the F4F-3 was discontinued, believing that the aircraft could just as easily be salvaged by the enemy as by friendly forces. In November 1942, due to the inherent lack of reliability of the automatic switches and introduction of the pararaft, the use of the Type-A raft system was discontinued as well. Production aircraft utilized either the Type-S (Standard) or Type-A (Automatic) MK-I life raft depending on manufacture date. With the use of the Type-S, the pilot would, upon ditching, release the compartment panel by the manual lever in the cockpit, then physically retrieve and remove it from its carrying case. This color image illustrates a pilot retrieving the raft by the grab handle provided. Ideally he would then unpack and inflate the raft on the wing, if still afloat. The zippered case signifies the raft to be a Type-S. (National Archives)*

By 1936, several developments had occurred that would affect future life raft development and integration for the USAAC. The standard life rafts at that time, Types A-1 and B-2, consisted of heavy duck rubber coated fabric bladders.

The Materiel Division had just prepared specifications for a new line of rafts that utilized latex rubber bladders encased in a heavy duck covering. Three new rafts, designated as Types A-2, B-3 and C-1, would be fabricated to conform to the new specifications, and were prepared for service testing. The Type C-1 raft was developed to supersede the Type B-2 for use in pursuit airplanes. Up to this date provision for carrying life rafts was required on all aircraft with a flight path over water, and the two-man, 400-pound capacity Type B-2 was provided for pursuit types, stored in the pilot's luggage compartment.

The Type C-1 life raft was a smaller version of the B-2, with a 250-pound rated capacity and an internal bladder construction. The Type C-1 was service tested from July 1936 to January 1937 by the 8th Pursuit Group, GHQ Air Force at Langley Field, Virginia. In their recommendations, it was found to be superior to the Type B-2 but not adopted due to the inability to properly carry it.

However, they did recommend that it be standardized for such use as it may be appropriate. In an alternate turn of events, the engineering division could not find suitable locations to install built-in flotation compartments in pursuit aircraft that would enable the plane to remain stable and right-side-up.

In January 1937, based on these findings, Chief of the Air Corps recommended that all combat aircraft be provided with flotation equipment except pursuit or basic combat. Rafts were to be omitted from pursuit types in the assumption that they would not be flying over water, therefore provisions for carrying a life raft would not be required. It was further stated that no provision for its accommodation was desired. While the recommendations were being debated, the Type C-1 life raft was standardized 13 April 1937, titled as Raft, Pneumatic Bladder, Type C-1 USAAF Specification No. 94-40241.

Air Corps Policy No.159 of 1937 directed the incorporation of a means of securing flotation for combat airplanes when forced down on water and for carrying life rafts. With the new policy, the Materiel Division recommended changes in the engineering of pursuit aircraft fuselages to incorporate the new Type C-1, since the current models did not provide suitable provisions, as it was impractical for the pilot to retain the life raft in the event of an emergency.

On an administrative level, details of the Type C-1 were prepared for inclusion in technical data for projected service use. Authority for procurement was withheld pending engineering revisions for aircraft, which would establish requirements and eventual allocation in tables of allowances.

Ultimately, for reasons unknown, compartments were not accommodated in future production of pursuit aircraft, resulting in no procurements of the Type C-1 life raft. Stations in Hawaii and on the west coast perceived it as a necessity, and as operations continued to expand in the Pacific Ocean, they established a requirement for the Type C-1, continually requesting quantities to outfit airplanes. The Materiel Division was forced to decline their requests since none was available, nor were any on procurement. In response to the continued demands, the Materiel Division began the investigation into a suitable one-man raft for use in pursuit aircraft.

*(Above and right) A torpedo shaped life raft compartment was adapted for this USAAC Boeing P-12 pursuit aircraft, mounted on top for use in emergency water landings, Wheeler Field Hawaii November 1931.*

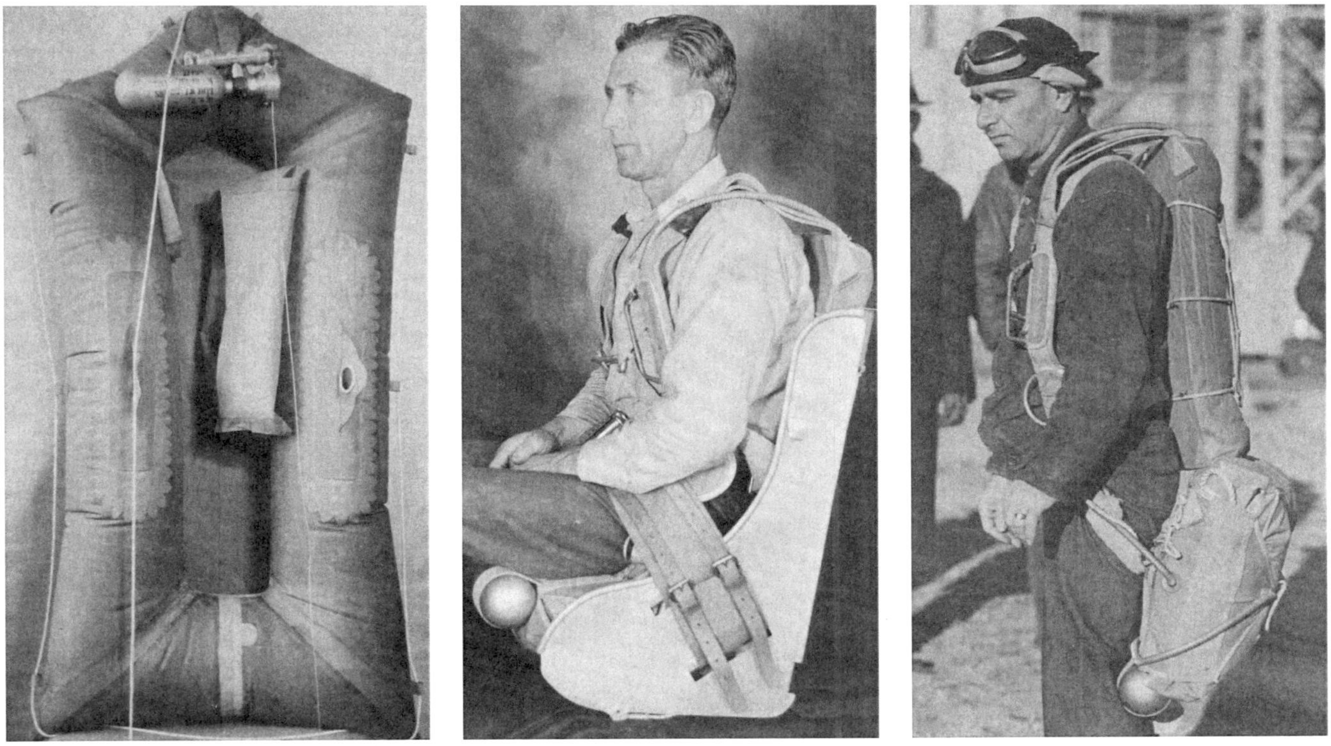

(Above) The MK-I life raft was the standard for US Navy VF class aircraft, but did not address the issue of having a raft stored separately from the user. Through the 1930s the BuAer attempted to experiment with parachute types that would be integral to the parachute harness and would remain with the user in emergency situations. Both one and two man configurations were tested. In two seater aircraft, the pilot would carry a parachute raft with a rated capacity for the addition of the gunner. A seat type system was developed and used in conjunction with a back type parachute. The seat type parachute life raft illustrated was tested at NAS Lakehurst in 1934. The concept was favorable, but the design was cumbersome and it was not adopted due to its bulk and discomfort. (National Archives)

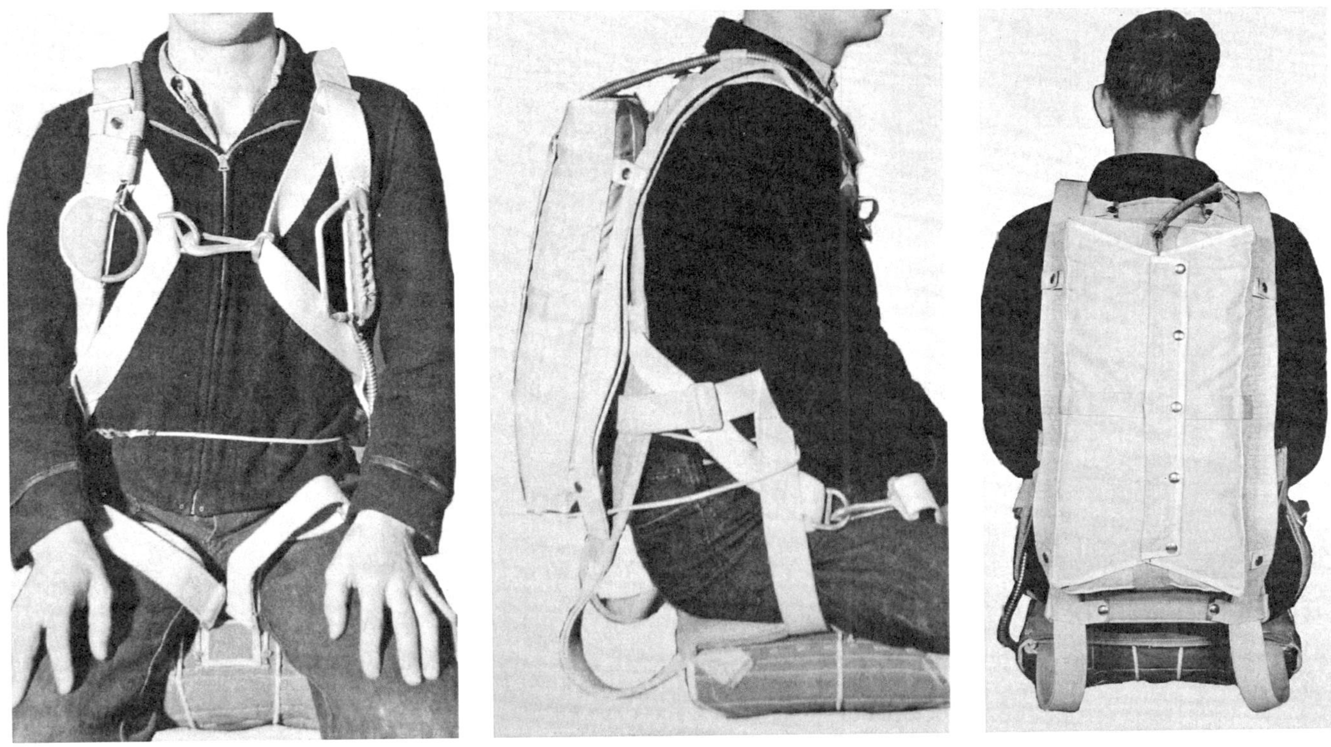

(Above) For the standard seat type parachute, the BuAer evaluated a back type parachute raft system. Like the seat type raft, a rip cord was provided to release the raft from the container, with an "O" ring located near the user's right shoulder. This configuration was tested at NAS Lakehurst in 1936. (National Archives)

Before the war, there was no urgent requirement for emergency provisions, especially for single seater aircraft. The USAAC had discontinued their requirement for rafts altogether, while the BuAer attempted to refine life rafts. The decision was based on operational priorities – the USAAC flying primarily over land and the BuAer conducting flights over water.

With war looming and the projection of conflict over and across oceans, certain entities had the foresight to initiate a program to develop individual life rafts that will most certainly become a military requirement. For the USAAC, the Materiel Division developed a one-man parachute raft in the spring of 1941 to satisfy pursuit group requirements in lieu of the Type C-1. Working in coordination with the US Rubber Co., a lightweight rubber-coated nylon raft was developed as per USAAC Dwg. No. 42B3447.

The raft was designed to be compact and fit in a cylindrical container 16" long x 7" in diameter, weighing 9 pounds. As part of its design emphasis, it was to be stowed in the cockpit and attached to the parachute harness prior to an emergency bail out or ditching. No emergency provisions were provided, being streamlined to include only a CO2 cylinder for inflation. A service test quantity of 100 were ordered.

In August 1941, the USAAC and BuAer received samples of the new British Type-K Dinghy in the Type-A Seat Pack and the Type-B Back Pack. Upon evaluation, both were found suitable for use in pursuit and VF class aircraft.

The initial reaction by the USAAC was that the stowage of the raft in the seat or back was objectionable because of the possibility of premature inflation, and that the raft should be stored on the cockpit floor like that of the 42B3447. Furthermore, they had no interest in the seat pack, having already experienced the deficiencies of such a concept with the development of a seat type jungle emergency kit in the mid 1930s. The possibility of integrating the raft in a backpack form was evaluated, since that configuration was acceptable with the jungle kit. On preliminary examination, it was found that the British raft was compact enough to be stowed in the jungle emergency kit 38J5271, with the sponge rubber insert removed. It was then recommended that service test quantities be procured for pursuit aircraft.

The BuAer were enthusiastically optimistic about the design concepts after satisfactorily field testing them at NAS Norfolk in September. They had attempted parachute designs in the past, but could not progress past the bulk issue. Of immediate interest was the raft design and its compact nature. Working in coordination with the US Rubber Co. and Fred Patten, the BuAer established preliminary specifications for a suitable design for VF class aircraft.

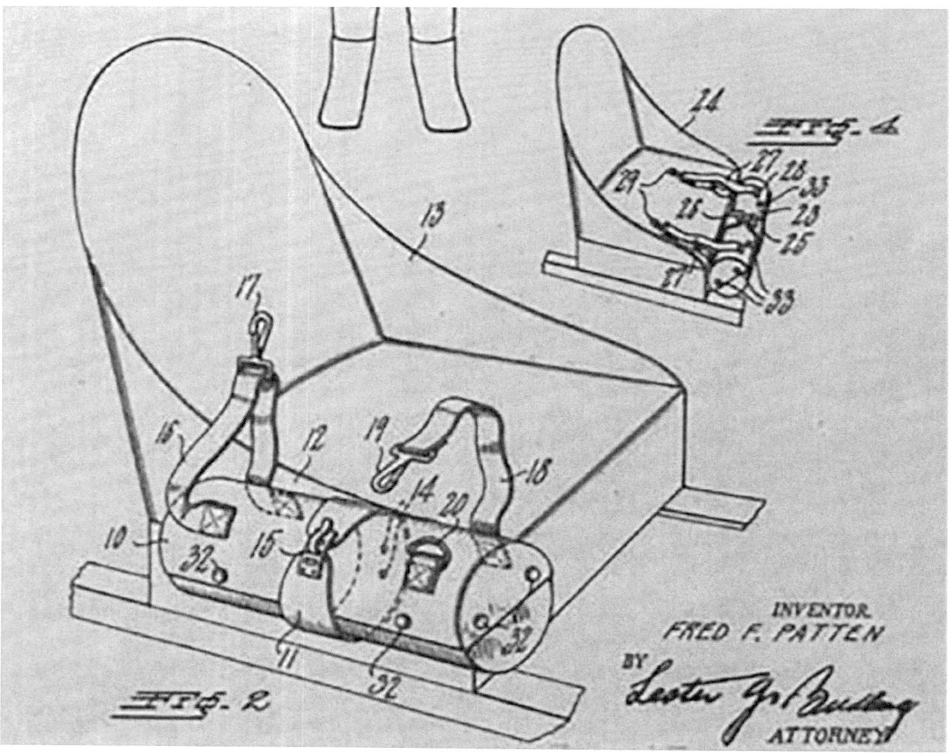


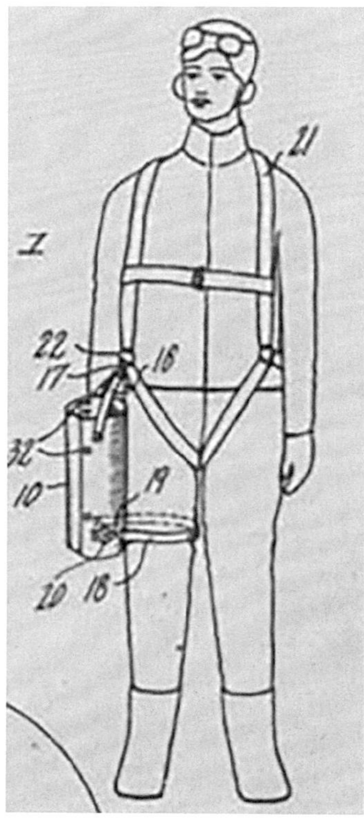

*(Above) Patent awarded to Fred F. Patten which offered the best look at the concept of the 42B3447. Fred Patten had substantial influence in life raft development for the military, dating back to 1939. He worked with the US Rubber Co. and would be a commissioned officer in the US Navy during the war. His passion for life-saving devices derived from the loss of his Naval aviator brother, who had died for the lack of an emergency life raft in 1934. Patten helped design the 42B3447 for the USAAC. In October 1942, he submitted the patent and it was assigned No. 2,366,455.*

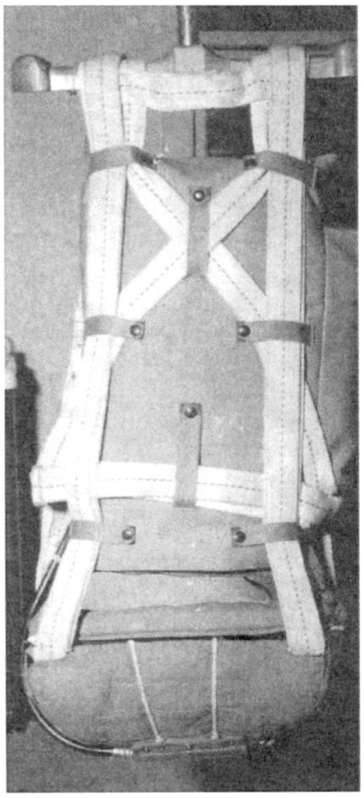
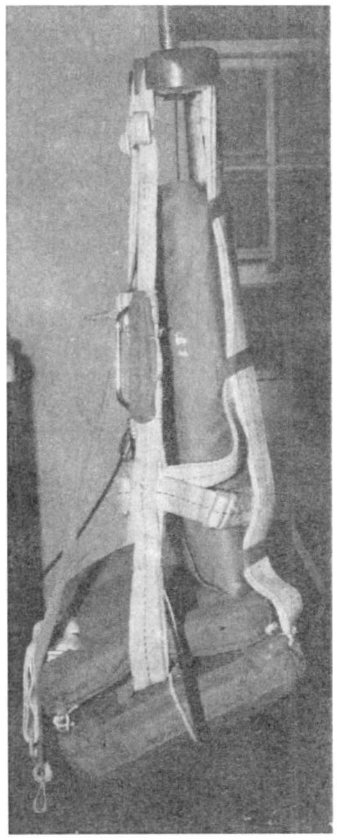
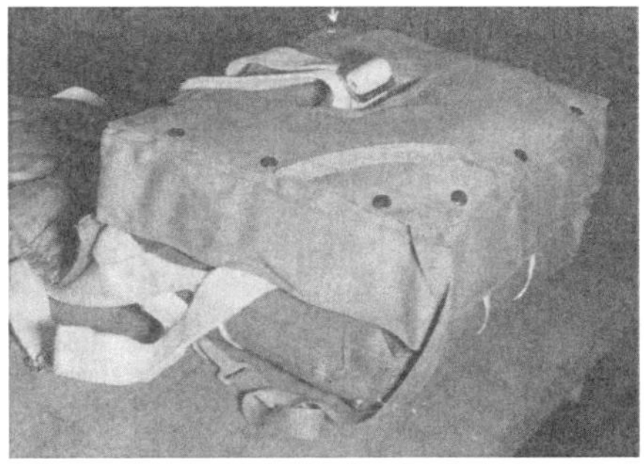

*(Left and above) Early British Type-K Dinghy packs field tested by the USAAC and BuAer in August-September 1941. Far left is the Type-B Back Pack attached to a standard USAAC seat type parachute. Center and above is the Type-A Mark-II Seat Pack. (National Archives)*

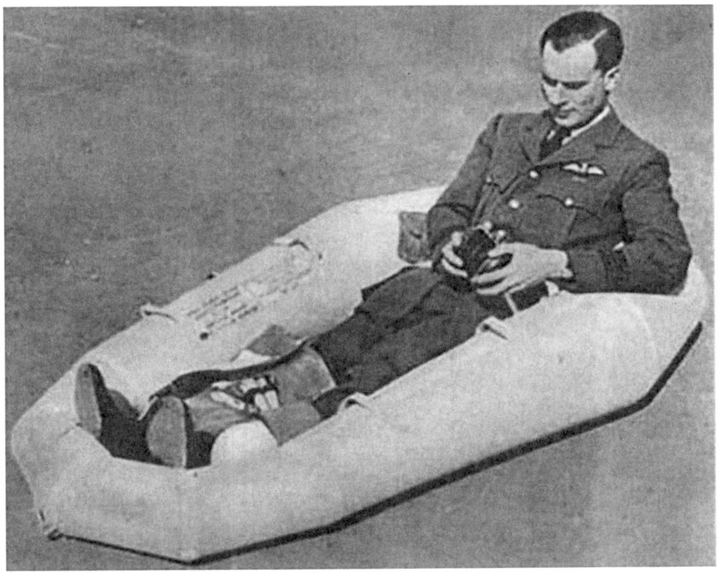

*(Left, above and below) The first-generation British Type-K Dinghy utilizing the Type-A Seat Pack, as featured in the Illustrated London News, 6 September 1941. This design was revolutionary at the time, being compact and containing a complement of emergency provisions. It could be worn comfortably with the parachute, adding minimal bulk increase. The Type-K would evolve to include additional features and accessories during its life span, but this was the principle influence for United States concepts.*

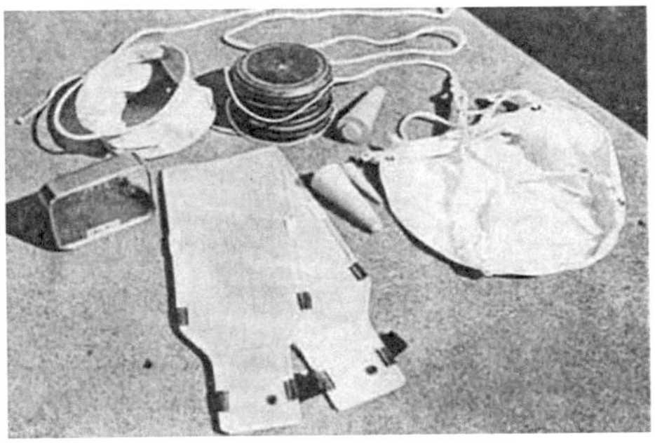
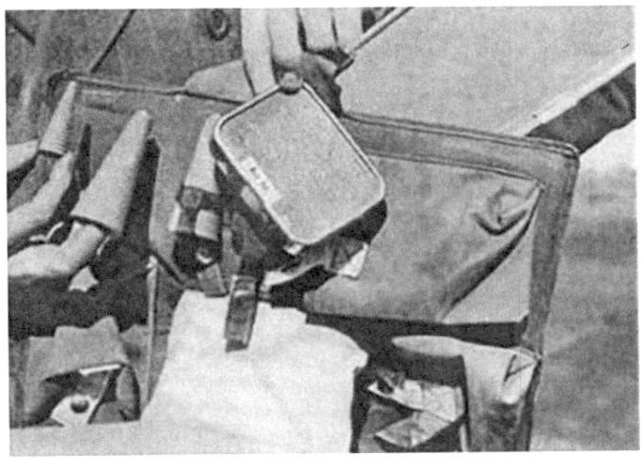

(Below) Utilizing the British Type-K Dinghy as a template, the BuAer drafted and approved specifications for a one-man parachute raft under NavAer Specification M-524, dated 24 November 1941, titled Rafts, Life, Pneumatic, One Man Parachute Type. The overall assembly was almost a direct copy of the British pattern, making only minor adjustments in design features. The life raft was constructed from orange-yellow rubberized fabric, 66" long and 40" wide at the bow, 30" wide at the stern, with a 12" diameter bladder tapering to a 6" at the stern. A service test quantity of 300 complete assemblies was procured from the US Rubber Co. in January 1942, awarded under Navy Department Contract No.98179. Shown is one of the 300 rafts manufactured in June 1942. At this date the M-524 had been approved for joint Army-Navy standardization, which was reflected in the nomenclature. Secured and folded with the raft is the sea anchor, bailing cup and concertina hand pump. (National Archives)

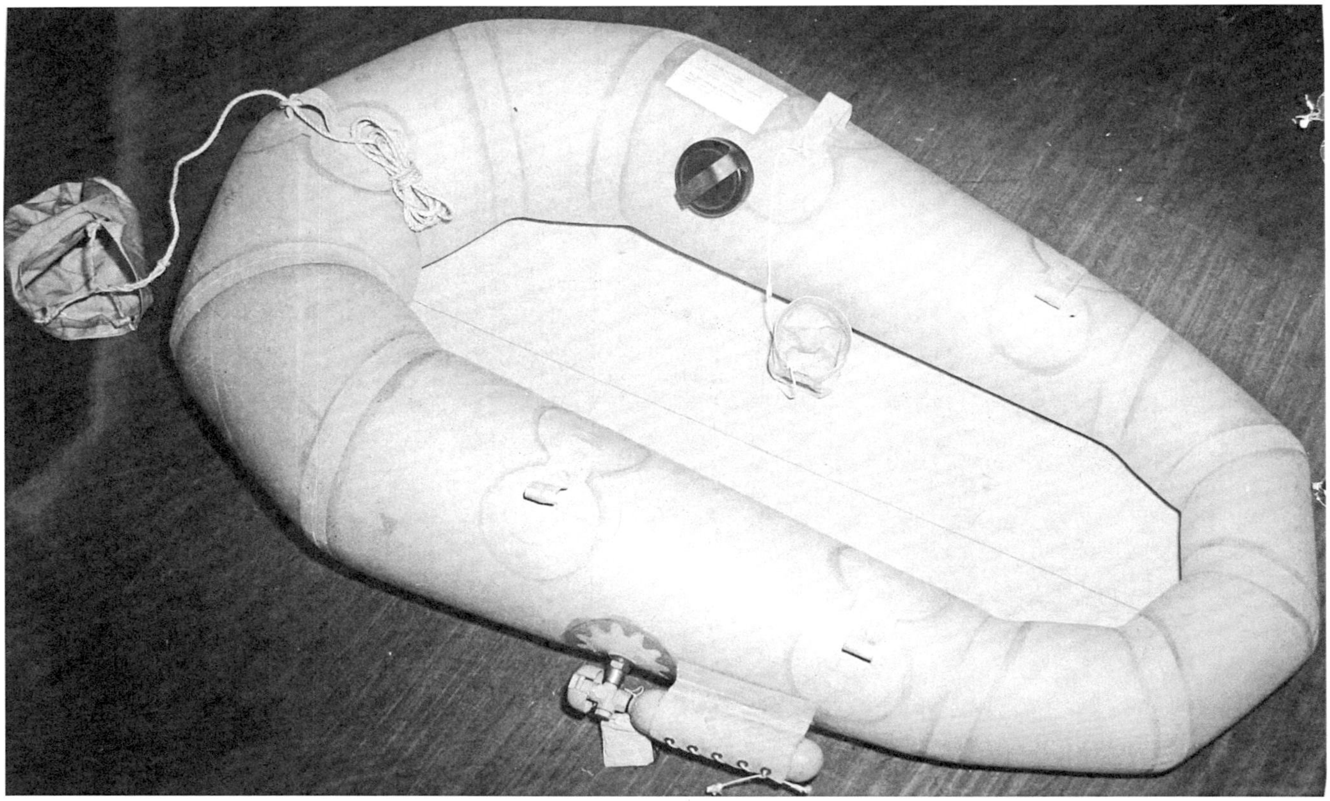

(Below) In this preliminary specification, two types of cases were outlined: the Type SP-Seat Pack and Type BP-Back Pack. Both designs closely copied the British patterns, with the seat pack being the preferred type. Illustrated below is the seat pack, and like that of the British Type-A, it included a centered slot for the leg straps of the parachute harness. Inside the case, provision for securing accessories was provided. Two hand paddles were secured to the bottom, with the repair kit, bullet hole plugs, drinking water and sea marker stowed along the edge. The complete assembly packed was 15" x 14" x 4" and weighed 14 pounds. As outlined, all external surfaces were to be orange-yellow to include both the life raft and case. (National Archives)

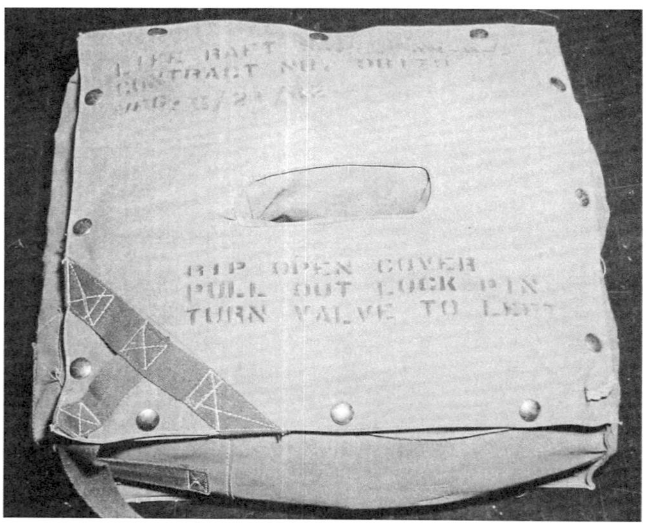


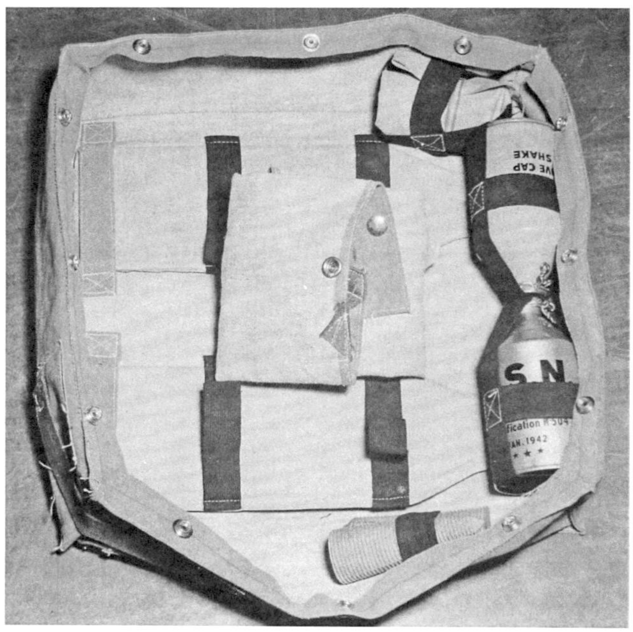

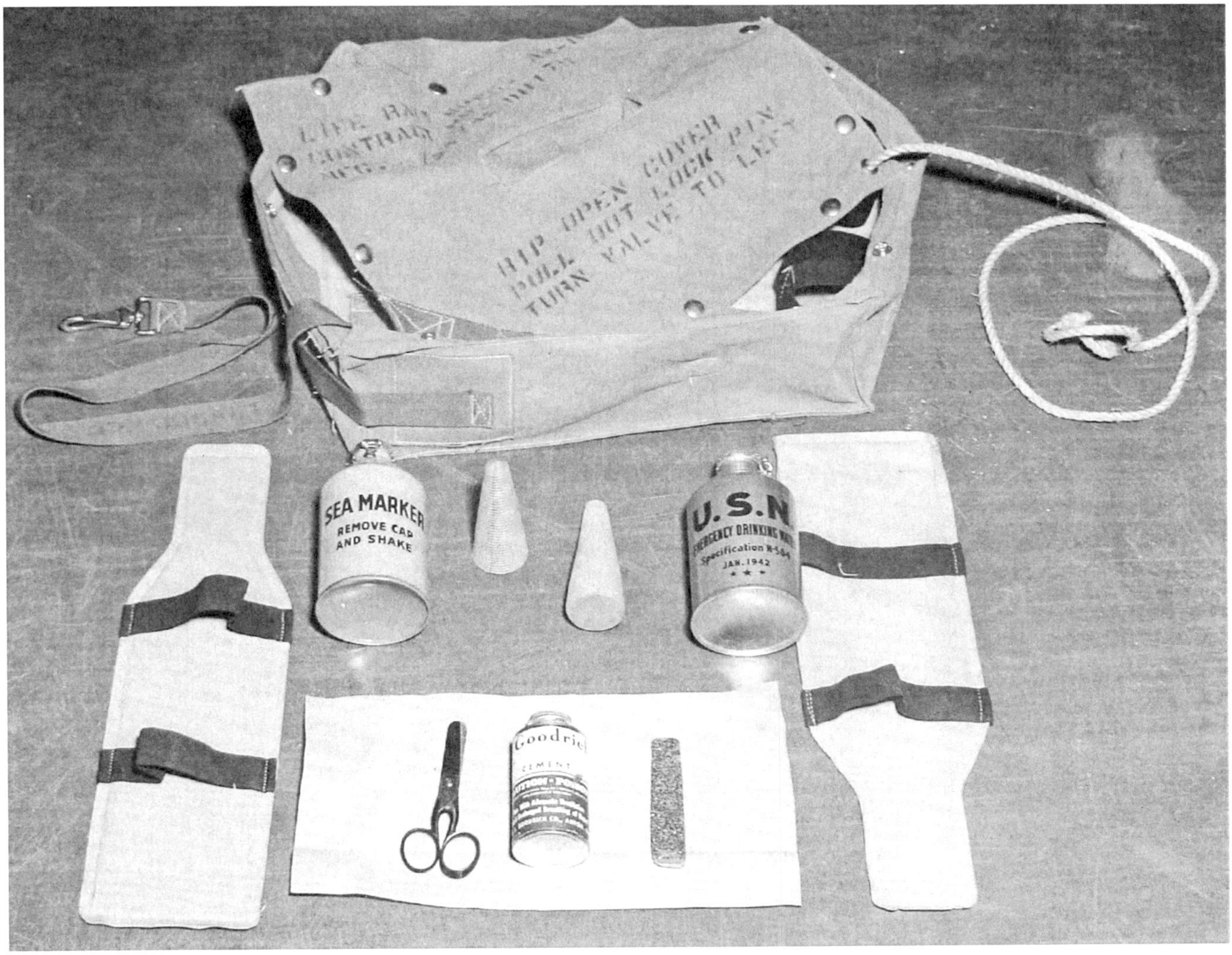


(Above) Accessories integrated into the M-524 assembly. The sea anchor and bailing cup were lashed to the grab handles of the raft and the concertina hand pump was screwed into an inflation inlet valve. Additional accessories included hand paddles, drinking water M-504, sea marker M-528 and a repair kit. The draft of M-524 pre-dated the adoption of the sea marker M-528, being based on the details described in M-524 of a cardboard mock-up 3" x 5". The repair kit included a 12" x 6" length of rubberized material for patching, Type-B rubber cement, pair of blunted scissors and a roughing tool. Everything was wrapped and rolled in the patching material and secured with an elastic band. Two bullet hole repair plugs were also to be provided, either rubber or wood, 4" in length and 1-1/2" wide tapering to ½", conical in shape and capable of floating. (National Archives)

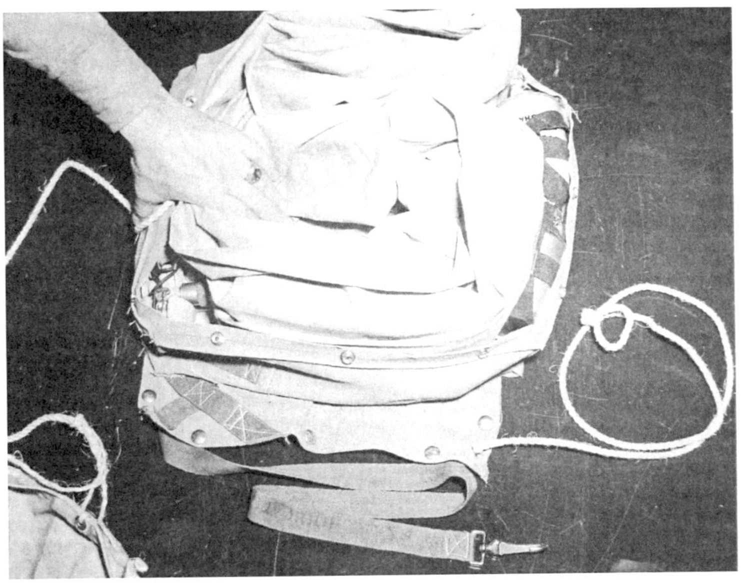

(Left) Packing the raft required it to be folded and coiled around the center slot. Picture shows the top cover removed, even though in normal use it was not completely removable (the sleeve of the slot was sewn to both the bottom and top, only the outer edge detached from the main body of the case). This design feature was later found to be problematic in removing the raft. The case was secured to the user by a lanyard with a snap hook on one corner. On the opposite corner a grommet was provided for securing the raft to the case. (National Archives)

*(Above) Ralph Douglas of the US Rubber Co. demonstrates the new one-man parachute life raft in preparation for testing it in the East River, New York, July 1942. These images were part of a publicity exhibit for US Navy fliers.*

*(Top left) Douglas illustrates the light weight of the unit at only 12 pounds.*

*(Left and below) As part of the demonstration, he conducts the proper procedure in an emergency by first inflating his life vest and then boarding the raft.*

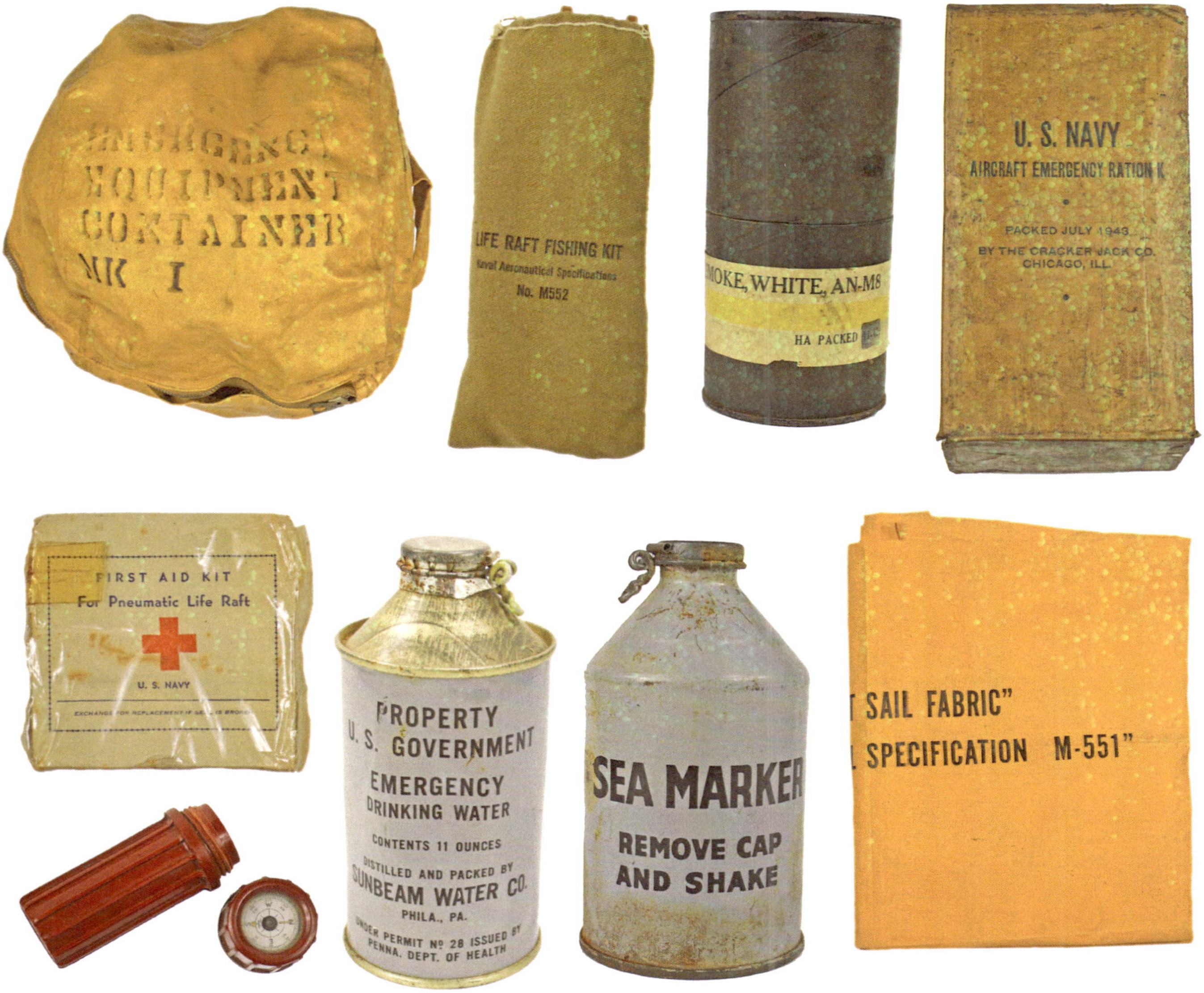


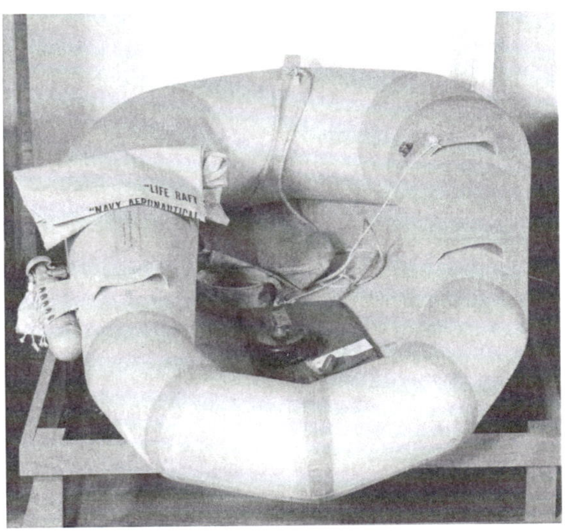

*(Above) An early production M-524/AN-R-2 pararaft in 1943. It retained the older features such as the all-yellow sail M-551, concertina pump and early pattern bailing cup. (National Archives)*

*(Above) The M-524 one-man parachute raft was to replace the standard MK-I life raft in VF class aircraft. When it became available, service personnel would be directed to remove the MK-I but retain its emergency equipment container. Stowage of the container was optional - either in the cockpit, or in the life raft compartment. Shown is a MK-I equipment container, 7" long x 12" in diameter and its accessories as outlined in Technical Order 1-43 for the inspection packing and maintenance of the one-man parachute raft. Not pictured is the smoke grenade clamp. Once the M-524/AN-R-2 entered service, local commands would amend the accessories in the container to suit their preference. The equipment container would be replaced by the emergency back pack kit in the spring, 1943. By late 1942, the parachute raft was in high demand. With the Wildcat and MK-I life raft being combat tested, multiple problems occurred: with the raft stored in the spine of the fuselage it was subject to bullet damage; during a ditching, structural damage distorted the fuselage, often jamming the compartment door making removal of the raft impossible. If the pilot bailed out, he left the life raft behind.*

Upon entering the war, an immediate requirement was established for the issuance and expedited standardization of life rafts for pursuit combat aircraft. This concern was generated from USAAF units preparing for deployment overseas as early as December 1941.

With a new precedence, a meeting was scheduled to discuss the issue by a Technical Sub Committee of the Joint Aircraft Committee, set for 20 January 1942, on the standardization of a joint Army-Navy pararaft. The USAAF cancelled the meeting as they felt no discussion was necessary since they were working to adopt the 42B3447.

As the scope of war became more serious, the vital importance of life saving equipment grew in magnitude, and the meeting was re-scheduled for 11 February 1942, to comprise representatives from the Army, Navy and British. It was the decision of the committee to adopt the M-524 for joint Army-Navy standardization. Tentative Army-Navy specifications were drafted and submitted to the Joint Aircraft Committee and approved 10 March 1942, assigned under joint Army-Navy Specification AN-R-2 and titled Raft, One Man Parachute Life Raft.

With it now being under joint Army-Navy cognizance, this decision would affect BuAer procurement, so authorization would be delayed until formal specifications were approved. Any submission of technical details had to be reviewed and approved by both services. The negotiation process took time to sort the necessary details. Most all materials detailed in M-524 were NavAer specifications, these then had to be converted to Army-Navy specifications and Federal specifications agreed for inclusion. The bulk of the work and input was conducted by the NAF and the joint Army-Navy Working Committee of the Aeronautical Board. Agreements on construction details were eventually agreed upon by the end of April with official approval made effective, 2 May 1942.

The significant detail in AN-R-2 was the deletion of the back-pack case, with the rest of the specification making clerical revisions to include updated Army-Navy and Federal specifications. Upon approval, funds were released quickly, followed with the BuAer awarding contracts for the manufacture for the urgently needed AN-R-2 pararaft.

The USAAF received samples of the M-524, 19 March 1942, for evaluation purposes. From this they expressed little interest in the design features and accessories. For expedition purposes, they agreed on technical features of materials. The Materiel Command began the revision of the AN-R-2 to suit their requirements. They found the case likely to interfere with the parachute harness and the life raft difficult to remove. They also felt the accessories were not properly secured and could be subject to loss.

Before recommendation of procurement, deviations needed to occur. With the approved deviations, the USAAF formally adopted the AN-R-2 pararaft as standard, 4 July 1942, awarding contracts the same month.

*(Right & below) For USAAF procurement, the slot was deleted. The new dimensions were 16" x 12" constructed from 8oz. Army grey duck fabric. The cover was secured by a perimeter of snap fasteners with an orange-yellow grab handle for quick removal and visibility. The case was secured to the standard seat type parachute by four tie tapes located at the corners on the back. Four 16" high tensile strength tapes were provided for alternate securing methods.*

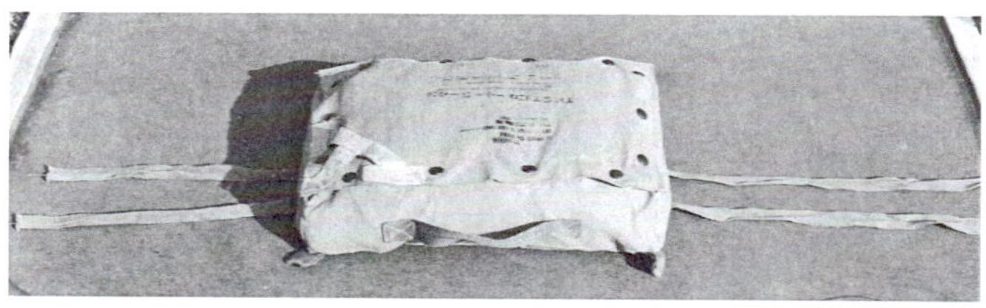

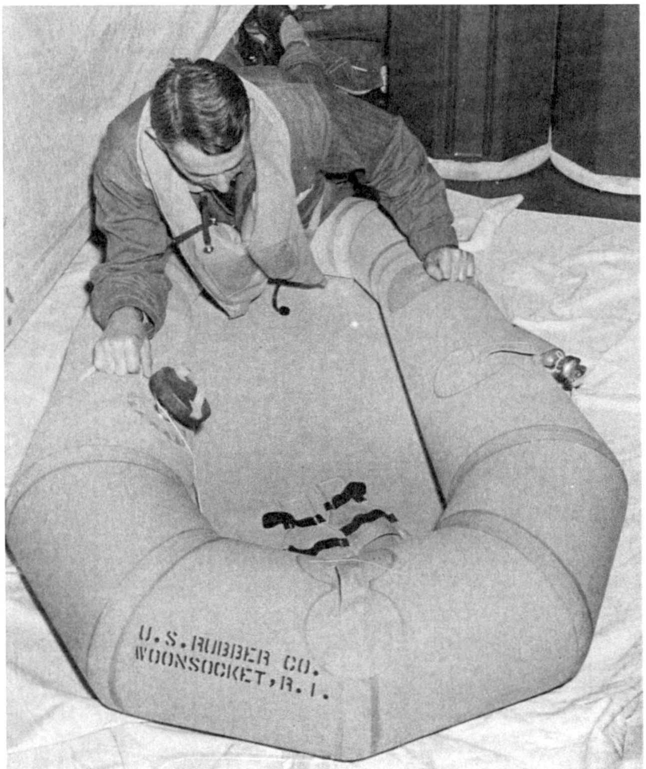


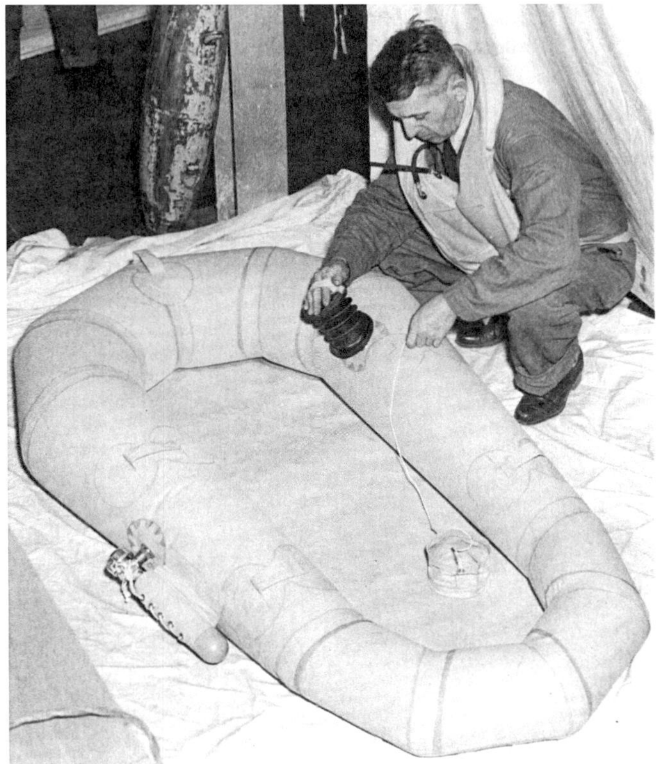

*Images from the USAAF Engineering Division evaluating the US Navy M-524 parachute life raft in April 1942. As part of the process they checked all design aspects of the inflation system, boarding, and the practical use of the complete assembly. Interestingly, these images were used in the first distributed USAAF technical order for the one-man parachute life raft, though this was not the exact design that entered service. (USAF)*

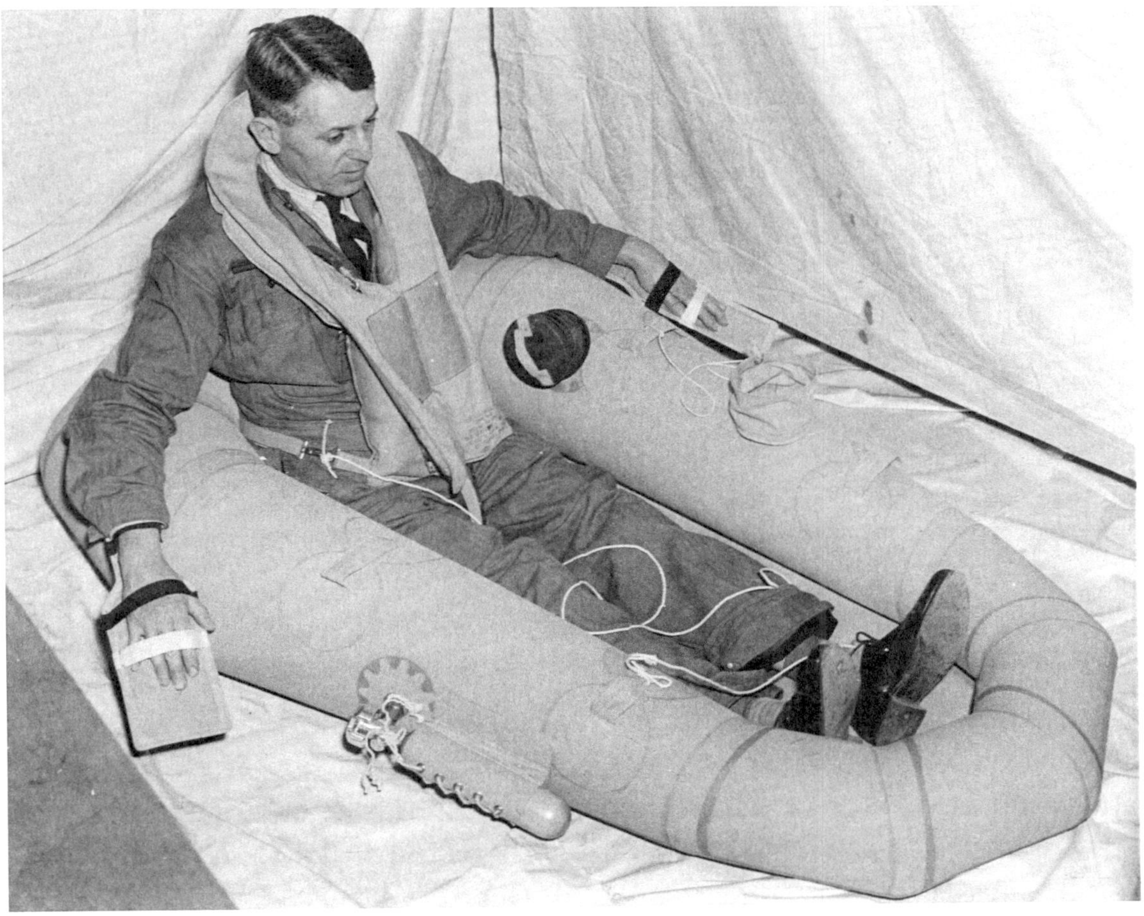

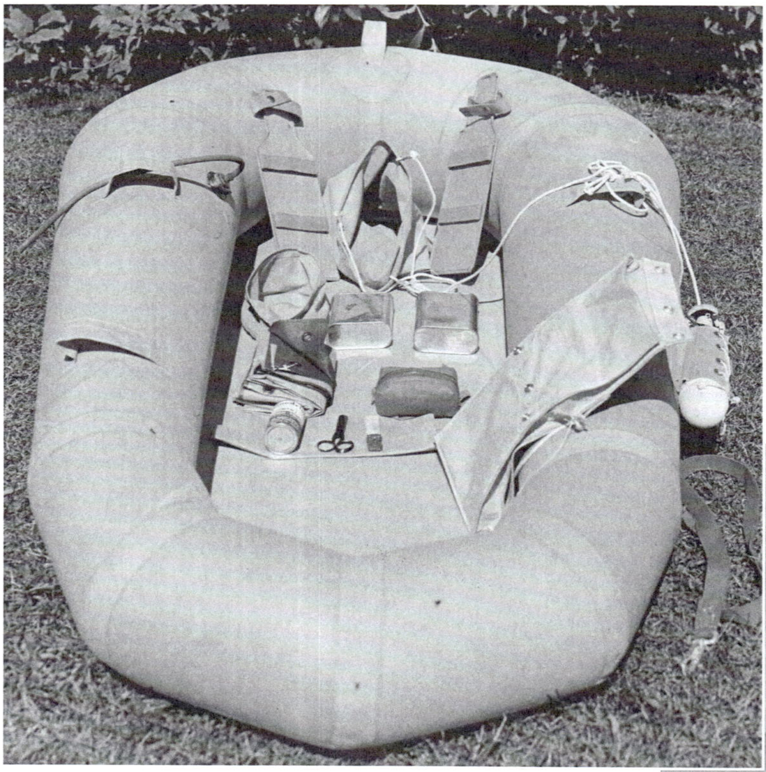

*(Left and below) The initial concept of the one-man parachute raft that was found optimal for USAAF use, August 1942. Changes included integrating a central accessory container with new type components, and reversing the installation of the inflation system. Below is the new type outer case. This design was accepted and placed on procurement, differing from the type being procured by the BuAer at that time. All changes were submitted to the Working Committee for suggested changes to Specification AN-R-2.(USAF)*

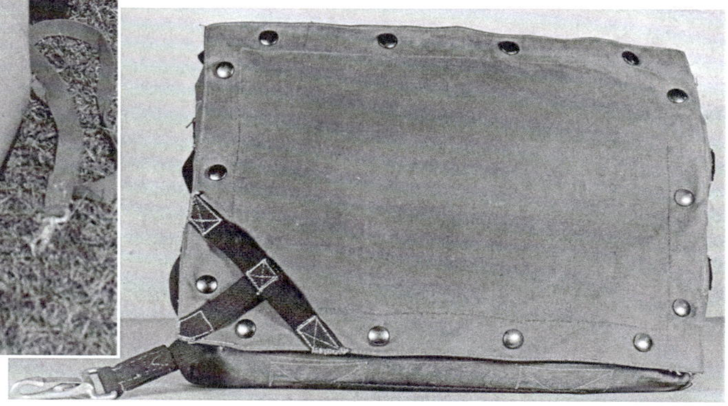

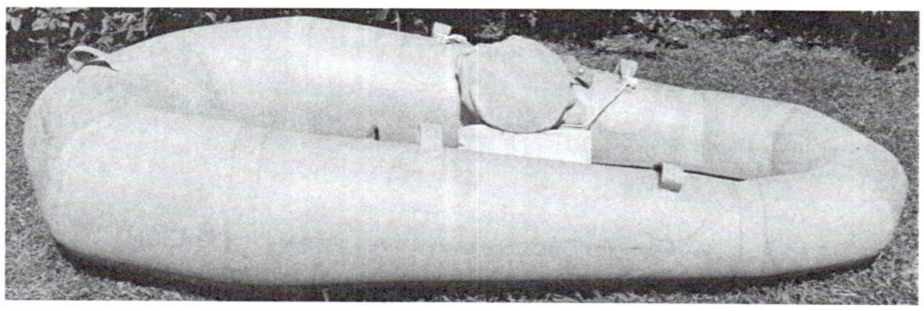

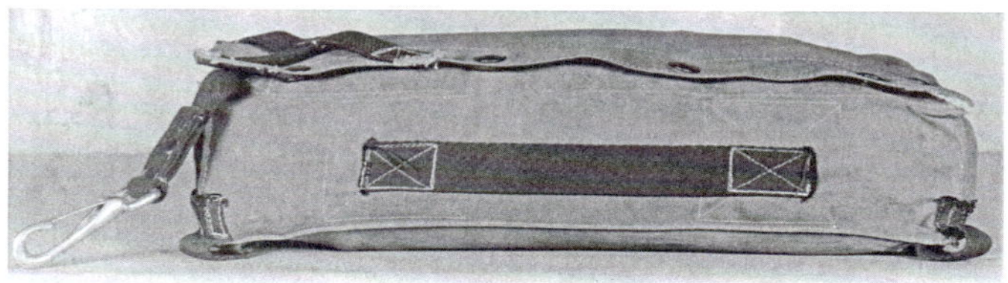

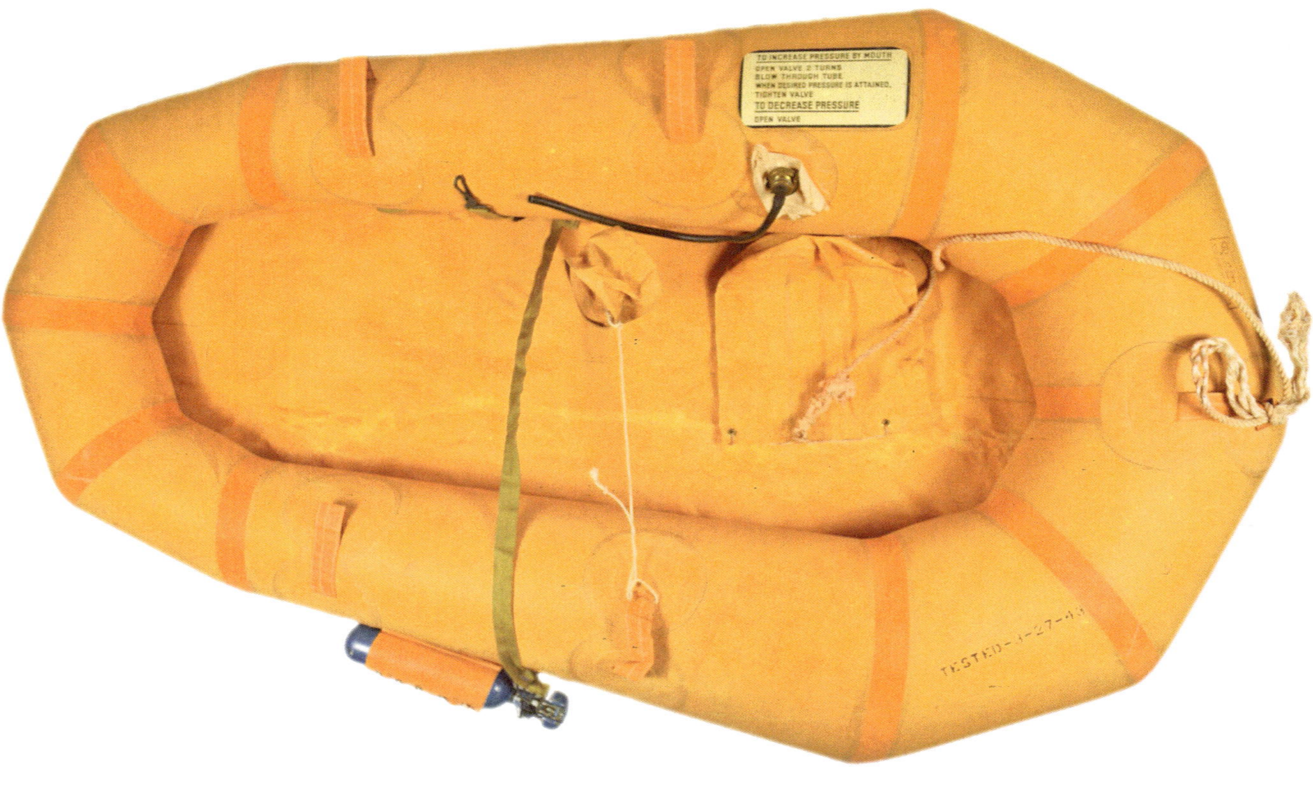

*(Facing page bottom and above) The overall construction of the life raft remained unchanged except the USAAF required the placement of the CO2 cylinder on the opposite side, and the bottom to be light blue, because it was believed that a darker color would be less likely attract sharks. Specifications M-524 and AN-R-2 had specified that all external surfaces be orange-yellow, including the bottom. For the USAAF the life raft would be directly secured to the user by a lanyard attaching the neck of the CO2 bottle to a D ring on the life preserver. The procedure for boarding the raft was over the stern, placing the CO2 and lanyard connection on the user's right hand side. The contract number for this raft indicates it was one of the originals from July 1942, though manufactured after the acceptance of specification AN-R-2A.*

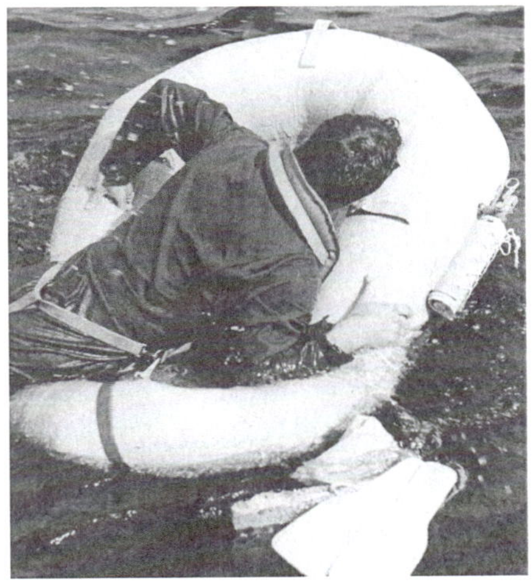

*(Right) Wartime image demonstrating the correct method of boarding. If the CO2 bottle was on the left of the user, they would risk getting entangled.*

*(Below) Accessories were stowed and folded with the raft with no discernible increase in thickness. When the user was released from the parachute harness, the top cover was completely removed. The raft and accessories remained with the user and the case was discarded with the harness. The concertina pump was deleted to reduce bulk when packing, replaced by a 16" oral inflation tube.*

*(Below) On 17 July 1942 the authorization for changing the water and sea marker cans to rectangular was distributed to contractors. In evaluation the USAAF found the round cans to be uncomfortable to sit on for long periods. The new rectangular cans for the AN-W-5A and M-528 would become the standard for both services. Other changes included the addition of 2" rubber patches and a metal roughing tool of commercial manufacture (the aggregate roughing tool was found to be useless after being submerged in water). By this date, the bullet hole plugs for all rafts were to be made from wood instead of rubber. For exposure and signaling purposes, a sail was added, per USAAF Specification No. 94-26756. In August 1942, the addition of the parachute first aid packet was approved. Illustrated below are examples of accessories for the first generation USAAF pararaft. The wartime image shows a complete USAAF AN-R-2 pararaft as procured in 1942 and manufactured into 1943 (note the paddles, which are an experimental wooden type).*

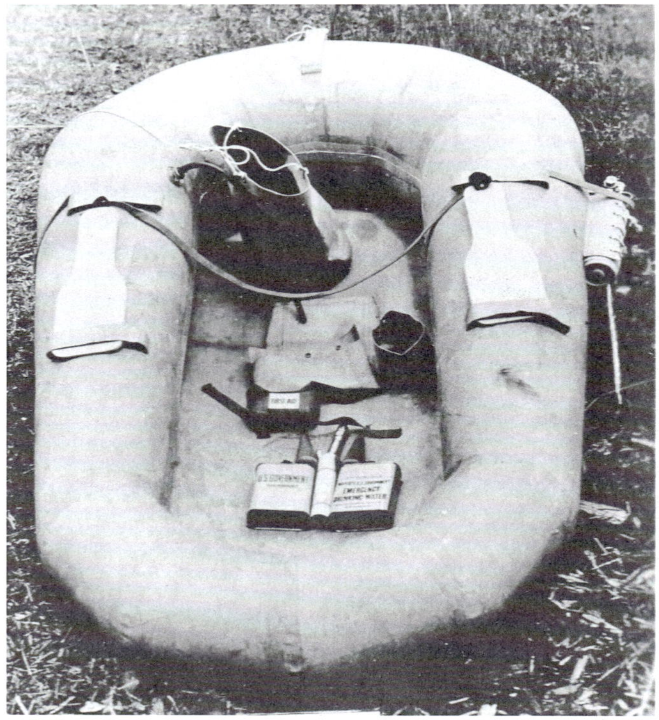

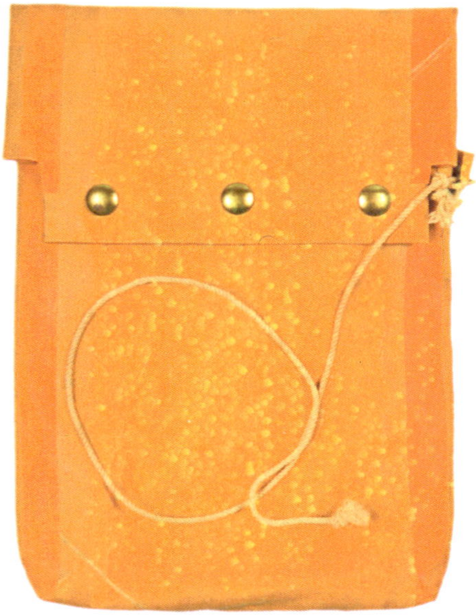

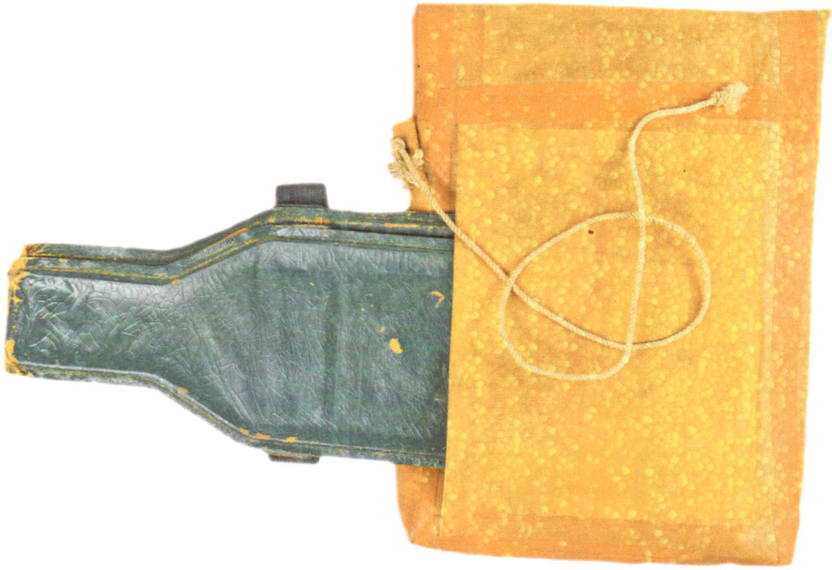

*(Above) Accessories were stowed in a rubberized container, 8" x 11". The hand paddles were tied together and the tips inserted into the sleeve on the back, then lashed to the tab using the grommet on the container. The complete assembly was secured with a short cord to one of the grab handles on the side of the life raft prior to folding. When more accessories were added, such as the S2-1055 first aid kit or signaling mirror, an alternative was to bundle all the accessories in the sail and fold into a tight, secure package.*

*(Below) Wartime color image showing the accessory container and paddles during inspection. (National Archives)*

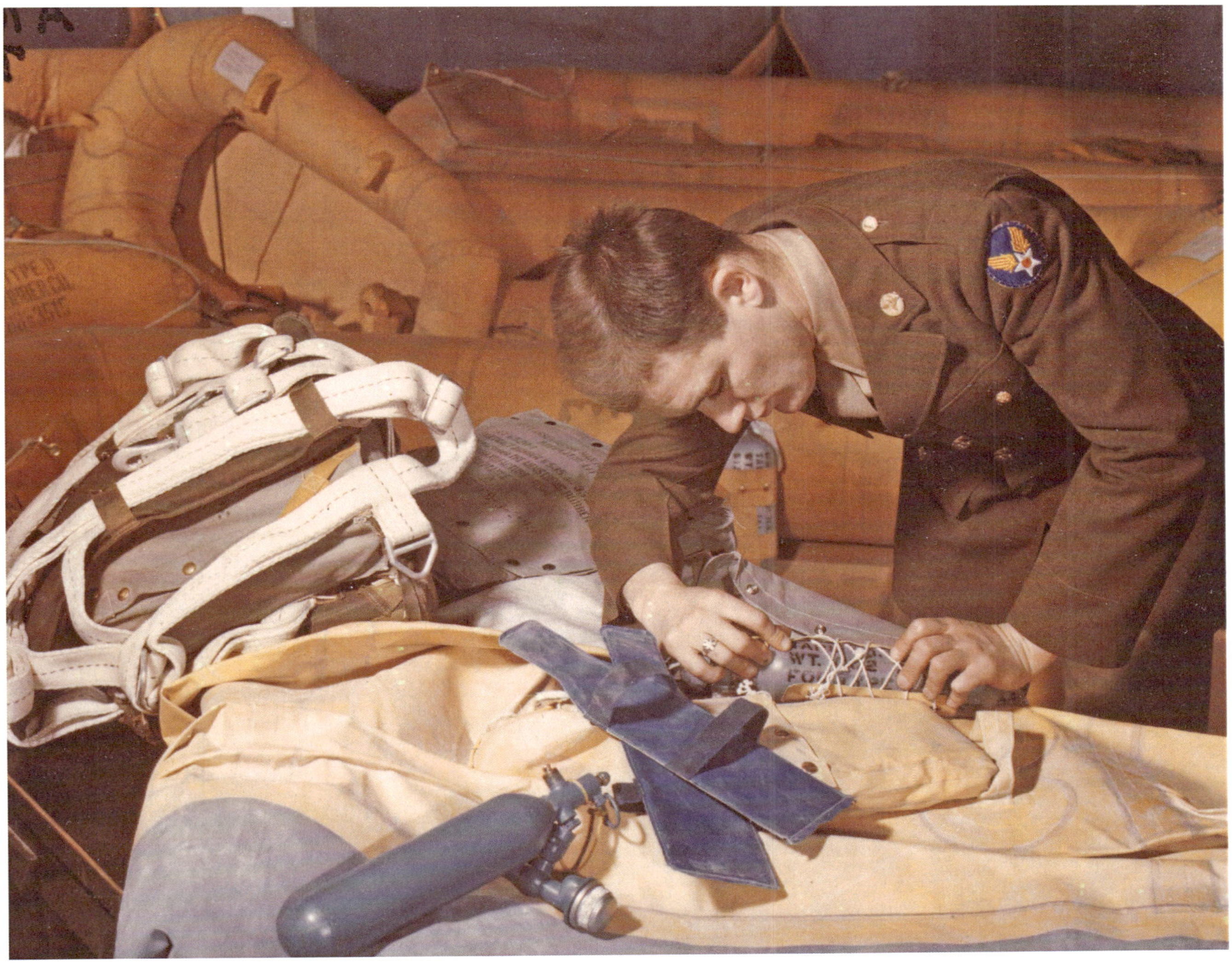

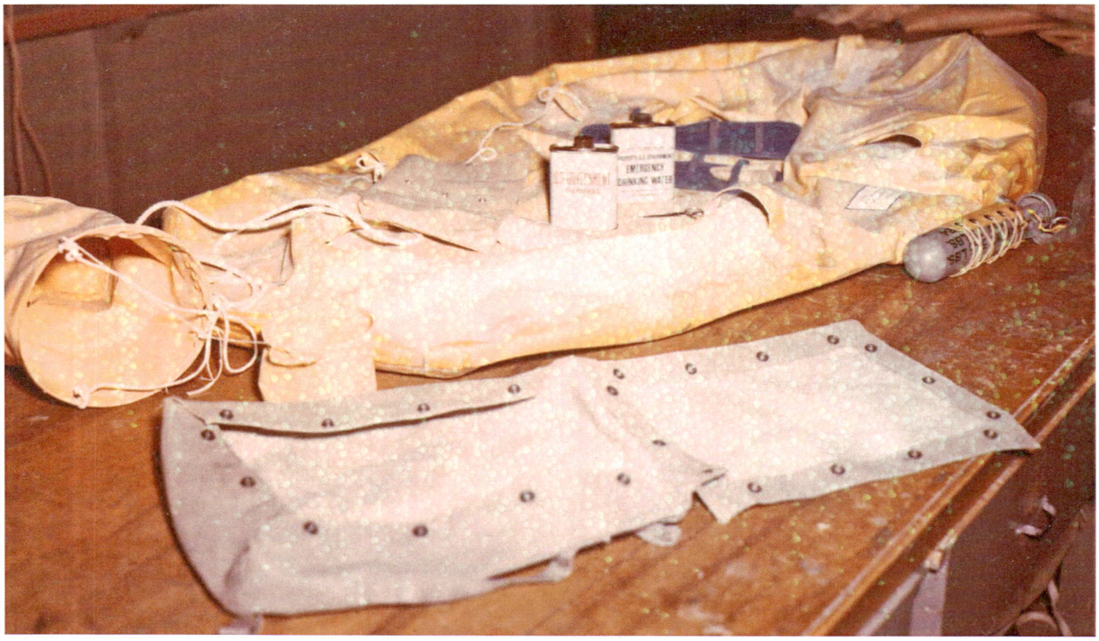

*(Right) Wartime color image showing the bulk of accessories for the first generation USAAF AN-R-2 pararaft. (National Archives)*

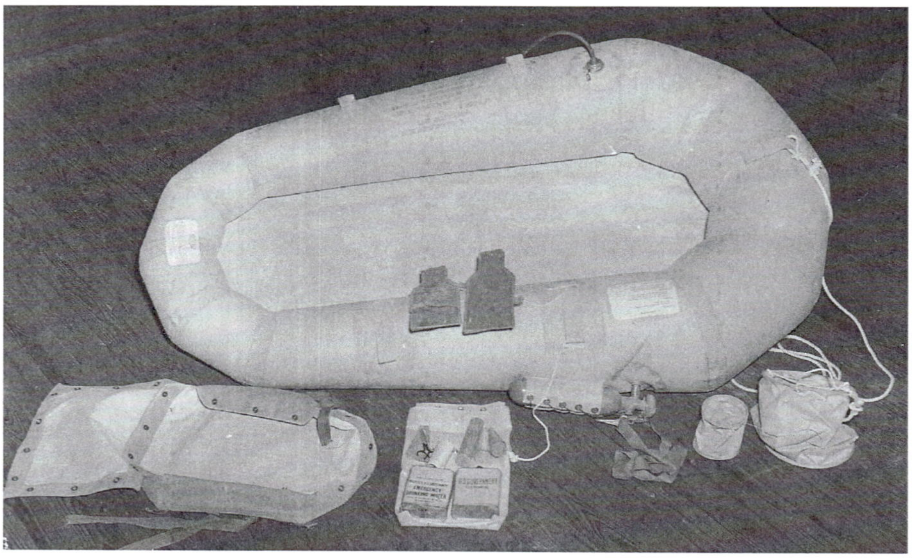

*(Left) This wartime image better illustrates the complete list of changes authorized by the USAAF for production of the AN-R-2. Not shown is the first aid packet. Note the accessories container lashed to the grab handle with appropriate articles to be stowed in it lying on top. (National Archives)*

The BuAer were the first to procure the AN-R-2 pararaft in quantity, awarding three contracts for a total of 20,000 units in May 1942. These were procured for use in VF class aircraft only, with sufficient quantity to supply all airplanes in service and on procurement.

Actual production was very slow, if not non-existent, for several months, as contractors needed to tool and organize their facilities for production. They also had to wait for materials such as the fabric, CO2 systems and accessories from sub-contractors. Production and delivery from any of the three contractors was, on average, only 50 per month until the winter of 1942.

The first quantity deliveries were reported as 180 in August, to west coast stations to outfit aircraft preparing for overseas embarkation. Additional quantity deliveries were not projected until January 1943. With the expected availability of the pararaft, contractors of VF aircraft were ordered to halt production the Type-A and Type-S MK-I life rafts as of November 1942, but to continue providing the equipment container. Once available, the pararafts could be requisitioned utilizing ASO stock number R83-R-15650.

With the approved changes, in 1942 the USAAF Materiel Command released funds for procurement of 15,000 pararafts for use in pursuit aircraft under seven purchase orders to five contractors. They could be requisitioned under stock number 6600-660450.

Only a few hundred had been delivered by the close of 1942, with the bulk going to units embarking to the PTO. At a meeting held in February 1942, the decision was made that the pararaft be reserved for use in aircraft of three persons or less. The BuAer delegated the use strictly for VF class, and it wasn't until November 1943 that the authority to issue pararafts and back pad kits to VTB, VSB and VC squadrons was officially approved.

In USAAF Military Requirement Policy No.33, dated 24 November 1942, the issuance of pararafts for aircraft carrying three persons or less became the official standard.

Distribution of the pararaft differed between the two services. For the BuAer, it was issued to the aircraft as Standard Aircraft Equipment, and when deliveries were made first with VF class it was the responsibility of the local authority to requisition the appropriate quantity for supply, plus 15% for spares. Distribution of the pararaft for the USAAF was only to stations from which aircraft may normally be required to fly over water, being a controlled item and distributed as needed. The procurement of 15,000 units was projected to meet these initial requirements.

In August 1942, the USAAF submitted the list of revisions to the working committee for evaluation and amendment of specification AN-R-2. This list was then distributed to the BuAer for approval with much of the proposed revisions acceptable.

With regard to the case, they found the packing of the raft to be more efficient but not the elimination of the slot. It was found that the non-slotted case did not allow proper adjustment of the leg straps creating a potential safety issue. After considering all proposed revisions, the NAF began the process of improving what had been presented.

In October 1942 the NAF introduced a newly-designed case that incorporated the best features of the center and non-slotted cases under NAF Dwg. No.69741.

*(Right) The new case as detailed under NAF Dwg. No.69741 retained the parachute leg strap slot, but moved it several inches forward so that the raft could now be folded similarly to the USAAF version, allowing ease of removal. It also incorporated four pockets for accessories: one for a first aid kit, one for the repair kit and plugs, one for the water and one for the sea marker. All components were properly secured to prevent loss. The cover on the case was completely removable. A lanyard from a grommet in the corner of case was tethered to the grab handle of the raft, and a lanyard from the opposite corner was secured to the D ring of the life preserver. Shown is the concept model. (National Archives)*

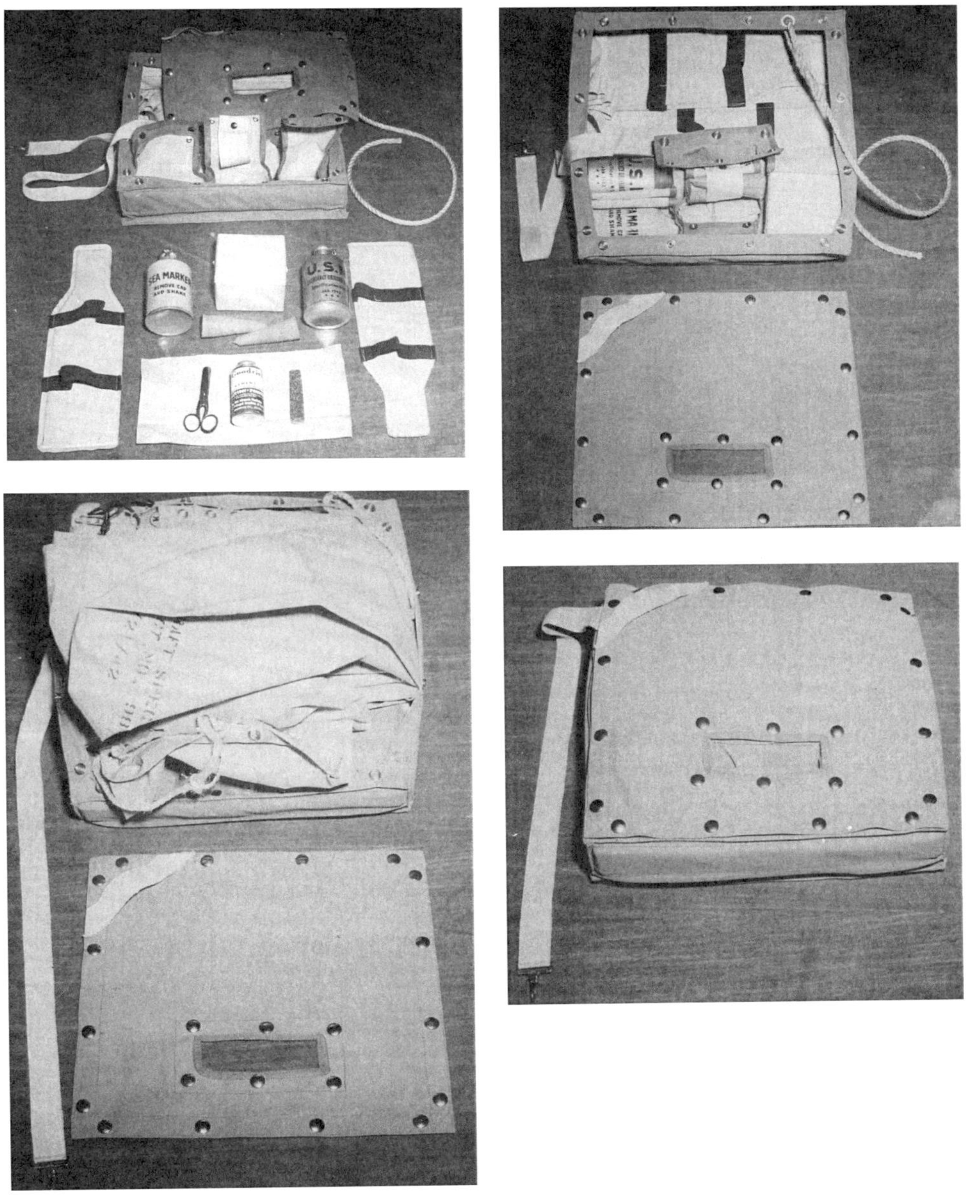

*(Below) The BuAer distributed change orders to contractors in November 1942 to switch over to the new case with forward slot, and for it to be fabricated using an aircraft grey color. It also ordered use of the concertina pump to be discontinued, replacing it with an oral inflation tube. In May 1942, upon acceptance of Army-Navy specifications, three contracts were awarded: one to the Firestone Tire & Rubber Co., contract NXS(A)-5785. Due to confusion and miscommunication, Firestone fabricated their contract per the original NavAer Specification M-524. In the example of the Firestone raft shown, the bottom of the raft is yellow as detailed in M-524 and AN-R-2. The nomenclature stenciled on the raft is as detailed in M-524 – this nomenclature and format were revised in AN-R-2. Like the other two contractors, the bulk of manufacture and deliveries did not occur until 1943 with orders delayed from May through December 1942.*

*(Bottom) Cases supplied by Firestone on this contract, conforming to NAF Dwg. No. 69741. For production, the two separate pockets for the water and sea marker were integrated into one, to accommodate the projected standardization of the new rectangular cans proposed by the USAAF. As prescribed, the new cases were to be fabricated using the aircraft grey. The Astrup Co., a sub-contractor for Firestone, initiated fabrication of the cases in error, utilizing orange-yellow material, which was later corrected after they were properly informed of the technical details.*

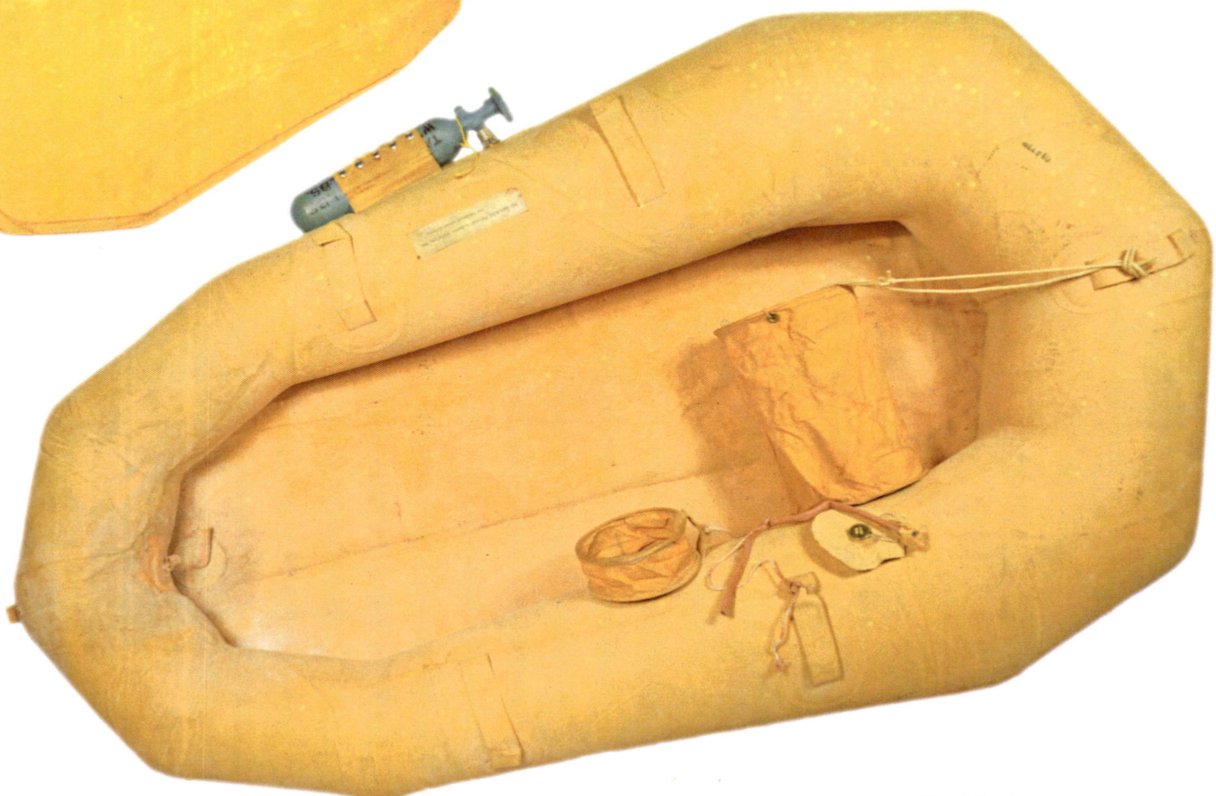

*(Paul Faulkner Collection)*

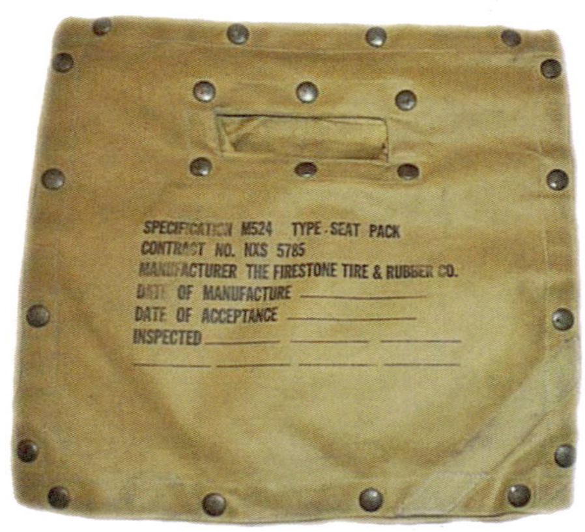

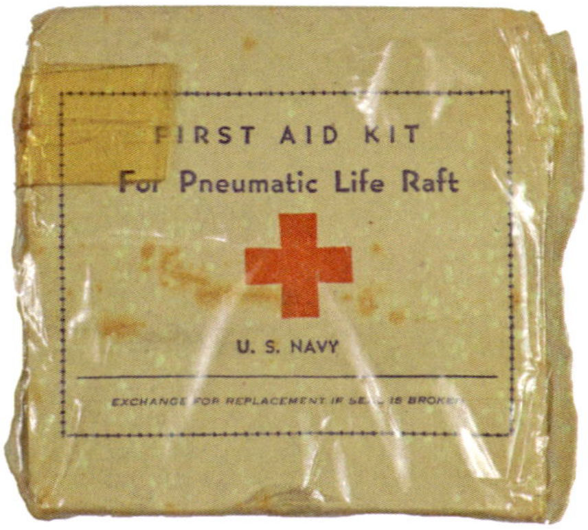


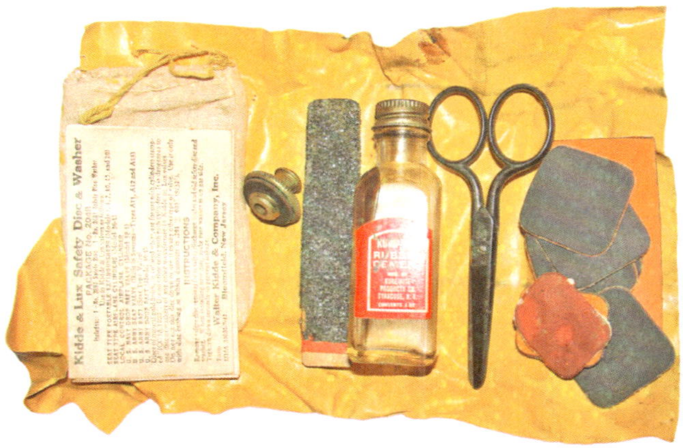

*(Above and right) Components supplied to early BuAer production pararafts. Adopting concepts from the USAAF, first aid provisions were included along with the new AN-W-5A round water can, replacing the M-504. The repair kit would also have included early generation, solid core bullet hole plugs and the addition of rubber repair patches of commercial availability, conforming to AN-R-2A.*

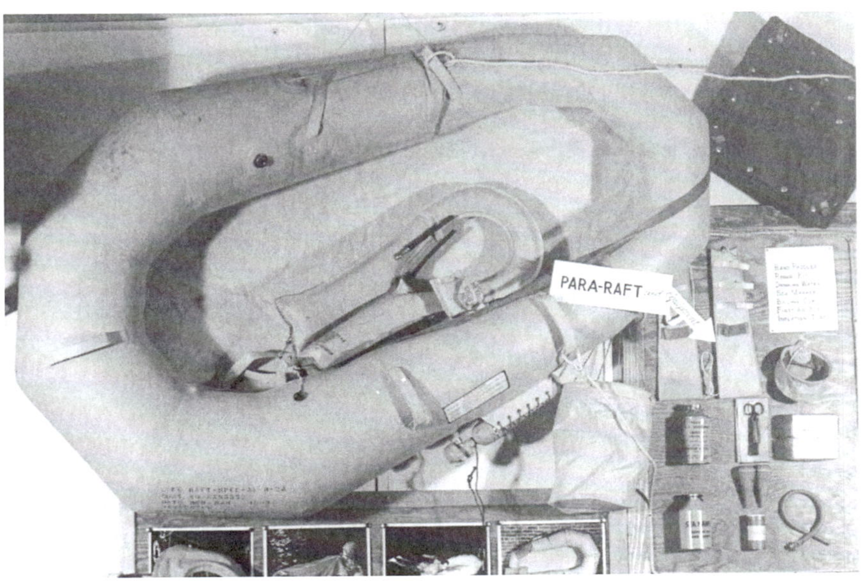


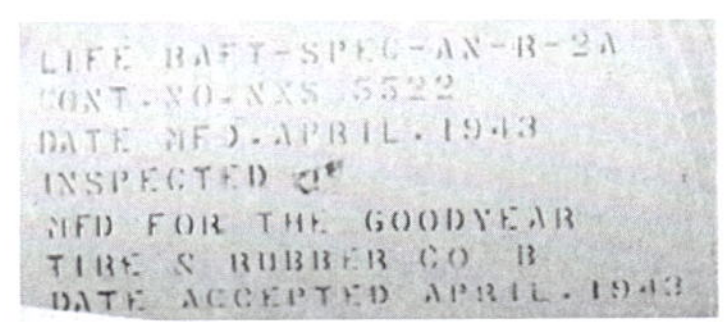

*(Above) Barr pararaft on an equipment display board at NAS Norfolk. Under sub-contract from the Goodyear Tire & Rubber Co. the Barr Rubber Products Co. manufactured AN-R-2 pararafts under contract NXS(A)-5522, awarded in May 1942. Barr had the worst struggle with delays, so it is unlikely that they made any deliveries in 1942, delivering the bulk of their product in 1943. These wartime images illustrating their life rafts reflect the manufacture after Amendment-A was made effective. (NNAM)*

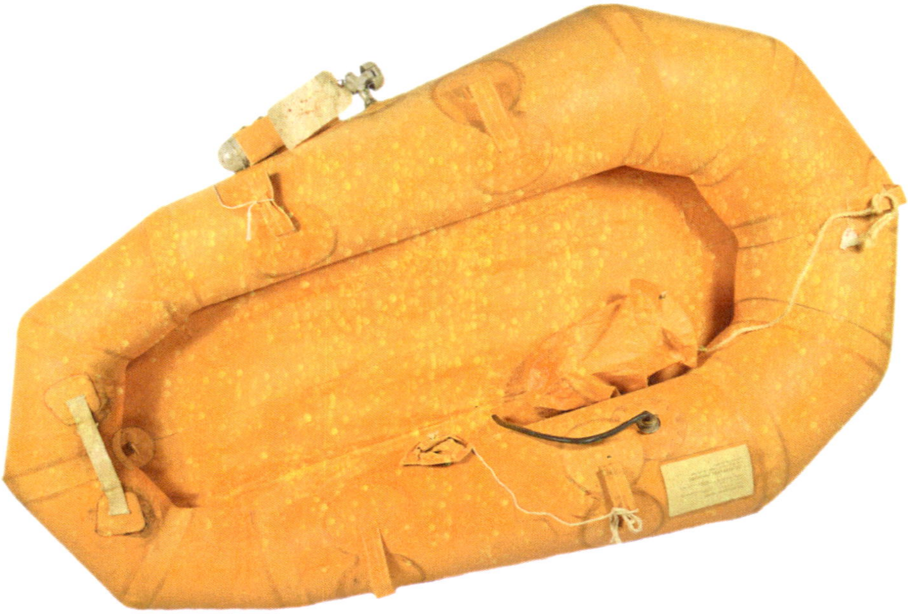

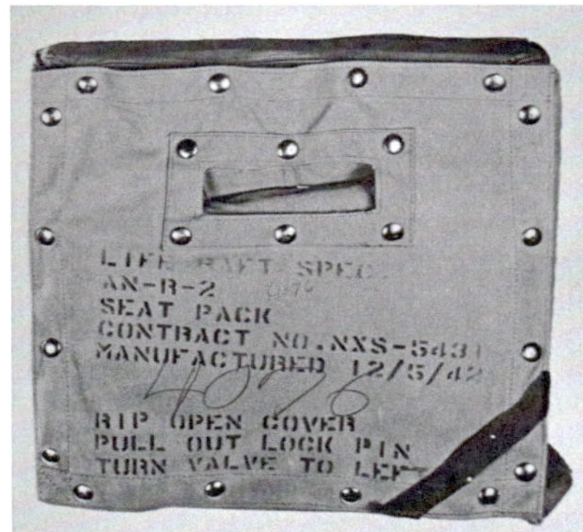


*(Above) The US Rubber Co. was the third contractor for the initial contracts, awarded in May 1942 under contract NXS(A)-5431. Being directly involved with the development of the pararaft from its inception, it would comply to all the changes through 1942. The example shown was manufactured in December 1942 and complies to the stenciling as outlined in specification AN-R-2. The raft retains the yellow bottom, as per specification, but updated with the oral inflation tube as per change order from November. It had been updated as per Technical Order 132-44 of November 1944 for the installation of the radar reflector MX-137/A on the stern.*

*(Below) The forward slotted case supplied on this contract. Still attached to the case is an inspection tag. The wartime image illustrates the components of an early production BuAer AN-R-2 pararaft.*

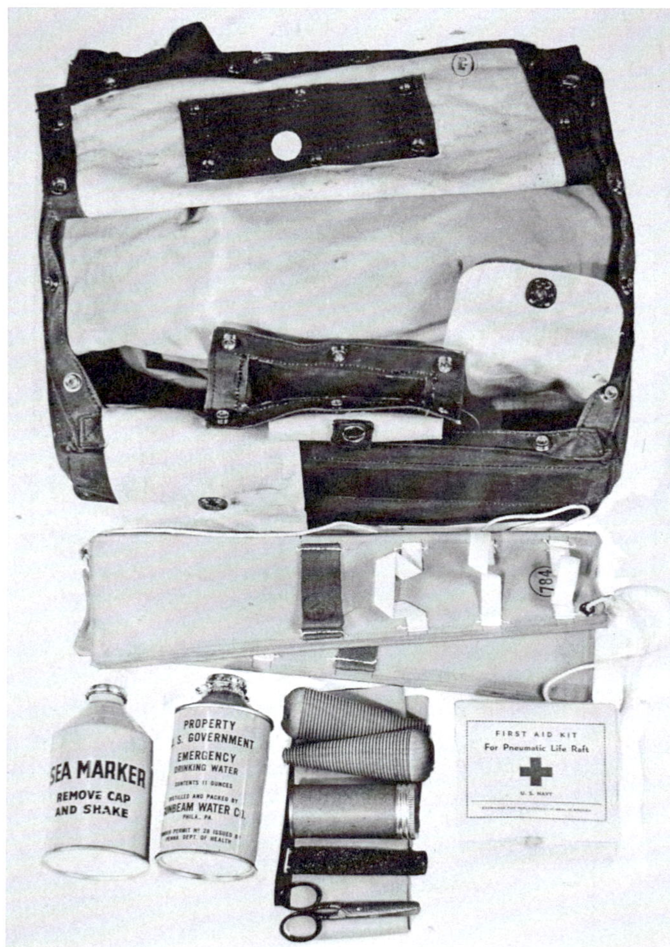

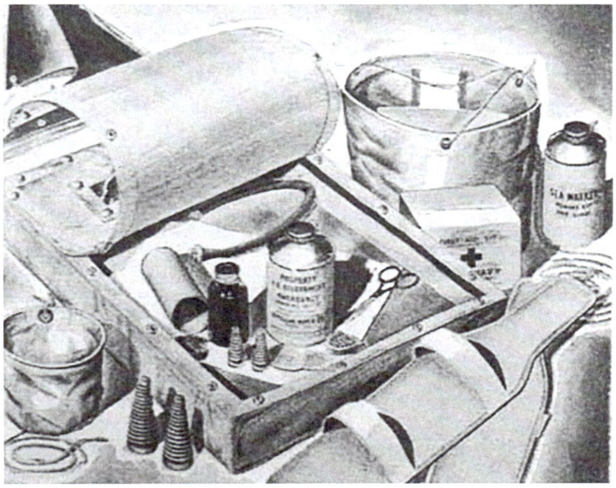

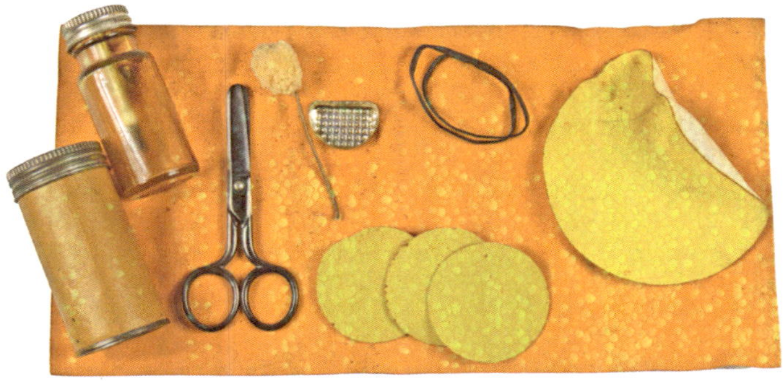

*(Right) Additional examples of components utilized in BuAer AN-R-2/A contracted rafts.*

*(Above) Detail from an advertisement for the US Rubber Co. c.-1944. Note the cone top sea marker can.*

The production of pararafts through 1942 could best be described as chaotic, with confusion between the services and contractors compounding the issue with production deviations.

By November 1942, contractors were receiving materials for bulk assembly, increasing the production. The most significant setback occurred when a large shipment of CO2 valves from Walter Kidde & Co. was found to be defective. The pararafts sat at the factory awaiting replacement valves, and in the interim, the BuAer distributed their change order to the new forward-slotted case. What should have been deliveries of center-slotted cases became NAF Dwg. No.69741 in early 1943.

With the new parachute type life raft standardized, it was decided to halt the manufacture of life rafts with a capacity of 1,000 pounds or less, and for them to be removed from specifications.

The USAAF authorized the deletion of the Type-B series of 500-pounds on 24 September 1942, with the BuAer following suit on 11 October, deleting the MK-1 from specification. This directive simplified issuance to bombardment and transport aircraft for the USAAF, with the BuAer retaining the MK-II, MK-IV and MK-VII life rafts.

By January 1943, there was a great deal of redundancy in equipment for life rafts between the services, and the joint Army-Navy Working Committee began the process to consolidate materials in Army-Navy form. Contractors expressed their concern regarding the inefficiency of having separate production lines for assembly of the pararaft for each service. To alleviate the issue, terms were agreed between the two proposed revisions for consolidation of manufacture being approved under Amendment-A to specification AN-R-2, effective 25 January 1943.

The new assembly would be referenced under Army-Navy Dwg. AN-6520. Prior to this amendment life rafts and cases were to be legibly marked with what is referenced as the Procurement Specification by which it is procured. Under Amendment-A it was directed that the part number, AN-6520-1, be used for procurement, and that the raft and case be legibly marked AN-6520-1. There were some intermediate variables in production. Procurements awarded in 1942 progressed with the changes, while assemblies manufactured after January 1943 reflected that they complied with Amendment-A.

Some early production contracts awarded after January still retained the old method reflecting the procurement specification. In time, all production would conform to the approved changes. Contracts awarded in 1942 were allowed to continue until completion.

*(Below) An existing example of a 1943 BuAer contracted pararaft conforming to standardized details as per specification AN-R-2A. As per the new approved guidelines, rafts were to include a light blue bottom, oral inflation tube and the legible markings of the part number. As per assembly drawing AN-6520, the NAF case Dwg. No.69741 was adopted, and alternately as per specification it too was to be legibly marked with AN-6520-1 (as illustrated). Included in AN-R-2A was the addition of tire patches and metal roughing tool, two sets nested bullet hole plugs, the rectangular water AN-W-5A and sea maker AN-S-10 cans, sail and first aid kit of a type specifically approved by the procuring agency. The bulk of the new approved details derived from the USAAF approved changes, with the case being the contribution of BuAer. The BuAer awarded their first contracts in February 1943.*

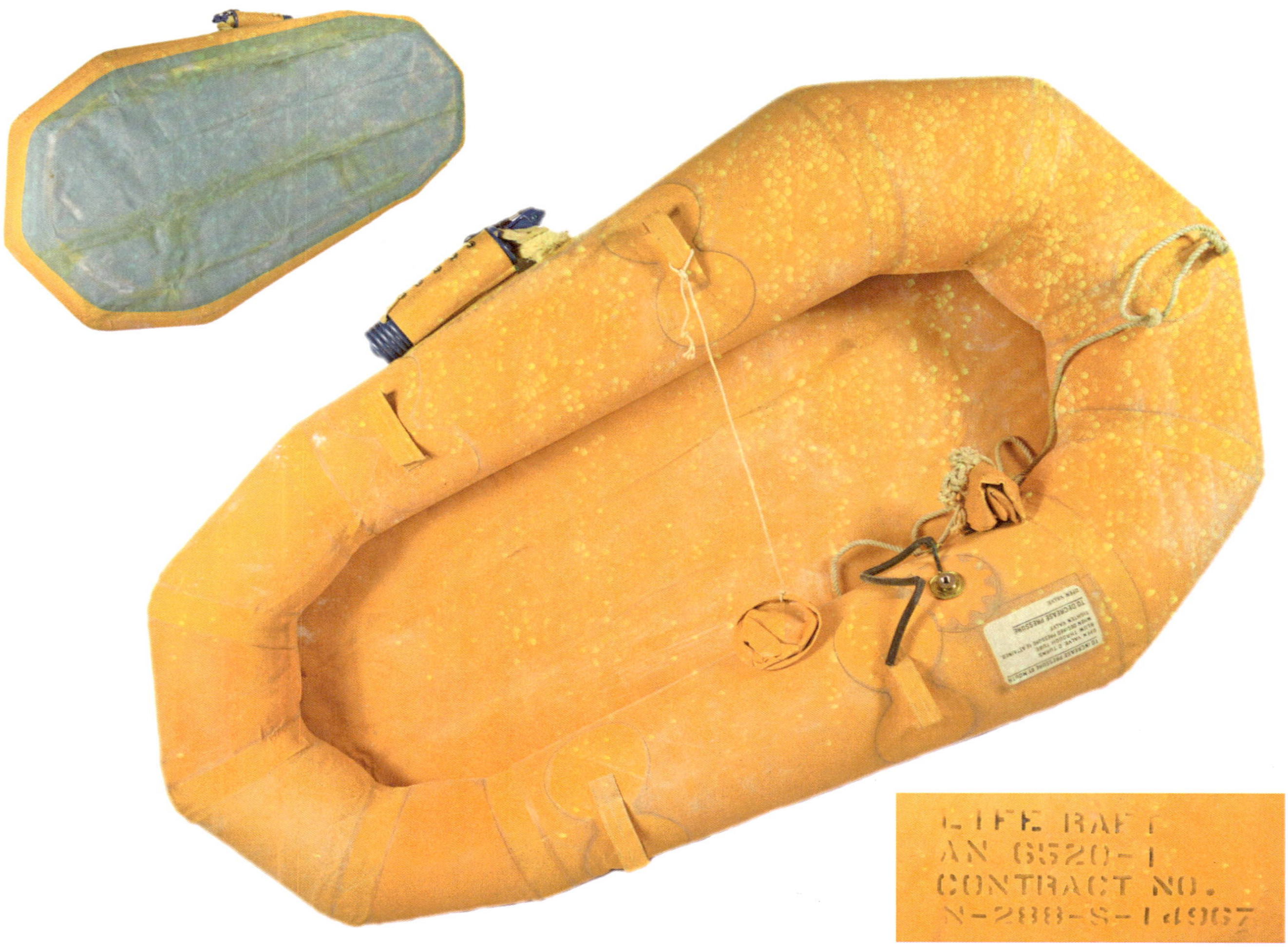

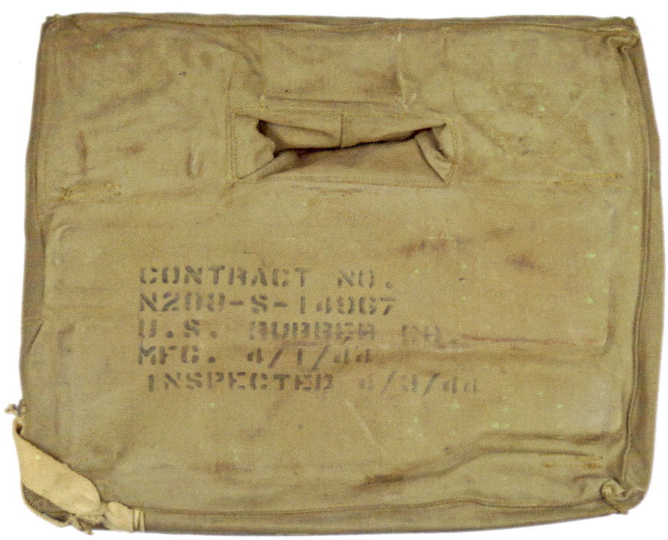

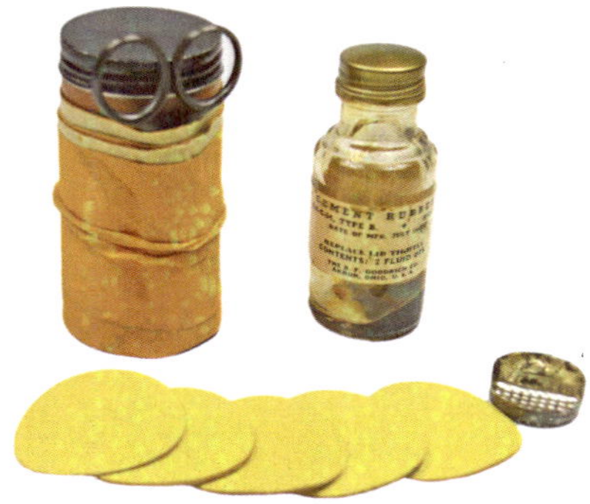

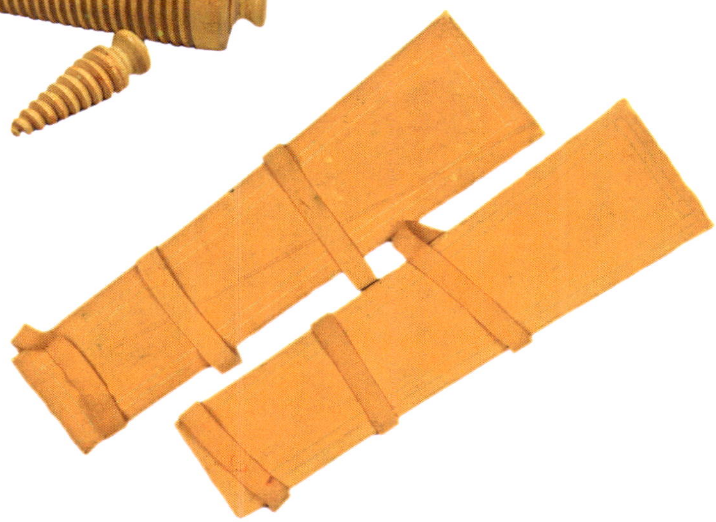

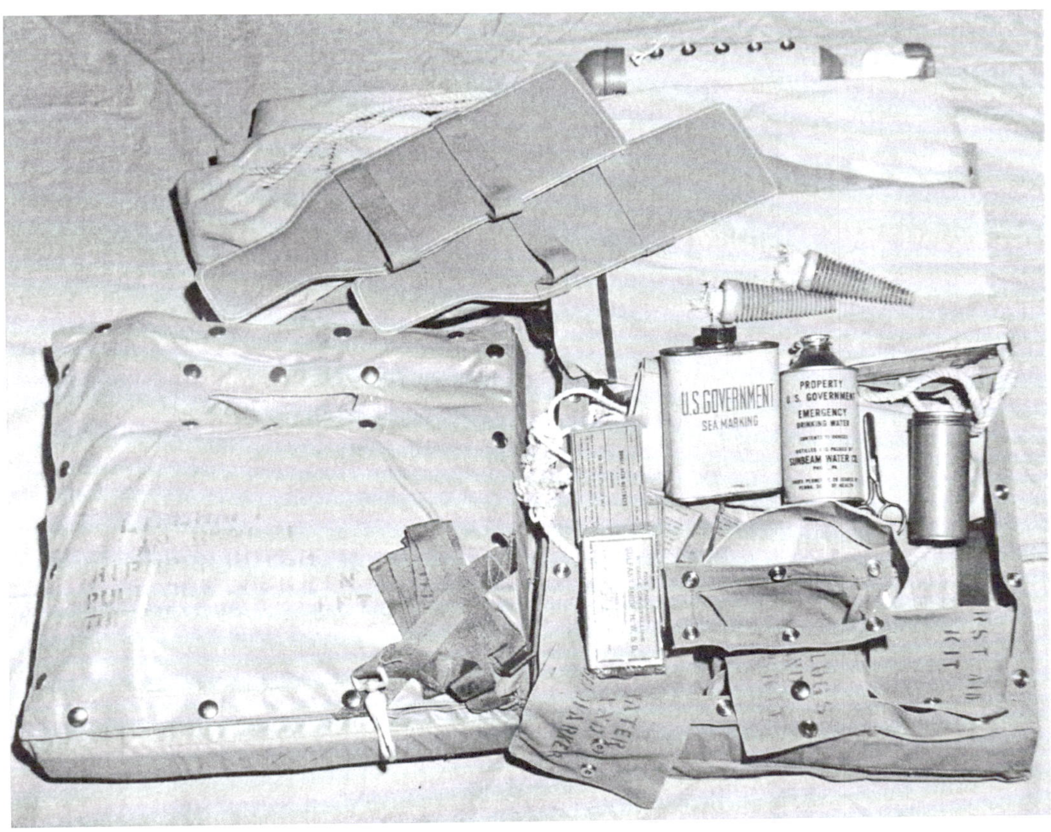


*(Above) Components supplied with BuAer pararafts as per specification AN-R-2A. This is more representative of a late 1943 packing with the new S2-1055 6 unit first aid kit.*

*(Left) Wartime image of two AN-6520-1 BuAer pararafts at NAS Lakehurst, October 1944. The opened unit exhibits many of its components and the round AN-W-5A water can remains in service. The nested bullet plugs are on top of the sail. (National Archives)*

*(Below) Complete assembly for both the BuAer and USAAF AN-R-2A pararafts as procured in 1943. Note that the raft is a USAAF AN-R-2, having the inflation systems reversed. The stenciling on the case reflects it to be a Navy contracted item. Production of the No.69741 case came later for the USAAF, here they utilized a Navy contracted AN-R-2A for testing the integration of their pararaft. All the components are conducive to both services. (National Archives)*

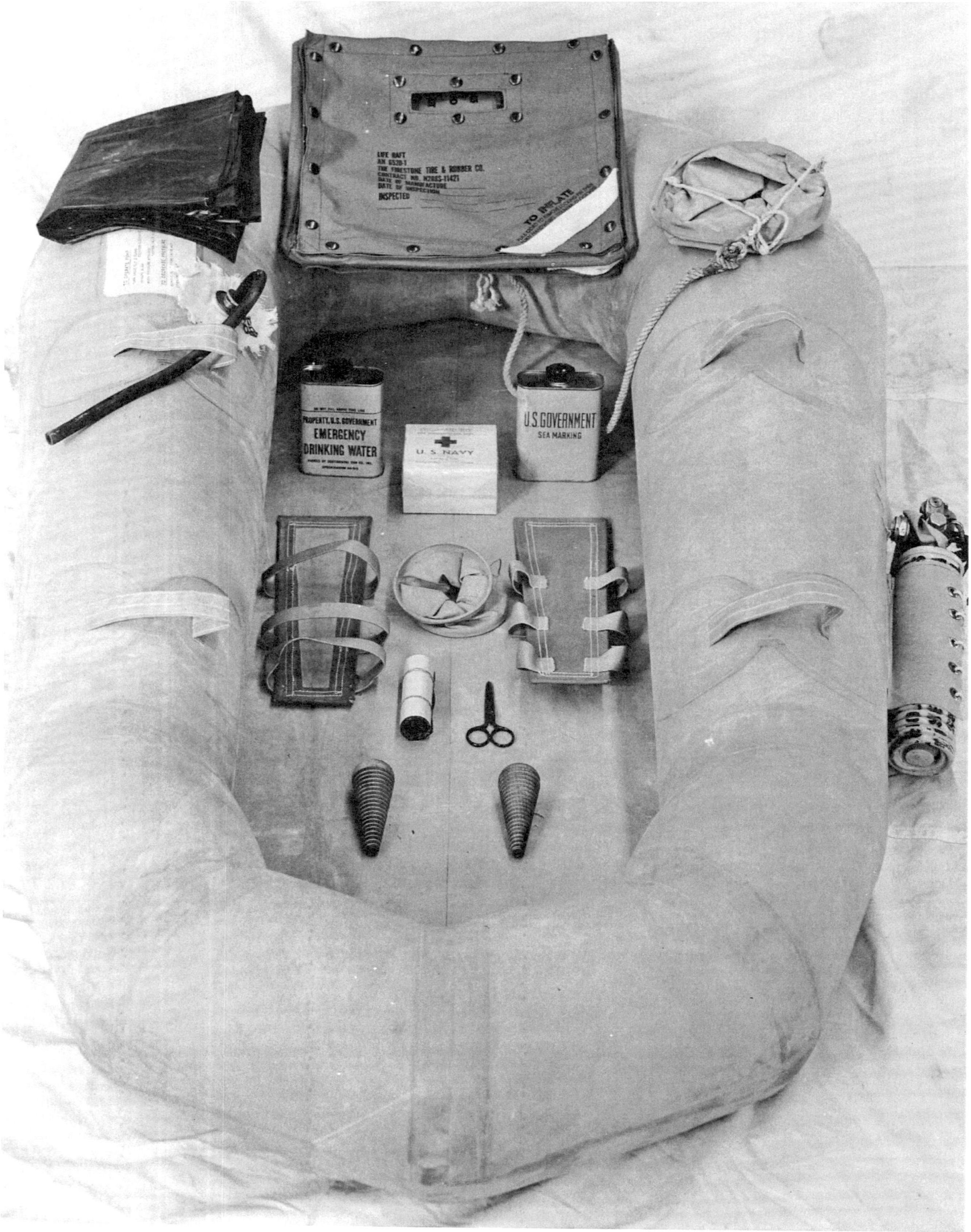

*(Below) For the USAAF, contracts awarded after January 1943 would conform to specification AN-R-2A. By this time, the rafts were indistinguishable between the services, aside from contract markings, increasing manufacturing efficiency. As of August 1943, it was reported that production was at 3,000 per month with a projected increase to 7-10,000 per month by the end of the year. Non-slotted cases continued to be delivered into 1943, until completion of 1942 awarded contracts. The USAAF drafted their first contracts in March 1943 under specification AN-R-2A. With significant outstanding deliveries due on the 15,000 procured in 1942, full procurement of NAF Dwg. No. 69471 cases was withheld pending an investigation in converting pararafts for quick attachment, as discussed in a meeting held in early April. After successful field tests, a method was adopted and applied to all future production of the No. 69471 cases for the USAAF.*

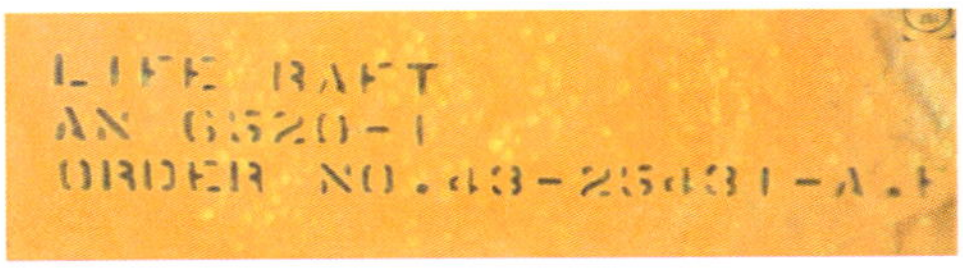


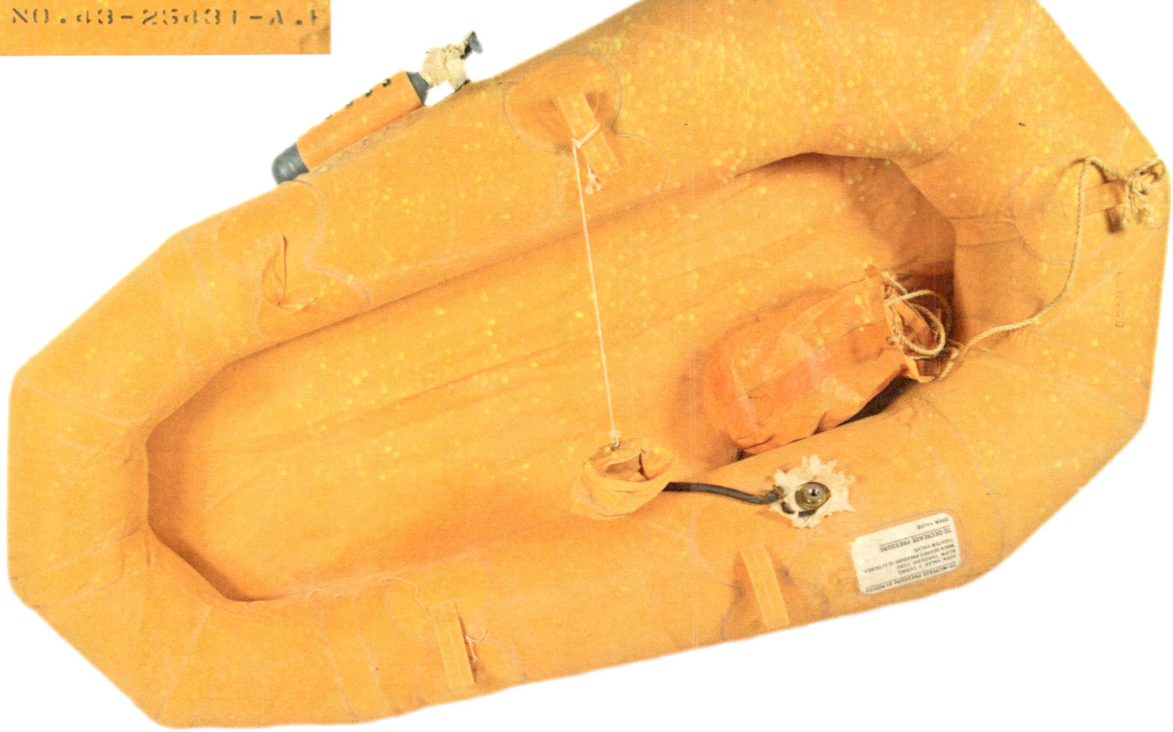

*(Below) An early example of the application of the quick attachment for use with QAC and back type parachutes. Also established at the meeting in April was the need, expressed by the 8th Bomber Command, to provide pararafts for all bombardment crews. This extended the requirements significantly. From the March 1943 approved contracts, Materiel Command had on order 40,000 pararafts to help meet the new projected requirements.*

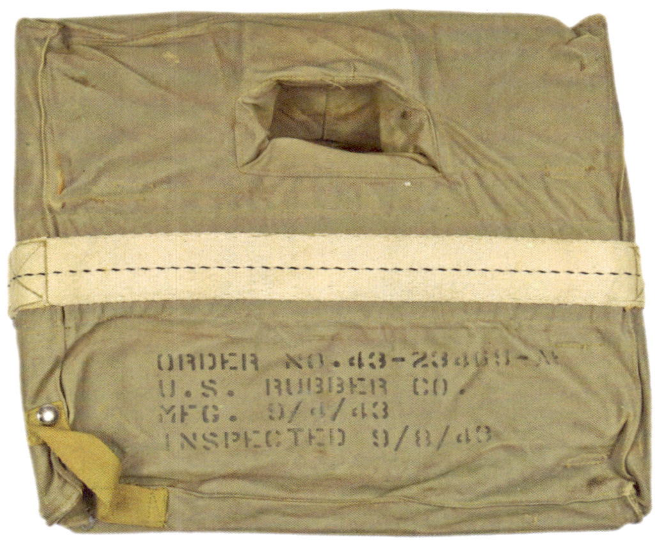

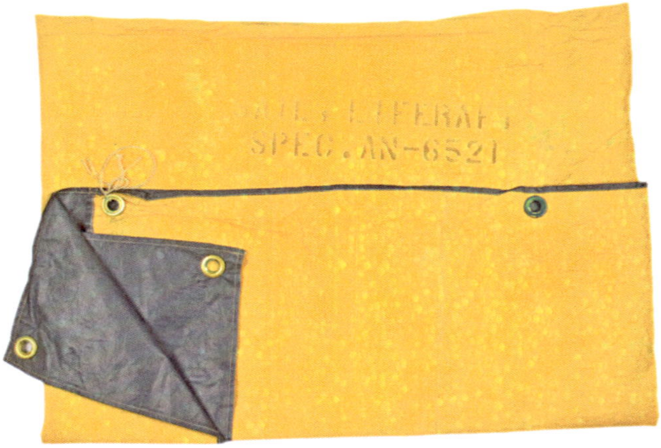

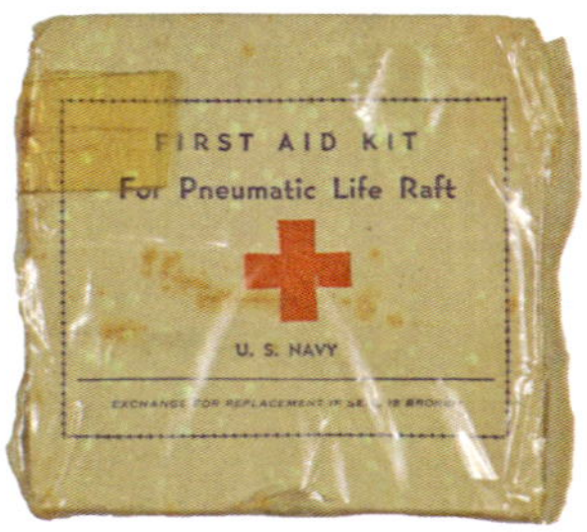

*(Left and above) Components for USAAF AN-6520-1 contracted pararafts. The USAAF did tentatively adopt the S2-1055 first aid kit, and requisitioned large quantities for inclusion in all life rafts including one-man types.*

*(Below) AN-R-2A pararaft on exhibit of the 351st BS 8th AF in England, December 1943. Just visible are the parachute first aid packet and the British MK-II hand held distress signals. (National Archives)*

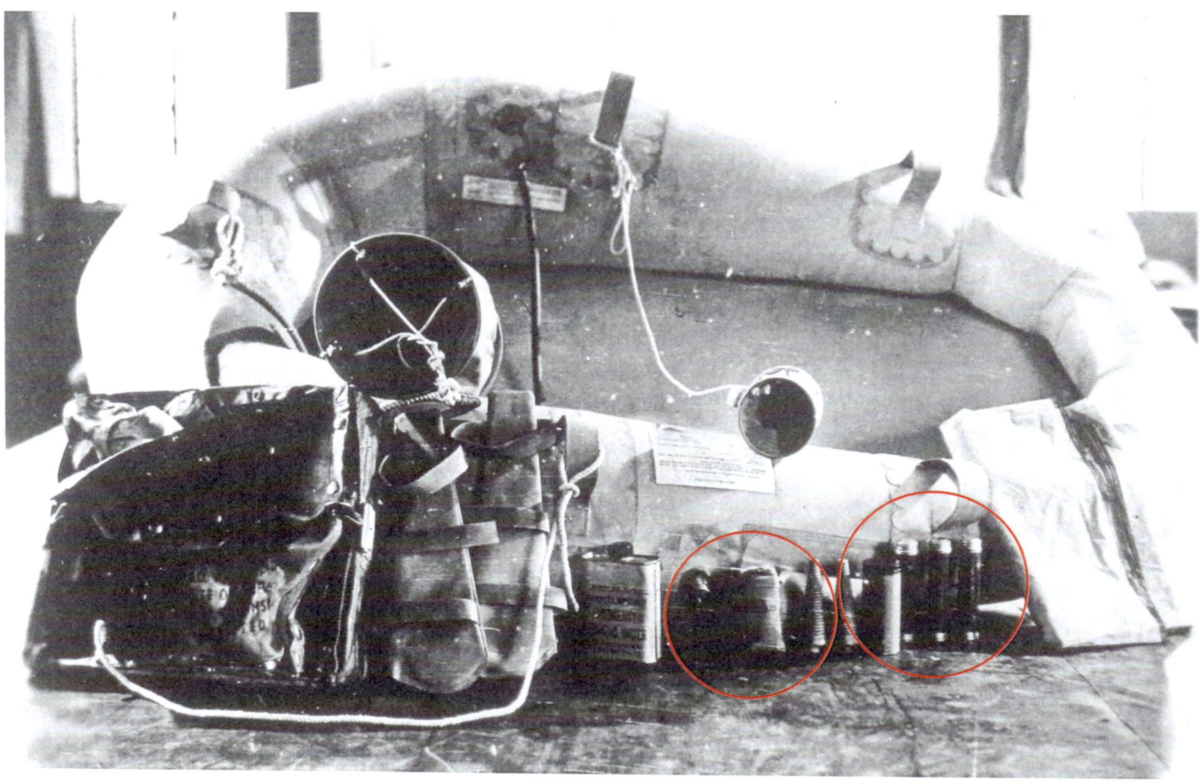

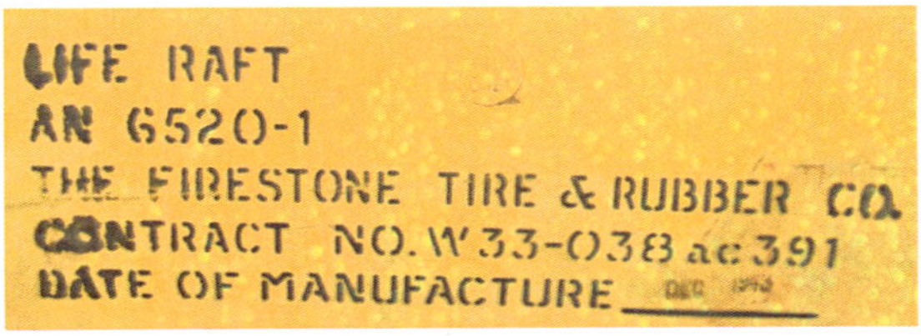


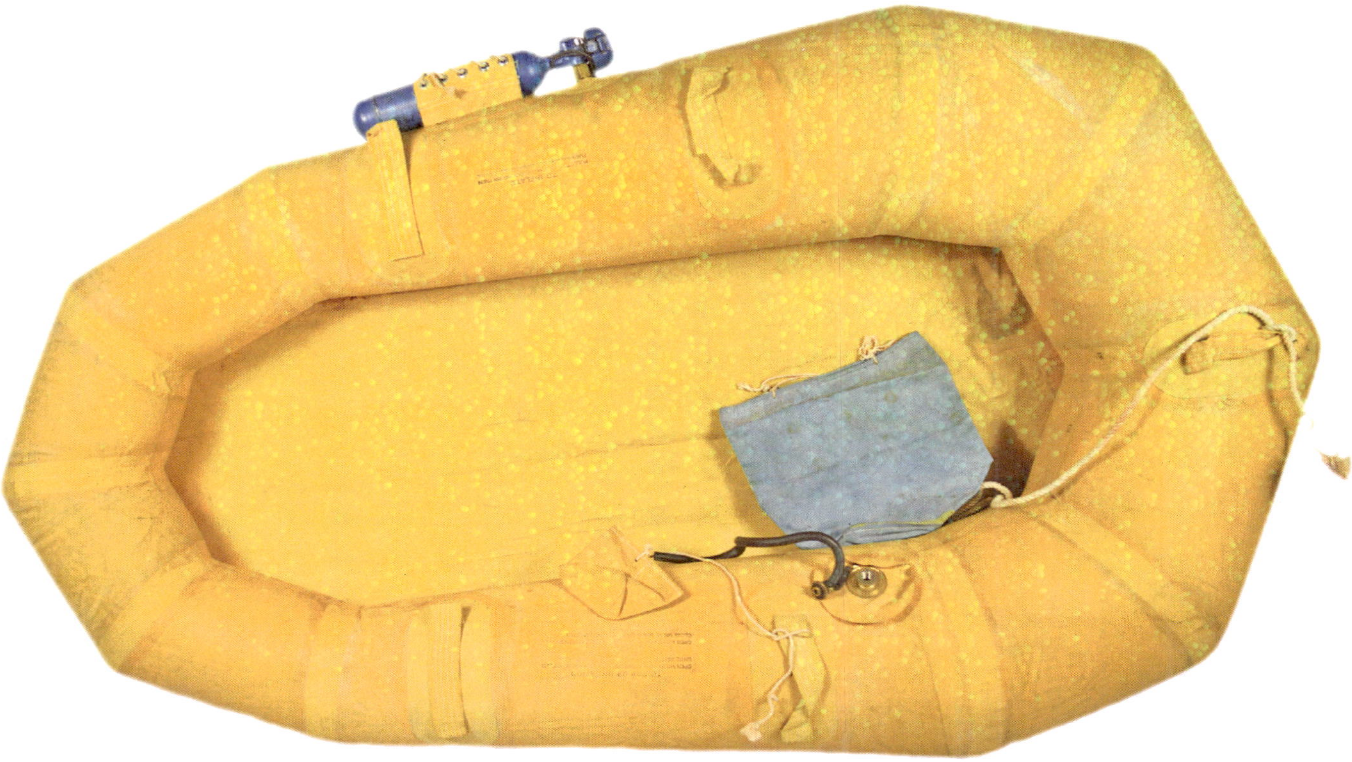

(Above ) AN-R-2A from one of the last contracts awarded. The first procurement of 15,000 were to meet fighter command requirements and priority for the PTO. As operations expanded, the 40,000 pararafts procured from March/April 1943 were principally to meet fighter command projected requirements for all theaters. With the new requirement established by the 8th Bomber Command, significantly higher quantities would need to be ordered to satisfy those demands. Six additional contracts were awarded in September 1943 for 80,000 pararafts to meet projections for both fighter and bomber commands for all theaters. These would be the last contracts awarded for the AN-6520-1 pararaft for the USAAF. In 1944 change orders for these six contracts would be issued to switch to the manufacture of the Type C-2 pararaft.

(Below) Case for the AN-R-2A pictured above.

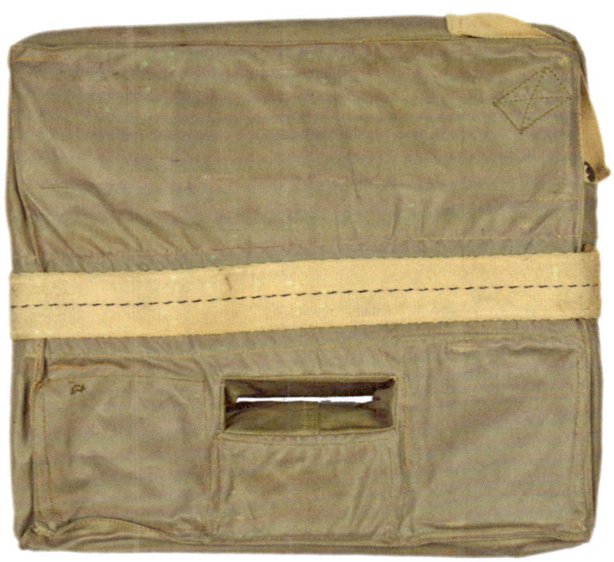

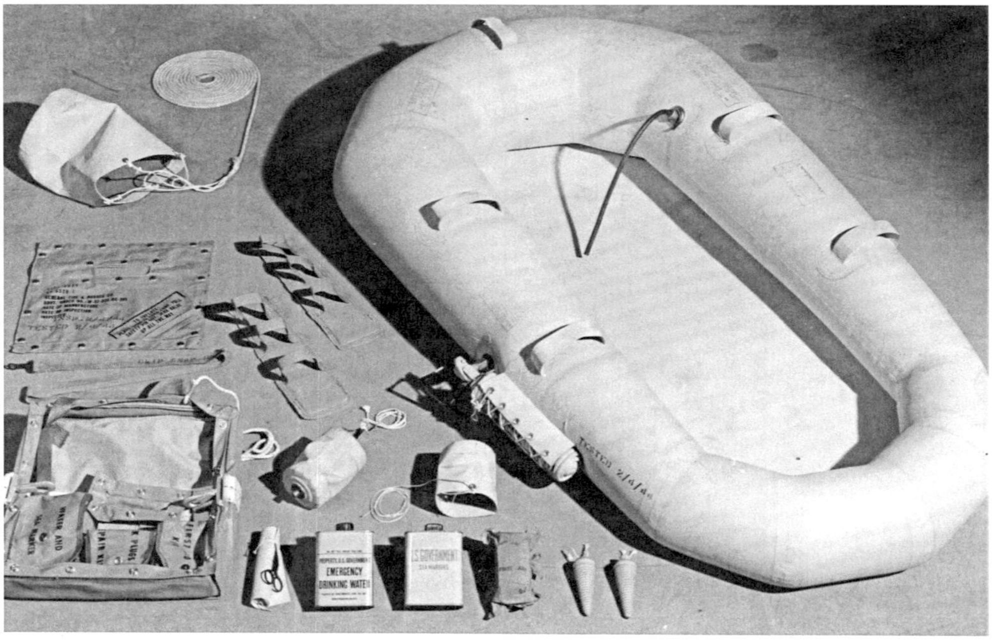


*(Left) AN-6520-1 pararaft as manufactured and assembled by the General Tire & Rubber Co. under USAAF contract 33038-AC-395. This assembly has the manufacture date of January 1944. This photograph was taken in December 1944, by which time the Type C-2 pararaft had superseded the AN-R-2A and it was instructed that all pararafts in service be updated to the Type C-2. The minimal updates applied here were the additions of a lever type valve for the CO2 and a hand bellows pump. All other accessories were standard as detailed in Specification AN-R-2A. (USAF)*

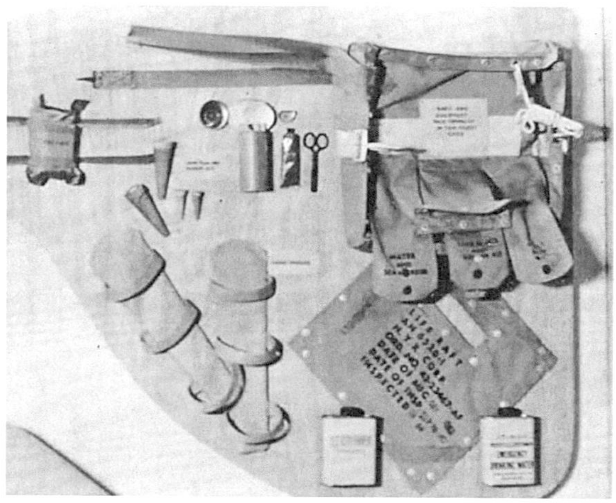

*(Right) The USAAF unanimously disapproved of the S2-1055 first aid kit, preferring the parachute first aid packet for pararafts. The first generation AN-R-2 units included the zippered pattern 42D11238. An attempt to transition to the use of the S2-1055, resulted in a mixed supply of both it and the parachute type in deliveries of pararafts through 1943. By the issuance of contracts for September, the parachute type was the approved standard, with a requisition of 80,000 to be supplied as G.F.E to contractors for inclusion. This wartime image is USAAF, but is equally representative of BuAer procurement, as all components were organic except the inclusion of the parachute packet.*

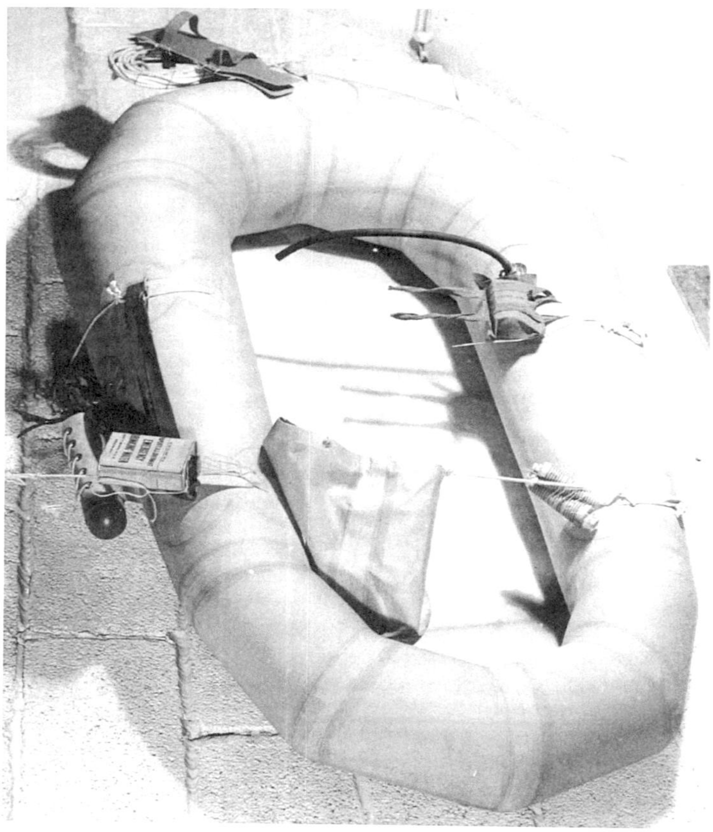

*(Left) AN-R-2A pararaft displayed in the equipment room of the 490 BG, England. Note lanyard attached to the neck of the CO2 bottle. This was a modification approved by the 8th AF Command in May 1944, to enable it to be used when the British Type-C packs could not be obtained. To improve speed of operation and to ensure the raft stayed with the user during a parachute descent into water, the cases were modified by eliminating the internal pockets and patching the leg slots, converting them back to the original design concept of the AN-R-2 non-slotted case and that of the Type-C. With this concept, the accessories would be packed in the sea anchor or secured to the grab handles and folded with the raft, August 1944. (Courtesy Gregory Pons)*

*(Below) In December 1943, the BuAer submitted recommended revisions to specification AN-R-2A to the working committee. The USAAF responded in kind, stating that no further coordination of subject specification with this command was necessary, as they were preparing specifications for a new pararaft to suit their requirements. This action allowed the amendment to be expedited and approved, 1 February 1944, as Amendment-B. The recommended changes were minimal to meet their service requirements and other concerns expressed by both services. With the introduction of the standard back pad kit M-592, a poncho raft cover was included, eliminating the need for a sail. It was recommended, and approved, that the sail therefore be deleted. During this time, the BuAer initiated the deletion of the canned sea marker, to be replaced by three dye marker packets M-566. Three packets were approved for inclusion in AN-R-2B. This would be formally addressed, service-wide, under Technical Order 53-44 of April 1944, for the replacement of the AN-S-10 with three M-566 dye markers in all emergency equipment. In specification it was detailed that protection against chaffing on any part of the raft, equipment and case was to be provided against. The AN-R-2B BuAer pararaft pictured has an anti-chaffing application the wheel on the CO2 cylinder and male coupling on the oral inflation tube. As prescribed in drawing below, the grab handles were to be moved forward towards the bow, but the example shown does not have that update applied. AN-R-2B specifications also outlines that the raft was to be legibly marked AN-R-2B, and the case was to be marked AN-6520-1. However, some examples of post-February 1944 contracts have AN-R-2B marked on the case and AN-6520-1 on the raft.*

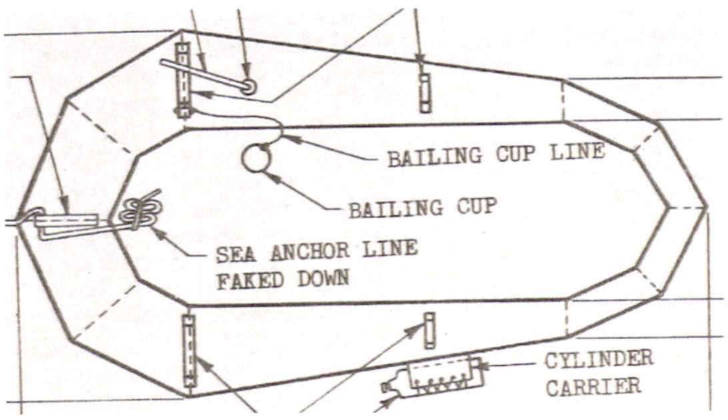

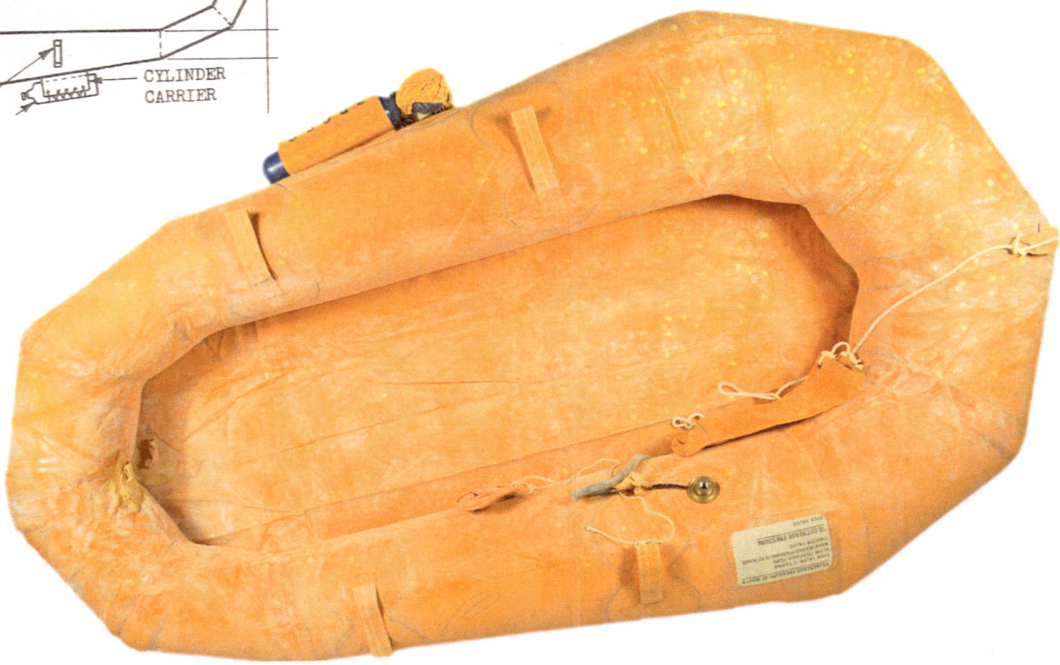

*(Above) Components for the AN-R-2B with the sail deleted and including the S2-1061 first aid unit. As per specification, three dye marker packets M-566 are provided (of any pattern, depending on packing date). The water can had friction tape applied to the cap for anti-chaffing purposes. With the USAAF transitioning to the procurement of the Type C-2 life raft, they made no further procurements of the AN-6520-1, and would not comply to these changes, making this amendment proprietary to the BuAer.*

*(Right and below) Wartime images illustrating updates to AN-R-2B specifications. Below is an early procured Barr Rubber Products AN-R-2 updated as per Technical Order 53-44 with three M-566 dye markers (NNAM). On the right is an AN-6520-1 as procured under specification AN-R-2B. Included in the display is the S2-1061 first aid kit.*

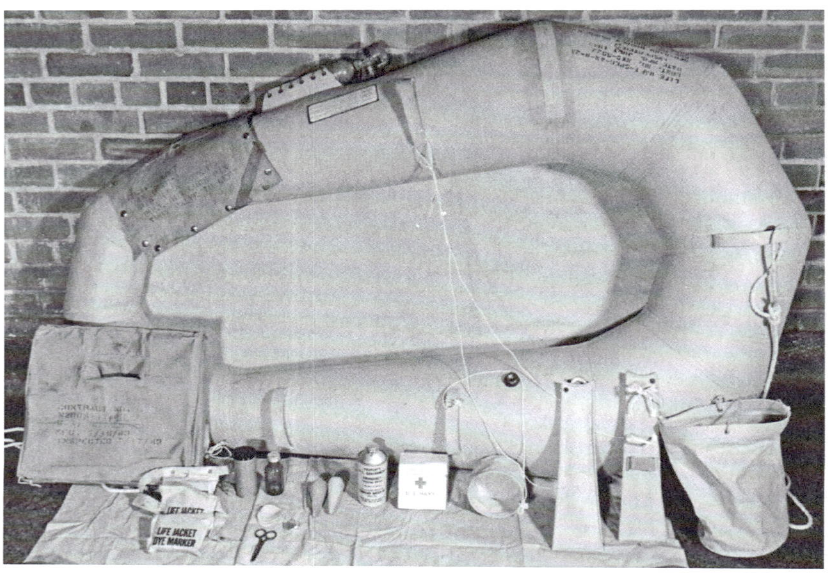

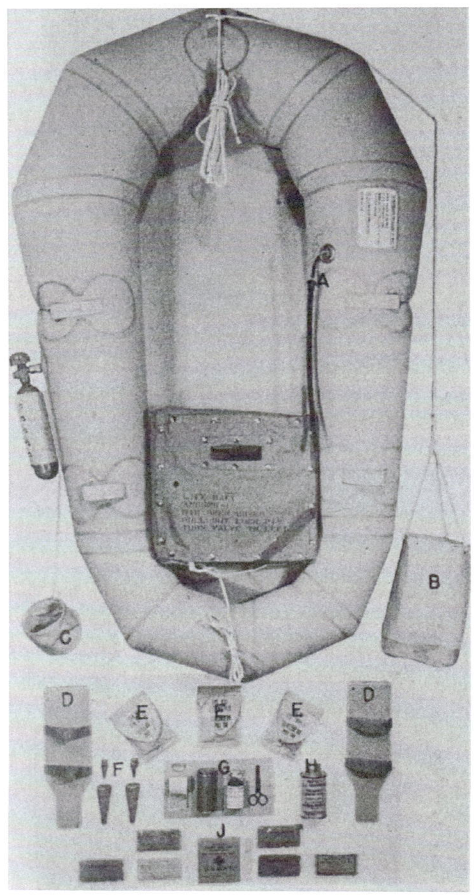

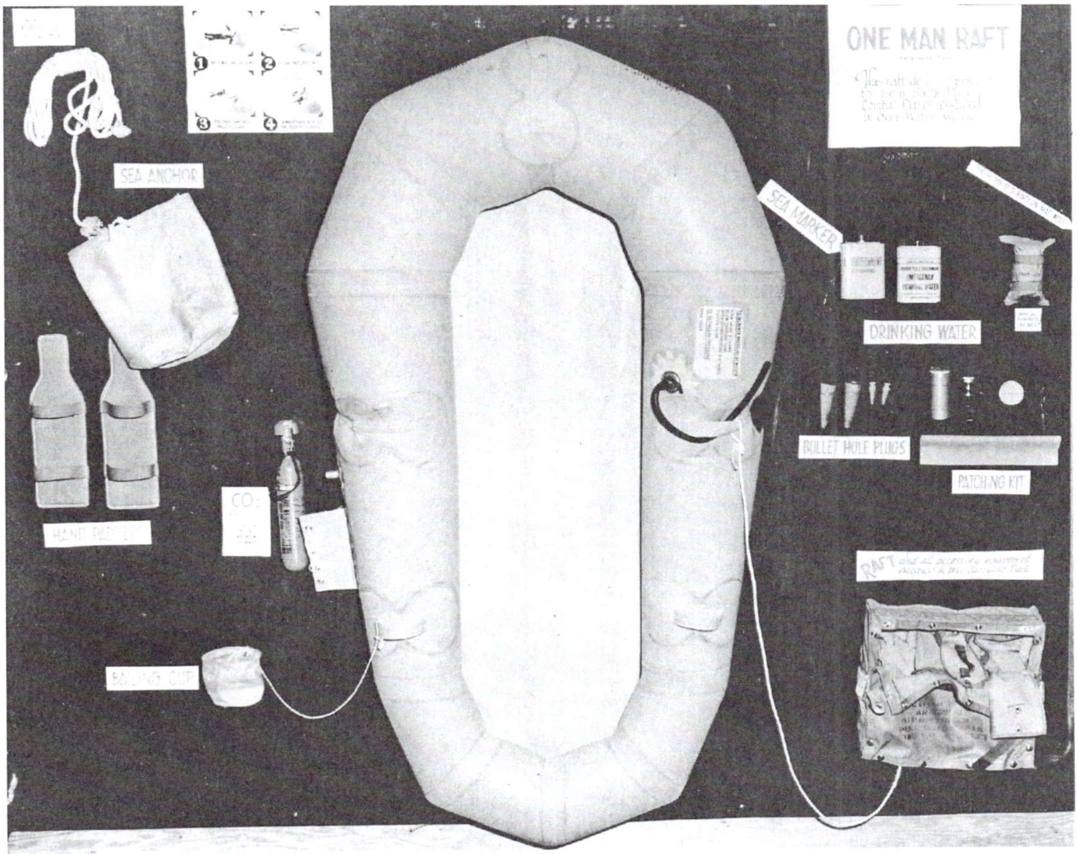


*(Left) A display at Chico Army Air Field in Chicago, November 1944. This illustrates the complete assembly of a AN-6520-1 pararaft orientated towards USAAF preference, but encompasses BuAer types as well, not shown is the sail. Of note is the correct attachment of the raft to the case, bailing cup, and the lanyard to the life preserver. The sea anchor would be secured to the grab handle at the bow. In keeping with the early concept of placing the CO2 bottle on the right side, the lanyard is secured to the grab handle to prevent entanglement while boarding. (National Archives)*

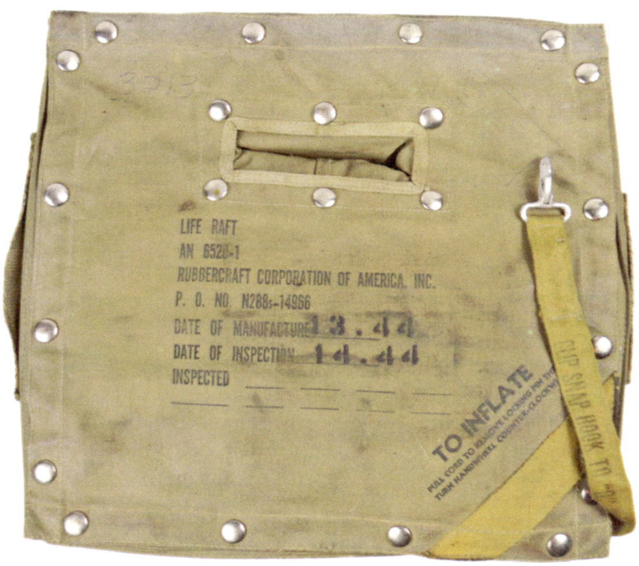


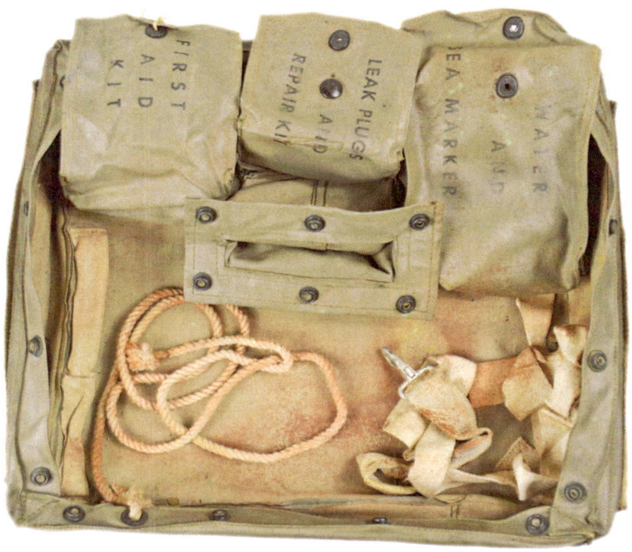


*(Above and right) Cases were fabricated by prime contractors and sub-contracted with subtle variations between them, but all conformed to Dwg. AN-6520, NAF Dwg. No. 69471. Known sub-contractors included the Kolonics & Company, American Lady Corset Co., and the Astrup Co.*
*Being a compact assembly, the packing and folding of the raft were accomplished with very little tolerance. The packing was essentially done by a layering system. Above left is a complete unit ready for issue, the lanyard ready for connection to a life preserver. Above right is a case with all components removed, the three internal pocket flaps legibly marked with its contents. The 5' lanyard for connection to the life preserver would be folded to allow only sufficient slack for attachment when seated on the unit, then secured by a strap and glove snap inside the case, as shown to the right.*

*(Above) Two examples of packed parachute raft cases with top cover removed. As can be seen, there is almost zero tolerance.*

*(Below) General placement of components. Paddles are placed in the bottom of the case secured by sewn straps, they are in turn lashed by a cotton cord to prevent loss. The sail was to be lashed to a convenient location by a 3' cotton cord, either folded up in the raft, or placed on the bottom of the case. After packing was complete, the raft would be folded and placed in the case. The updated BuAer Technical Note 76-43 of November 1943 for the one-man parachute raft included the back pad kit, and packing instructions stated that, since the M-592 was provided with a first aid kit, the unit provided in the pararaft was to be removed and additional equipment as desired placed in the first aid pocket. This was evidently optional – the first aid kit was still provided as G.F.E to contractors to be packed prior to shipping.*

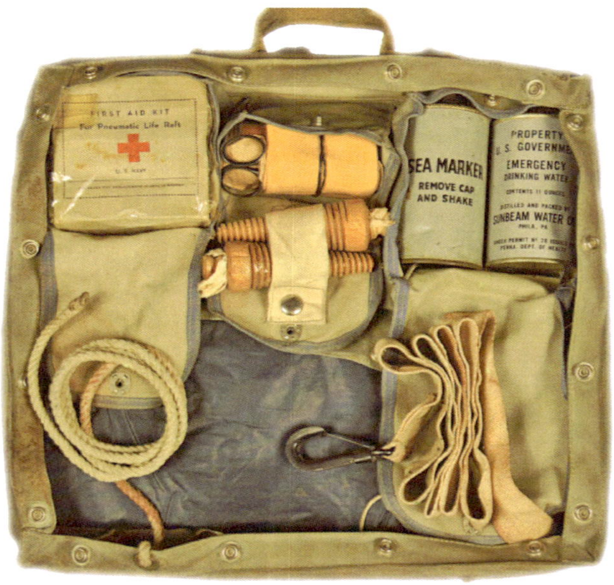


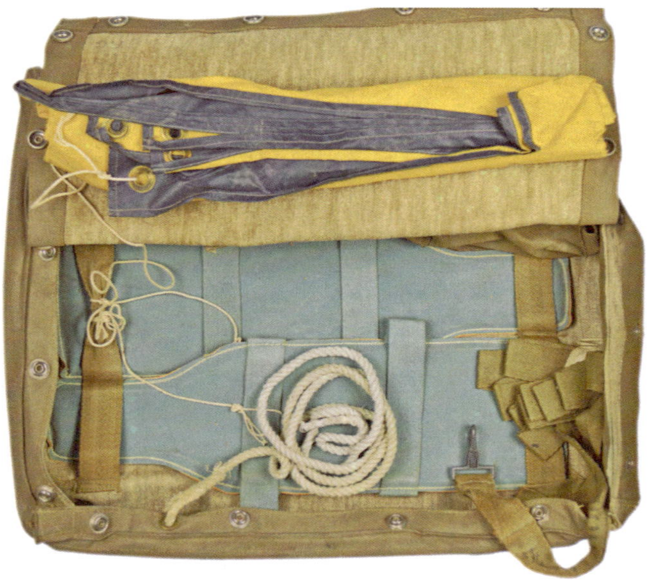

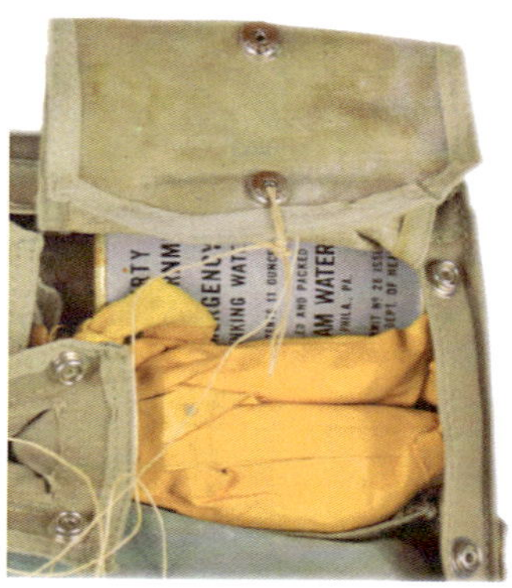

*(Right & far right) Alternate packaging of the rectangular water and sea marker cans to the right. On the left is the AN-W-5A round water can with three M-566 dye marker packets as per Technical Order and specification AN-R-2B.*

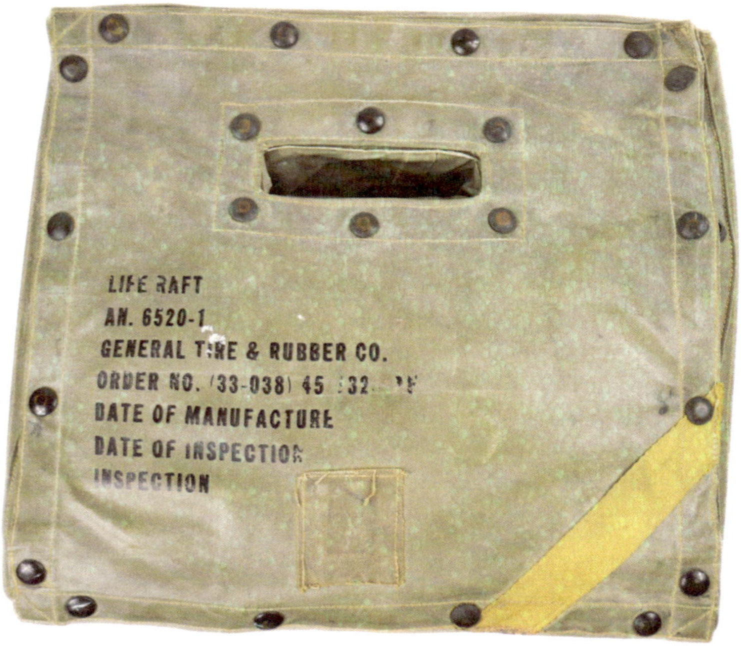


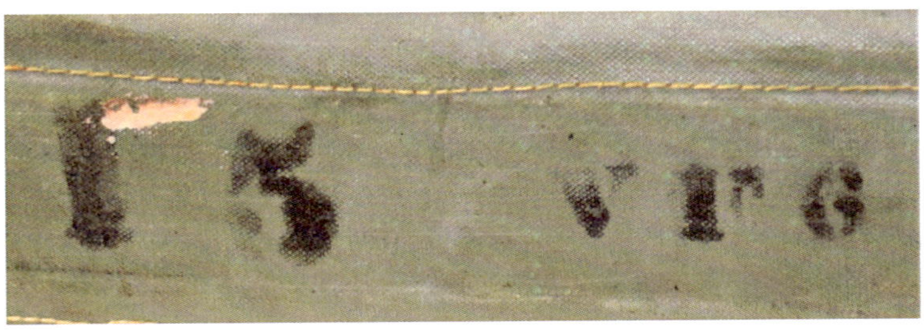

*(Left) As per BuAer Technical Note 1-43, dated 6 January 1943, for the packing inspection and maintenance of the pararaft, sufficient quantities of the new NAF Dwg. No. 69471 cases had been procured to replace the old center-slotted types in service. Pararafts, like life preservers and parachutes, would be inspected prior to and after operational use, and would receive 30-day inspections and mandatory semi-annual inspections. Exposed surfaces of the case would be the first component to receive damage and would have to be repaired or replaced. By 1944, extended service use of the pararaft took its toll on the cases, requiring replacements. In July 1944, the BuAer had negotiated with the USAAF for the supply of replacement cases under USAAF procurement. The USAAF allocated 3,918 from the New York Rubber Corp. under contract 33038-AC-392; 5,030 from American Lady Corset Co. under P.O. 45-1321; 1,957 from the US Rubber Co. under P.O. 45-1322; 2,010 from General Tire & Rubber Co. under P.O. 45-1323; and 2,084 from Hood Rubber Co. under P.O. 45-1324. These cases would meet BuAer requirements and not be equipped with the quick attachment. Shown is an example of the allocation from the General Tire & Rubber Co. The case is property marked to VF-6 and has had an inspection pocket added on the cover.*

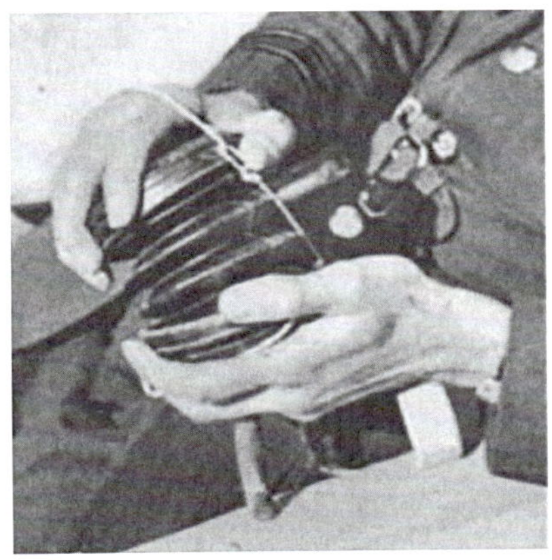

*(Above) The concertina pump, copied from the British model, was an early inflation feature.*

*(Left) The pump incorporated into the M-524 and AN-R-2 was 4" in diameter and 2" high with a full extension of 8". The top and bottom were fabricated utilizing a hard rubber, with a pliable rubber for the bellows. The USAAF deleted this item from their requirements to reduce bulk. BuAer followed suit on their change order of November 1942, switching to the oral inflation tube as they found the pump susceptible to breakage. In Technical Note 1-43, concertina pumps in service were to be replaced by the oral inflation tube. (National Archives)*

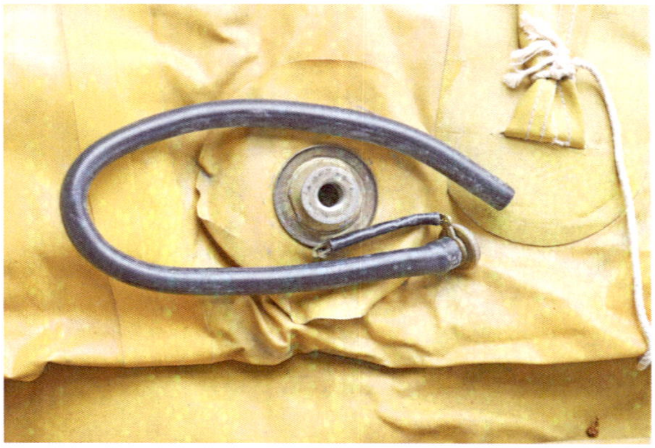

*(Left) Replacing the concertina pump was an oral inflation tube, 16" in length with a threaded male end for connection to the manual inflation inlet valve located in the bladder of the raft. This would remain the standard feature in the AN- series. Replacements were available from the ASO under stock number R83-T-704150 and 4500-920100 from USAAF stock.*

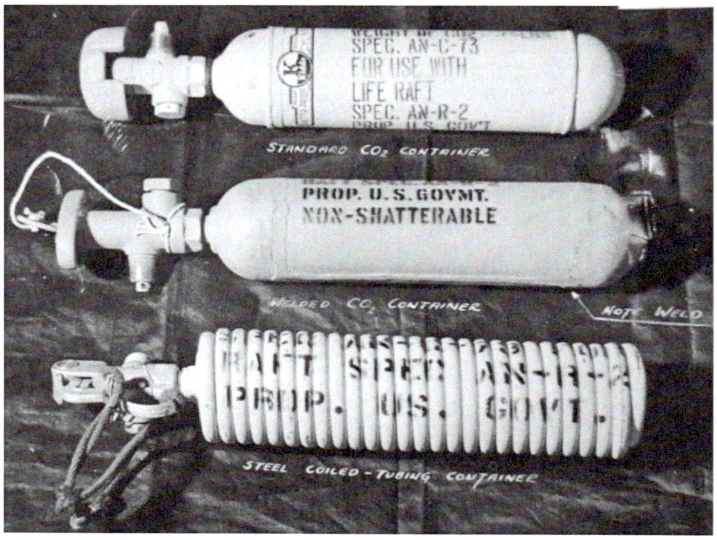


*(Right) Wartime image of the three principle variants of CO2 cylinders utilized with the AN-series pararaft. They were originally designated under NavAer Specification M-108 Type-A, Non-Shatterable. With joint Army-Navy standardization, they were re-designated under Specification AN-C-73 Dwg. No. AN-6025 23 July 1942. The top cylinder is wire wrapped for reinforcement, the middle cylinder is a welded seam type and the lower bottle is a coiled tube type cylinder. They were available for replacement from the ASO under stock number R51-C-10735 (bottle only). The assembly with valve could be requisitioned under stock number R-83-C-94670. The USAAF supplied the complete assembly, bottle and valve, under stock number 4500-341052. (National Archives)*

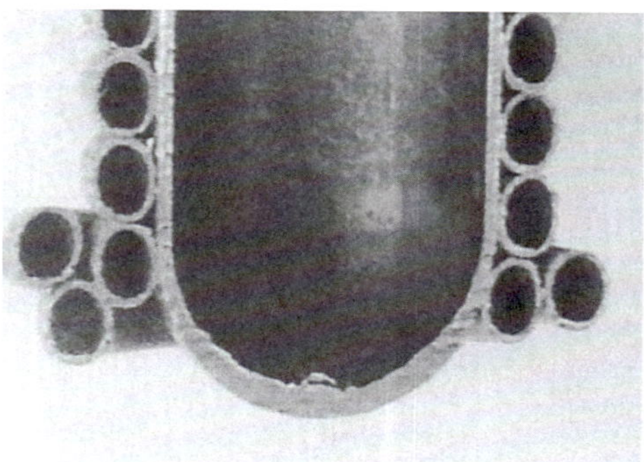

*(Above) The coiled tube type cylinder met the requirements of AN-C-73 and was approved in September 1942.*

*(Far left) Its unique design featured a small diameter cylinder integrated with an external coiled tube system, both of which held the CO2. These coiled cylinders were manufactured by the Andrews-Alderfer Processing Co. and were provided with a lever type ring pull valve manufactured by Jenkins Bros. The valve was approved by the USAAF but not the BuAer, who required the Walter Kidde & Co. wheel type valve on their deliveries.*

*(Left) The Andrews cylinder and Jenkins valve seen in a wartime image of a US-AAF AN-R-2 pararaft. (National Archives)*

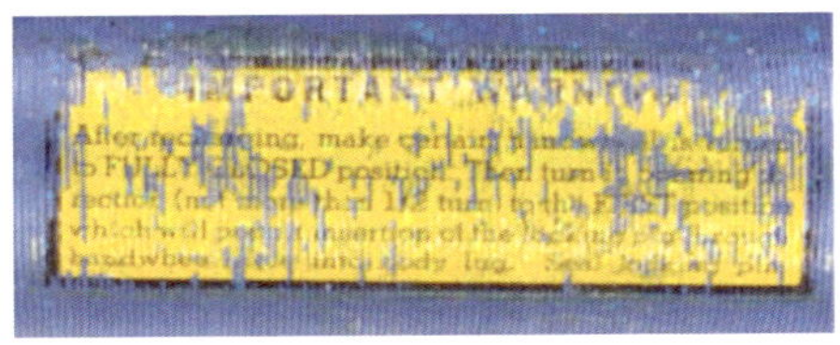

(Above) AN-C-73 CO2 cylinders and valves as provided with the AN-series pararafts. As per original specification, the bottles were to be painted grey, but for reasons unknown, contractors transitioned to painting the bottles blue. Originally, they were provided with wheel type valves with the transition to lever types in 1944. The third bottle from the left has been updated with a Walter Kidde & Co. lever valve. To discharge the CO2, a safety pin had to be removed, then the valve turned, rupturing a disc to discharge the gas. Due to problems encountered with corrosion and failure of the safety pin and wheel valve to function, the lever type was adopted requiring only a swift pulling action. Due to improper adjustment and recharging of CO2 bottles, an instruction label addressing the issue was required as of September, 1943. Two examples of this application are shown. Principle suppliers of bottles were Walter Kidde & Co., Turner Brass Works and Andrews-Alderfer Processing Co.

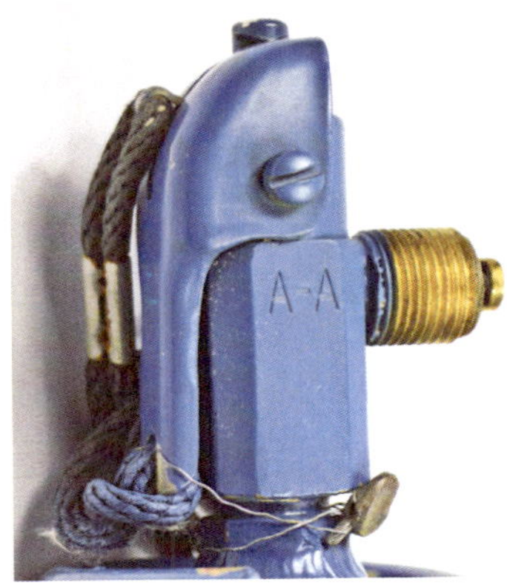


(Above left) Premature inflation became a concern after several isolated reports stated that while in the cockpit, the pararaft inflated, (causing a death in one incident). Shown is a demonstration of premature inflation. The BuAer addressed this issue, first in Technical Order 61-43 of September 1943, discussing the proper adjustment and setting of the wheel. The valve in question was from Turner Brass Works. Multiple notches were available for adjustment, and when not set properly, the wheel could turn and seep air into the raft. Labels on this procedure were recommended and available for application from the ASO. Safety wire was applied in some cases for tying the valve, but if the wire was too thick, removal could be problematic.

(Above center and right) BuAer Technical Order 27-44 of April 1944 directed that a copper wire witn a breaking strength of 17-pounds be applied. Both valves pictured have the safety wire applied.

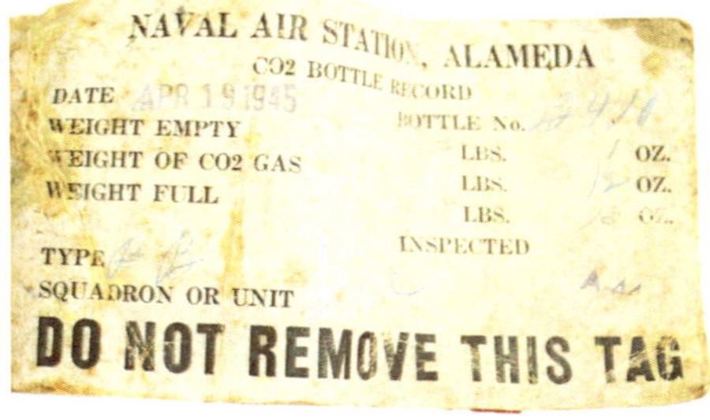


*(Above) BuAer Technical Order 91-44 of July 1944 prescribed the 17-pound sealing wire with the addition of a red tag or similar indicating that the cylinder had the proper CO2 charge. Shown is an example of the tag. This Andrews valve has been painted blue, indicting that it is not interchangeable with the Walter Kidde & Co. inlet check valve assemblies, as outlined in Technical Order 14-45 of January 1945.*

*(Above) BuAer Technical Order 130-44 of November 1944 stated that the Rego Type 4397-wheel valve (Turner Brass Works) be replaced with the Andrews-Alderfer lever type A-39 or the Walter Kidde & Co. wheel type. Top is an example of the Andrews lever valve and, as noted by Technical Order 14-45, non-painted valves were universally interchangeable. Lower picture shows a Walter Kidde & Co. wheel type.*

*(Below) Chaffing of the life raft fabric was an additional concern, and various remedies were implemented to prevent wear from exposed metal surfaces. The USAAF did not circulate any official directives, but recommended friction tape be applied over the safety pin and the coupling nut on the side of the valve stem. Another solution was the application of a gauze wrapping over the coupling nut. The BuAer addressed the issue first through Technical Note 75-43 of November 1943, which directed that an interior reinforcing patch be applied on the back of the inlet valve. This was revised in Technical Note 29-44 of March 1944 to include either an internal and/or an exterior reinforcing patch. Both services recommended the rounding of the edges of the coupling nut. The purpose of the coupling nut was for recharging the CO2. Specification AN-R-2B made it a requirement for the contractor to apply anti-chaffing measurers prior to delivery. Bottom center is an Assembly & Repair Depot application; wrapping the valve with felt, then a scrap piece of rubberized fabric.*

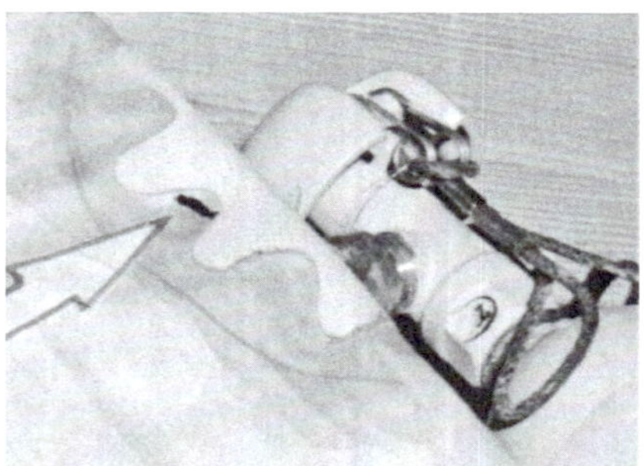

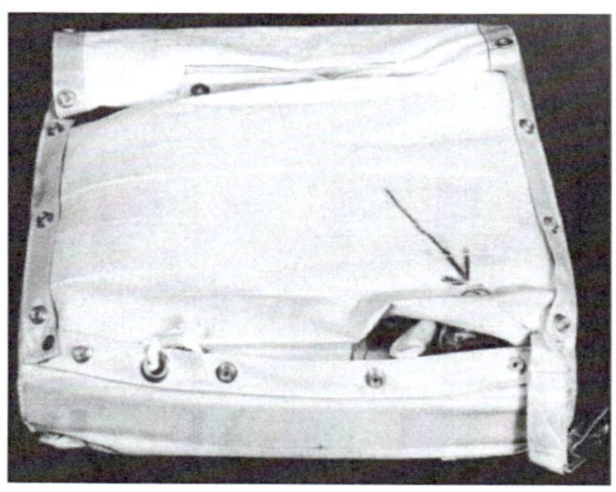

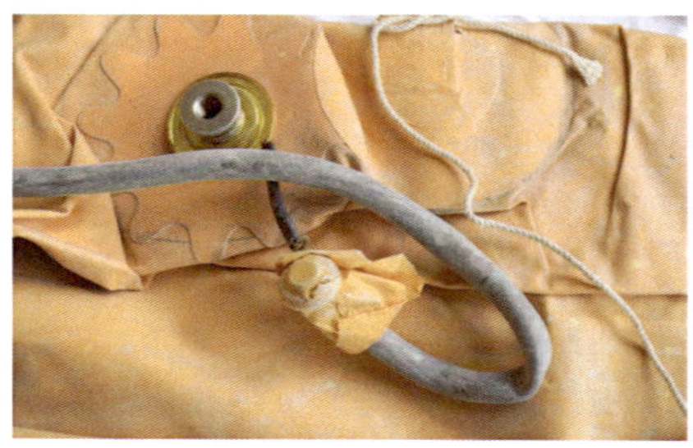

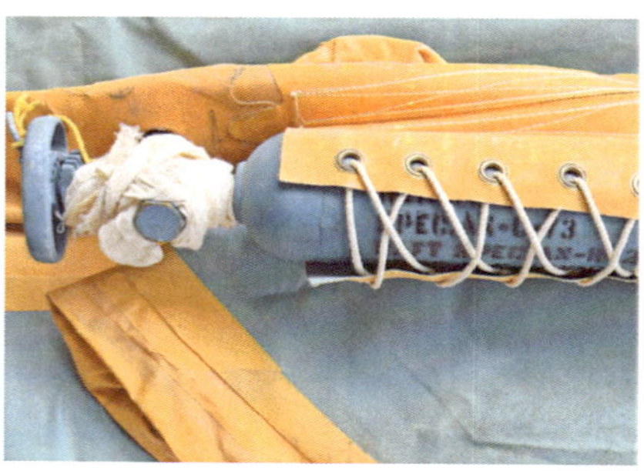

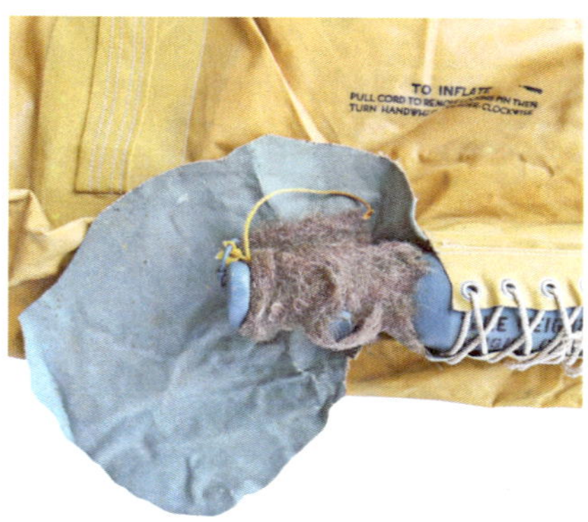

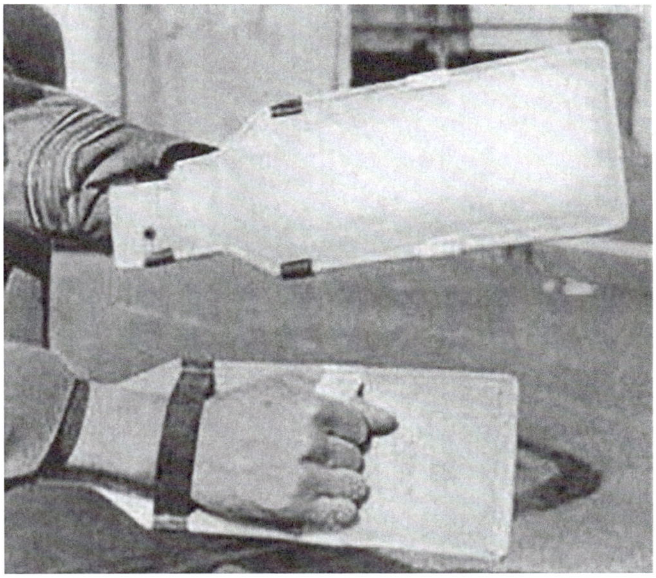

(Left and above) Hand paddles as provided with the British Type-K Dinghy. Constructed from rubberized fabric over a wire frame. This pattern would carry over to production of the AN-R-2.

(Right) As part of the revising effort by the USAAF, hand paddles made from wood were tested. Their concern was that the wire framed types did not float. A simple wrist strap was provided with the user grasping the tapered end as illustrated right. These were field tested in June of 1943 as part of a series of proposed revisions to the AN-R-2A.

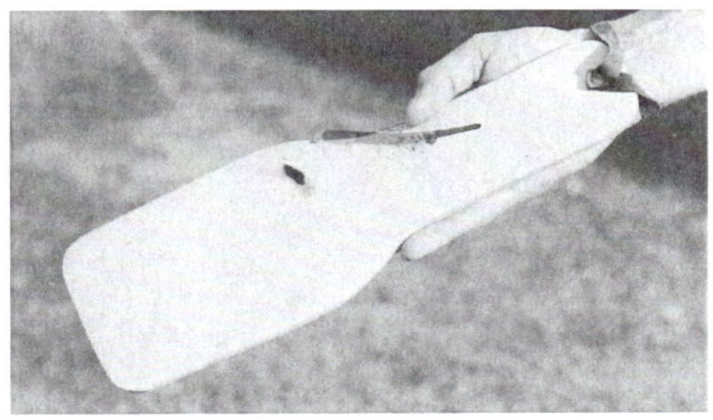

(Left)) Consistent details from specification M-524 to AN-R-2B stated that paddles were to be fabric over a wire frame, 14" in length, 4" at the finger tips and shaped to fit the wrist. Three 1" non-elastic bands were to be provided for securing the hand, wrist and forearm.

(Below) A selection of paddles. As can be noted from era photographs, not all conformed to specifications! Each was organic to both services and could be replenished by ASO stock number R83-P-4800 and 4500-621580 for the USAAF.

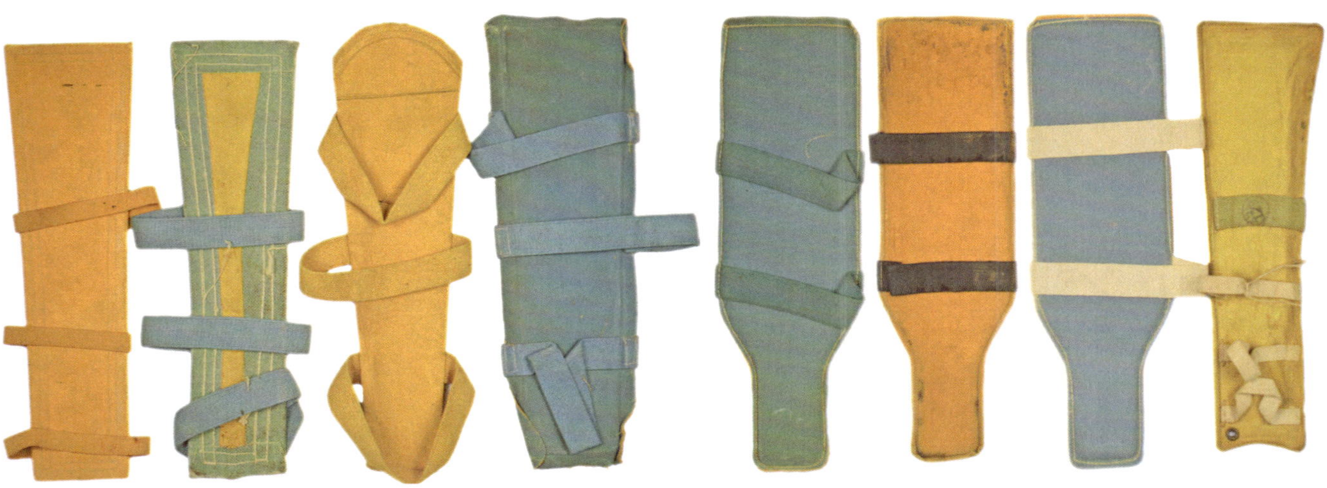

*(Left) Sea anchors provided in the AN-series pararaft consisted of rubberized fabric in a bucket shape 8" in diameter and 12" deep that could double as a water storage bucket. A 10' rope was provided and secured to the grab handle at the bow of the raft. Early buckets included heavy gauge wire to maintain rigidity at the top and bottom, transitioning to a narrow-gauge wire or doubling of material at the top only. These are just some examples. They could be replenished under ASO stock number R6-A-1950 and 4500-705000 for the USAAF.*

*(Below) Bailing cups were also provided and were of similar design to the sea anchor but smaller. Early examples were shaped like drinking cups, evolving to the same construction as the sea anchor, 4" in diameter and 4" deep. These would be secured by a 3' cotton cord to a raft grab handle. Replacements could be requisitioned utilizing ASO stock number R-83-C-81505 and 4500-331000 for the USAAF.*

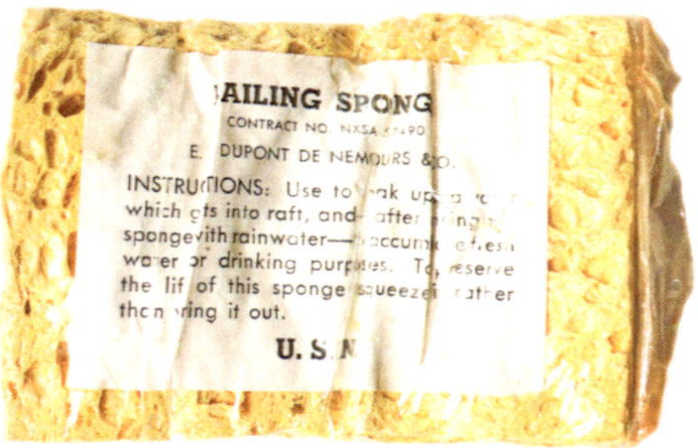

*(Above) In September 1943, the Army-Navy Working Committee approved the adoption of a sponge to replace or supplement the bailing cup. By this date it was determined necessary to keep the inside of the raft dry to prevent salt water immersion wounds. This feature was adopted and implemented in future specifications for both services under Federal Specification L-S-626, Type-I, Class A, Size 4 for all life rafts. The BuAer procured 200,000 under contract NXSA-57490 11 April 1944 for stock purposes. Life raft contractors were required to supply sponges as C.F.E. The primary supplier was E.I. DuPont and Co. of the Size 4 measuring 4-1/2" x 3". For BuAer contract NXSA-57490, sponges would be provided with a paper label and wrapped in cellophane, as shown above left. Life raft contractors often supplied with commercial versions as illustrated center. Upon availability, pararafts in service would be updated with the sponge. The sponge without label was a service upgrade for an AN-R-2 retrofitted with a 3' length of cotton cord. Replenishment was by utilizing ASO stock number R83-B-648500 and 6700-868980 for the USAAF.*

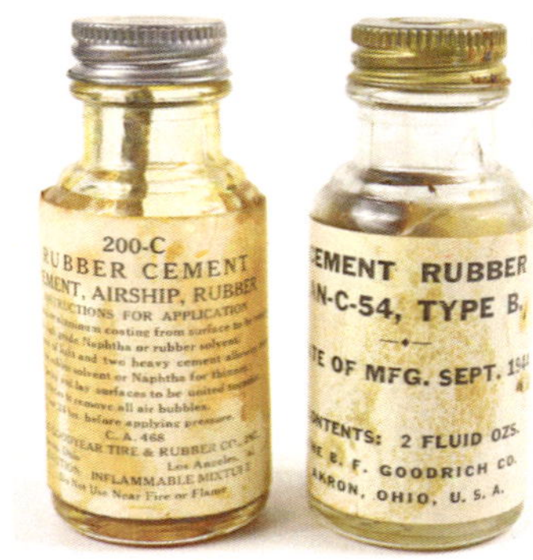


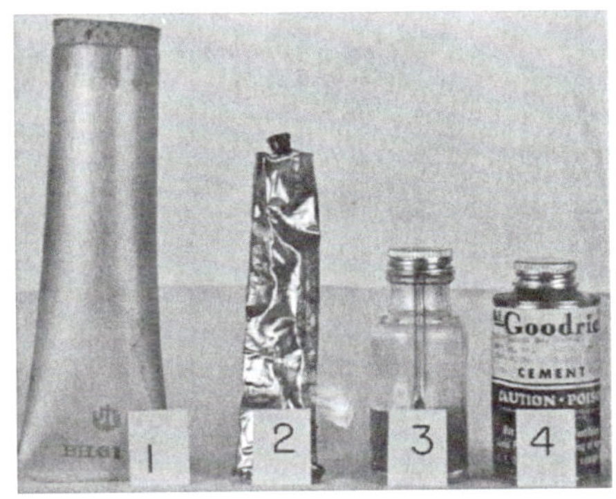

*(Above and left) Two ounces of Type-B AN-C-54 rubber cement was to be provided in the repair kit, packed in what was referenced as "a substantial container" to resist damage. Glass, tin and flexible tubes were provided by contractors, making for an almost endless variety. A typical feature was the supply of a cardboard tube, with screw top, for storage of the glass and tube cement containers.*

*(Right) A roughing tool was to be provided. The original tool for inclusion comprised wood sticks with an aggregate applied or sand paper provided in BuAer life rafts of all types. Two variations are illustrated. These were found to be useless after long periods of submersion. As part of USAAF modifications, a metal roughing tool was included in AN-R-2A. The standard metal roughing tool, as shown, was integrated into all life raft production.*

*(Above) Specifications M-524 and AN-R-2 required patching material 6" x 12", rubber cement, roughing tool and scissors. Tire patches were added in specification AN-R-2A. Pictured are examples of the tire patches, patching material and scissors. Also illustrated are two typical methods for the complete packing of the repair kit. Replacements were available from the ASO under stock number R83-K-710165 and 4500-526600 for the USAAF.*

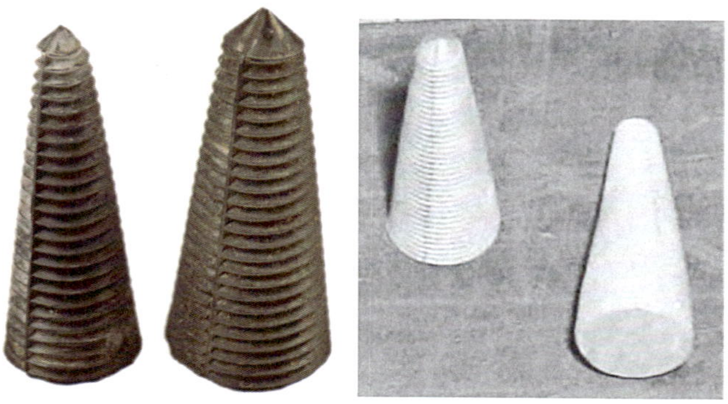

*(Left) Early bullet hole plugs provided were solid molded rubber. The British pattern is illustrated far left, but were hollow and included a smaller plug that would nestle into the larger plug. This feature was not initially incorporated into US manufacture.*

*(Left) Conservation of rubber caused the transition to fabricating the plugs from wood, with the initial patterns being a solid core. Two plugs were provided for each pararaft.*

*(Left) By late 1942, the design had changed and plugs were provided in sets for all life rafts. These were not covered by specifications, but were provided as a C.F.E. Two sets of nested, shellacked wooden plugs were supplied as a threaded cone. Variations were available with some principle examples illustrated. Two sizes were issued: 4" for pararafts and 5" for multi-person rafts. For replenishment purposes, the 4" plugs were requisitioned under ASO stock number R-83-P-408550 and 44500-526600 for the USAAF.*

*(Right) USAAF wartime image illustrating how the plugs are to be applied for the repair of punctures in the bladder of a life raft. (National Archives)*

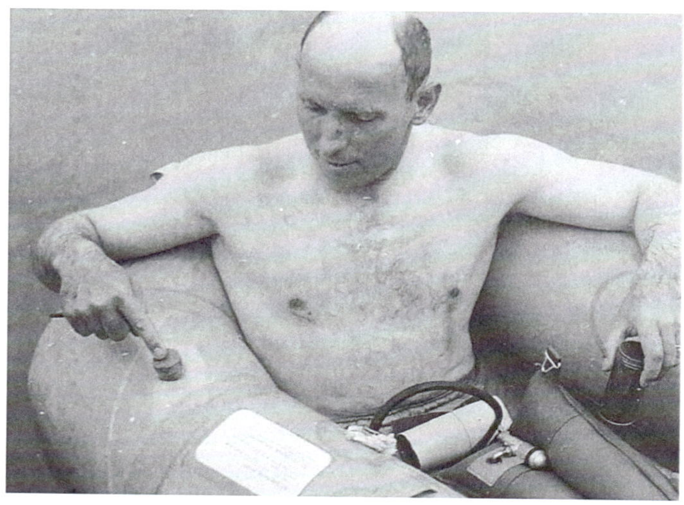

*(Right) Attempts to improve existing designs were always under consideration. The pararaft in this image incorporated two additional features: a spray shield and rudder. The BuAer investigated the concept of a spray shield, based on the updates of the Type-K Dinghy, but found it to be problematic, adopting instead the poncho raft cover as contained in the M-592. To improve sailing characteristics, both services experimented with ballasts and keels for all life rafts. A sailing apparatus was applied to this pararaft, but it is not clear exactly how it functioned. (National Archives)*

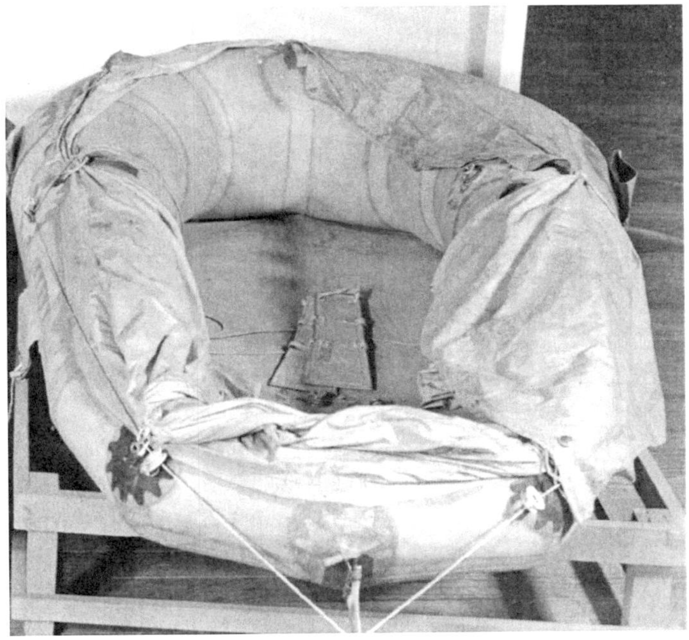

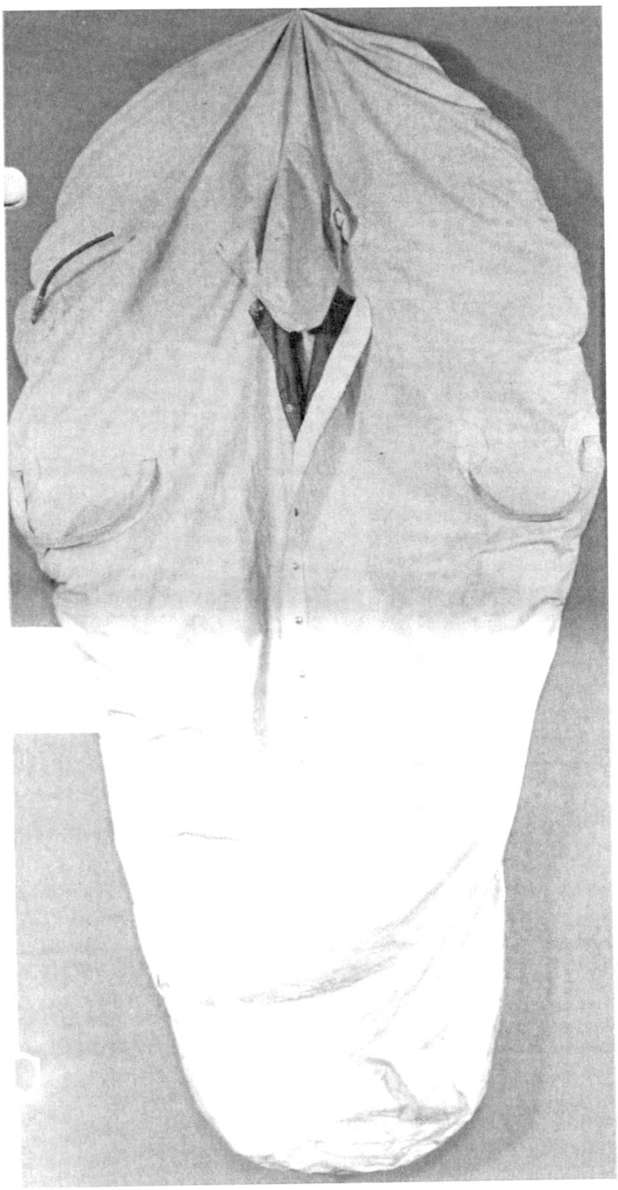

*(Left) Exposure was always a concern, hence the adoption of protective headgear, the addition of the sponge and use of sails or paulins. To address concerns for Arctic waters, insulated rafts were devised. Shown is a standard AN-R-2 incorporating an inflatable bottom for insulation and an insulted raft cover.*

*(Below) A completely new concept incorporating an inflatable bottom and side walls acting as a shelter from winter conditions. Both models were tested in 1943. Interestingly, the inflatable sheltered concept was adopted in the LRU series as the LRU-16/P in the early 1960s. (National Archives)*

*(Below) Equipment was often modified to suit specific needs. Illustrated here is an AN-series pararaft modified by the 343rd fighter group, 11th AF, USAAF. A case has been fabricated for use as a back type, incorporating both the pararaft and emergency back pad kit, making for a complete single unit. In the display of components, BuAer articles, typically contained in the M-592 back pad kit such as the hand projector, table rations, mag lens, and some first aid items are all exhibited. Also included are articles from the standard USAAF emergency parachute kit, including the folding machete, pocket compass, ammunition and mosquito headnet. (National Archives)*

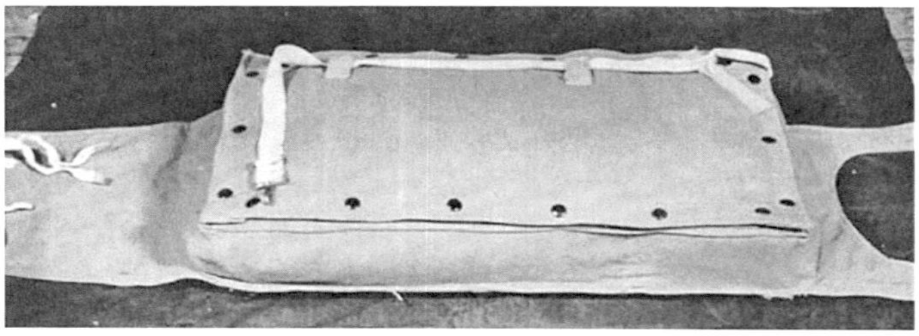

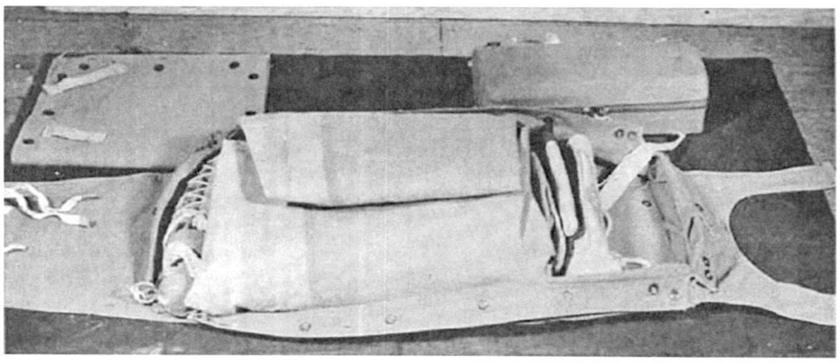

(Above and below) A series of photos showing the use of the AN-series pararaft by USAAF personnel in the PTO with standard seat type parachutes, c.-1944. A typical application, as prescribed by technical order, was to remove the snap hooks for added comfort, by cutting the webbing loop enough to remove the hooks. In the image above right the reinforcing strap has been completely removed. For additional comfort, if practical, the seat cushion was often used. (National Archives)

*(Left) Pararafts used in the ETO were predominately British types, principally due the unavailability of American models, as most of the early deliveries were sent to the PTO. After the revision of modifying the case for quick attachment it became acceptable for use. By late 1943, significant deliveries began to arrive in England and filtered into use amongst the air forces in the region. A stack of AN-R-2A pararafts can be seen in the equipment room of the 91st BG, England c.-1944. (National Archives)*

*(Below) The requirements in the ETO were for issue to both fighter and bomber commands. On the left, Maj. Herschel Green is utilizing and AN-R-2A pararaft with his standard seat type parachute, Italy c.-1944. Pararafts were distributed to all crew positions on bombardment aircraft. On the right are several, readily accessible for B-24 gunners, Europe c.-1944. (National Archives)*

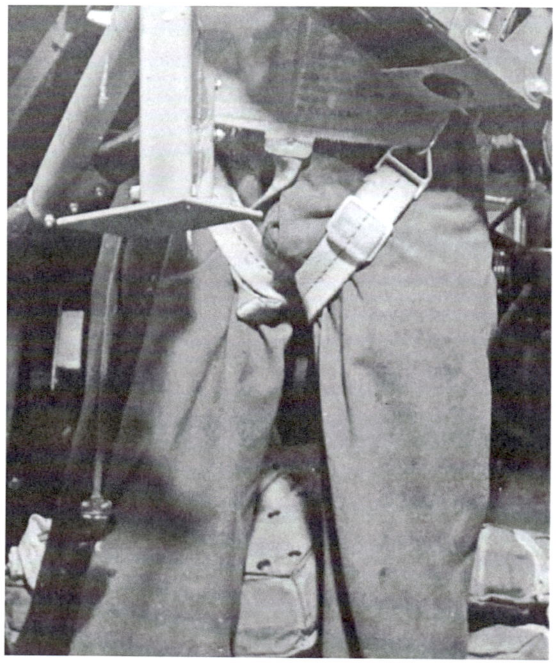

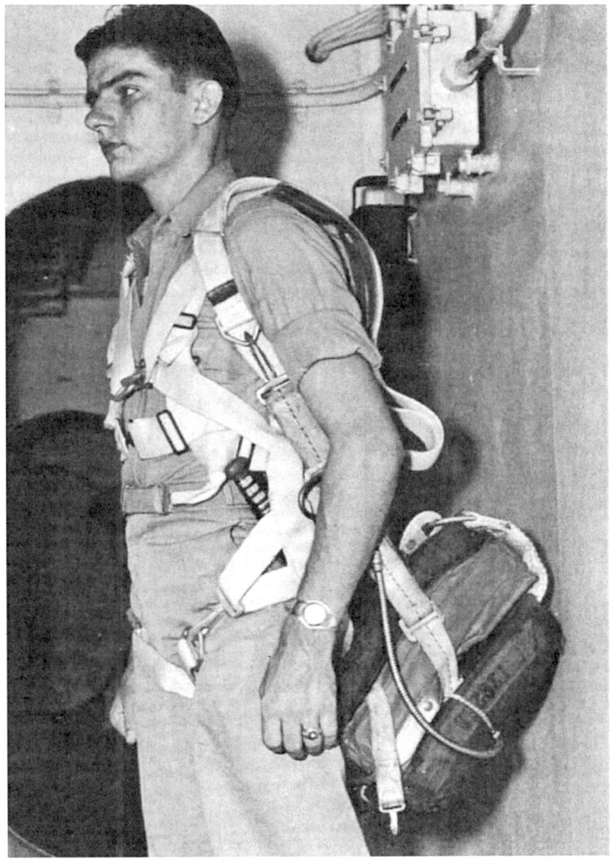

*A series of photographs of the pararaft in use with the BuAer and USMC. Above left, USMC 2nd Lt. Thaddeus Czarnecki of VMF-211 is utilizing a standard AN-series seat type parachute, Bougainville February 1944. Above right, an unknown individual aboard the USS Hornet demonstrating the QAS parachute with pararaft. Below Left, Ens. Raymond Klingerman of VF-82, August 1944. Below right, A naval aviator undergoing survival training on the proper use of the parachute raft, NAS Corpus Christi, May 1945. (National Archive)*

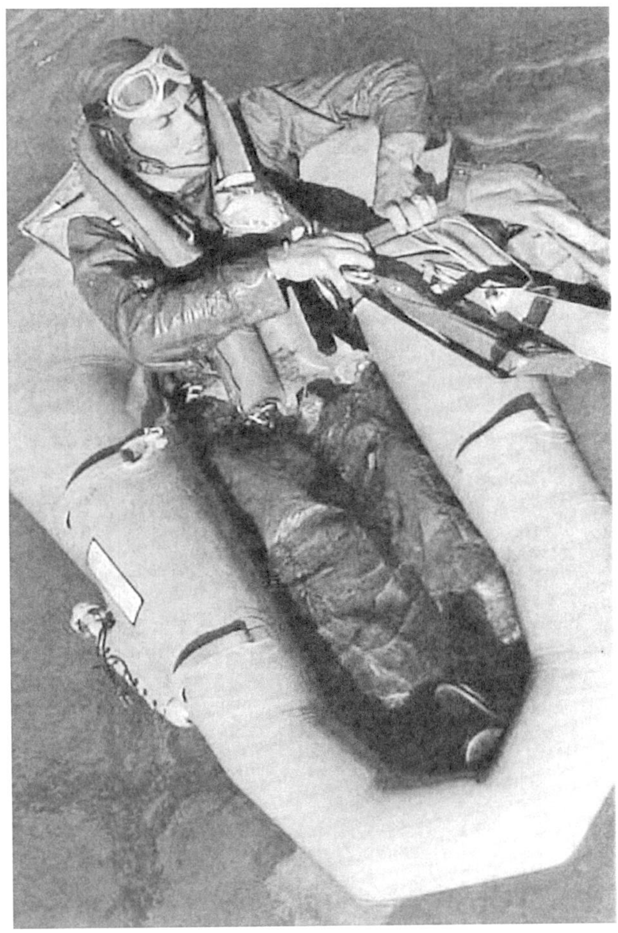

*Additional views of service use of the pararaft with USMC units. Top left, an unknown USMC fighter pilot of VMF-222 has been transported to the flight line preparing for strikes on Rabaul, he is utilizing the original back pad kit, incorporated with a standard seat type parachute, Bougainville April 1944 (National Archives). Top right, USMC gunners and radiomen of the 2nd MAW on Okinawa being transported back to their quarters after a long night of flight operations, April 1945 (National Archives). Below, In the equipment room of VMO-155 are parachutes ready for issue that include the complement of emergency back pad kits and pararafts, Kwajalein c.-1944. (Courtesy Gregory Pons)*

*Known contracts and quantities of the one-man parachute raft for the Bureau of Aeronautics. The quantities listed are for the original award, documentation suggests that extensions were approved on some contracts but cannot be substantiated. The total quantity below is 59,000 with a more realistic production figure near 70,000. Note: this list may or may not be complete.*

| CONTRACT | AWARD DATE | CONTRACTOR | QUANTITY |
|---|---|---|---|
| No.98179 | January 1942 | U.S. Rubber Co. | 300 for R&D |
| NXS(A)-5431 | May 1942 | U.S. Rubber Co. | 6,000 |
| NXS(A)-5522 | May 1942 | Goodyear Tire & Rubber Co. (Barr Rubber Productss) | 9,000 |
| NXS(A)-5785 | May 1942 | Firestone Tire & Rubber Co. | 5,000 |
| 288s-11421 | February 1943 | Firestone Tire & Rubber Co. | 4,000 |
| 288s-11408 | February 1943 | U.S. Rubber Co. | 3,000 |
| 288s-14939 | June 1943 | New York Rubber Corp. | 1,500 |
| 288s-14964 | June 1943 | Goodyear Tire & Rubber Co. (Barr Rubber Products) | 3,000 |
| 288s-14965 | June 1943 | Firestone Tire & Rubber Co. | 5,000 |
| 288s-14966 | June 1943 | Rubbercraft Corp. of America Inc. | 2,000 |
| 288s-14967 | June 1943 | U.S. Rubber Co. | 6,000 |
| 288s-23650 | June 1944 | New York Rubber Corp. | 3,000 |
| 288s-23651 | June 1944 | U.S. Rubber Co. | 7,000 |
| 288s-24214 | June 1944 | Anderson Rubber Co. | 3,000 |
| 288s-20455 |  | Astrup Co. | Cases only |
| 288s-22633 | May 1944 | U.S. Rubber Co. | Cases only |

*(Below left) Factory workers at the Barr Rubber Co. undergoing final inspection of one-man parachute rafts, c.-1943.*

*(Below right) Factory worker at the assembly plant for U.S Rubber Co. with a USAAF AN-R-2 pararaft, c.-1943. (National Archives)*

*Known contracts and purchase orders with quantities for the USAAF for the one-man parachute raft.*

*The purchase orders and contracts in the March 1943 bracket extend into April (most being drafted in March, but not approved until April).*

*At the time of awards, contractors still had outstanding deliveries due from 1942 contracts, delaying production under March/April awards.*
*In this bracket, the total quantity procured was 40,000.*

| CONTRACT | AWARD DATE | CONTRACTOR | QUANTITY |
|---|---|---|---|
| P.O. 42-21278-P | June 1942 | U.S Rubber Co. | 750 (Service tests) |
| P.O. 42-21279-P | June 1942 | U.S Rubber Co. | 250 (Service tests) |
| P.O 43-541 | July 1942 | U.S Rubber Co. | 3,000 |
| P.O. 43-542 | July 1942 | Firestone Tire & Rubber Co. | 3,000 |
| P.O. 43-543 | July 1942 | Armstrong Rubber Co. | 1,000 |
| P.O. 43-544 | July 1942 | Hood Rubber Co. | 1,000 |
| P.O. 43-545 | July 1942 | General Tire & Rubber Co. | 1,000 |
| P.O. 43-9495 | October 1942 | General Tire & Rubber Co. | 1,250 |
| P.O. 43-9496 | October 1942 | Firestone Tire & Rubber Co. | 1,250 |
| P.O. 43-9497 | October 1942 | Armstrong Rubber Co. | 1,250 |
| P.O. 43-9498 | October 1942 | Hood Rubber Co. | 1,250 |
| P.O. 43-23467 | March 1943 | New York Rubber Corp. | 3,000 |
| P.O. 43-23468 | March 1943 | Hood Rubber Co. | 2,500 |
| P.O. 43-23469 | March 1943 | U.S Rubber Co. | 4,000 |
| P.O. 43-23470 | March 1943 | Armstrong Rubber Co. | 5,000 |
| P.O. 43-23471 | March 1943 | Kastar Inc. | 750 |
| P.O. 43-25428 | March 1943 | Unknown | unknown |
| P.O. 43-25429 | March 1943 | Unknown | unknown |
| P.O. 43-25430 | March 1943 | Master Tire & Rubber Corp. | 1,500 |
| P.O. 43-25431 | March 1943 | U.S Rubber Co. | 1,250 |
| P.O. 43-26257 | March 1943 | Firestone Tire & Rubber Co. | 2,500 |
| P.O. 43-27412 | March 1943 | International Latex | 1,500 |
| 535-AC-37926 | March 1943 | New York Rubber Corp. | 6,000 |
| 535-AC-37927 | March 1943 | Hood Rubber Co. | 5,000 |
| 535-AC-37930 | March 1943 | Unknown | unknown |
| 535-AC-390 | September 1943 | U.S Rubber Co. | 18,000 |
| 535-AC-391 | September 1943 | Firestone Tire & Rubber Co. | 18,000 |
| 535-AC-392 | September 1943 | New York Rubber Corp. | 15,000 |
| 535-AC-393 | September 1943 | Hood Rubber Co. | 15,000 |
| 535-AC-394 | September 1943 | Air Cruisers Inc. | 7,000 |
| 535-AC-395 | September 1943 | General Tire & Rubber Co. | 7,000 |

*Note: this list may or may not be complete.*

*BuAer Technical Notes pertaining to the one-man parachute raft.*

| | | |
|---|---|---|
| 1-43 | One-Man Parachute Type Life Raft –Packing , Operation, Inspection and Maintenance of. | 6 January 1943 |
| 41-43 | Parachute-Raft Assembly Quick Attachable Chest Type. | 11 June 1943 |
| 61-43 | One-Man Parachute Life Raft –Premature Inflation of. | 11 September 1943 |
| 75-43 | One-Man Parachute Type Life Raft-Damage by Chaffing. | 1 November 1943 |
| 76-43 | One-Man Parachute Type Life Raft and Back Pad Kits –Packing , Operation, Inspection and Maintenance of. | 12 November 1943 |
| 29-44 | One-Man Pararaft Protection Against Chaffing. | 21 March 1944 |
| 74-44 | Parachute Harness Quick Attachable Seat Type. | 30 August 1944 |

*BuAer Technical Orders pertaining to the one-man parachute raft.*

| | | |
|---|---|---|
| 27-44 | Life Raft-Carbon Dioxide Cylinder Valves-Safety Wiring of. | 6 April 1944 |
| 53-44 | Substitution of Life Jacket Dye Markers Packets for Specification AN-S-10 Dye Sea Mark ng Cans. | 17 April 1944 |
| 91-44 | Life Raft Inflation Gear Safety Sealing Wires. | 12 July 1944 |
| 114-44 | Attachment of Pararaft to Quick Attachable Chest Parachute | 1 September 1944 |
| 130-44 | One-Man Parachute Type Life Raft Cylinder Valves-Inspection of. | 10 November 1944 |
| 132-44 | Life Raft Corner Reflectors-Installation of. | 13 November 1944 |
| 9-45 | Pararaft and Equipment Container Case-Model-A. | 13 January 1945 |
| 14-45 | One-Man Parachute Type Life Raft Inflation System-Non-Interchangeability of. | 22 January 1945 |
| 17-45 | Method of Attaching the Pararaft Kit to the Navy Standard Seat Type Parachute. | 31 January 1945 |
| 18-45 | Navy Standard Back Type Parachute. | 6 February 1945 |
| 22-45 | Parachute and Harness Quick Attachable Seat Type with Method of Attaching Pararaft Kit. | 26 February 1945 |
| 24-45 | Pararaft Kit, Model PK-1 and Supplementary Air Sea Rescue Items of Personal Equipment Description, Inspection and Maintenance | 2 March 1945 |
| 37-45 | Pararafts-CO2 Cylinders-Replacement of. | 12 April 1945 |

*USAAF Technical Orders pertaining to the one-man parachute raft. After Revision A, the pararaft was included in Technical Order 04-15-1 which encompassed all life rafts and would remain detailed in all subsequent revisions through the duration of the war. The pararaft was first included in the revision of May 1943, later revisions would include the Type C-2 pararaft.*

| | | |
|---|---|---|
| 04-15-2 | Use, Maintenance and Inspection of One Man Parachute Type Life Rafts. | 18 September 1942 |
| 04-15-2A | Revision to include rectangular water and sea maker cans AN-W-5 and AN-S-10 only | 28 December 1942 |
| 04-15-1 | Handbook, Operation and Service Instructions with Accessory Parts List– Life Rafts | 26 May 1943 |

Procurement of the pararaft for the BuAer was similar to the USAAF in that new requirements for their issue to VPB, VTB and VSB squadrons were recommended.

Priority for distribution was to VF class, then carrier squadrons of planes of three persons or less, and if available, for crew members of PBY and PB2Y type aircraft (but due to their size they were more heavily equipped with other emergency provisions). Purchases and deliveries were maintained at a steady rate from 1943 into 1945.

As discussed in the chapter on emergency back pad kits, they were developed to be used in conjunction with the one-man parachute raft. Together they would provide a complete complement of survival provisions at sea and on land. The pararaft was for VF class only retaining the MK-I emergency equipment container, which was replaced by the back pad kit in the spring of 1943, when available. It was during this time that the pararaft was recommended for all carrier aircraft.

A new allowance of Standard Aircraft Equipment was authorized in November 1943 for the distribution of the pararaft to all crew positions in aircraft of three persons or less. VTB and VSB aircraft had and maintained the standard complement of multi-person life rafts in a designated compartment.

The back pad kit would remain authorized for the pilot only, they were however requisitioned and distributed among aircrew. Each pilot and aircrew member would also have the complement as detailed under allowances provided in the authorized list of Pilot's Personal Equipment.

Multiple deficiencies were reported through service activities, principally regarding bulk, and attempts were made to streamline the combination of pararaft and back pad kit. On 15 May 1944 the BuAer Director of Military Requirements distributed a memorandum to the engineering division stating that, after appraisal of combat reports, survival experiences, plane design, and physiological research, data indicated the need for certain changes in present survival equipment. It was desirable to eliminate the back pad kit and incorporate basic equipment into a seat pack, supplementing it by issuing other articles as personal equipment.

The design emphasis of the seat pack was towards signaling. After its principle components: the life raft, paddles, bullet plugs, bailing cup, sea anchor and line for lashing equipment came signaling equipment to include signaling mirror, sea markers, pyrotechnic kit and radar reflector.

Survival items were separated into two categories of "primary" and "secondary." Primary articles would include one can of water and two desalting kits, which together would provide water for 12-14 days, plus a water storage bag and three tablet rations. Secondary items included sunburn ointment, poncho, safety pins, matches or burning lens, survival booklet and simple first aid kit.

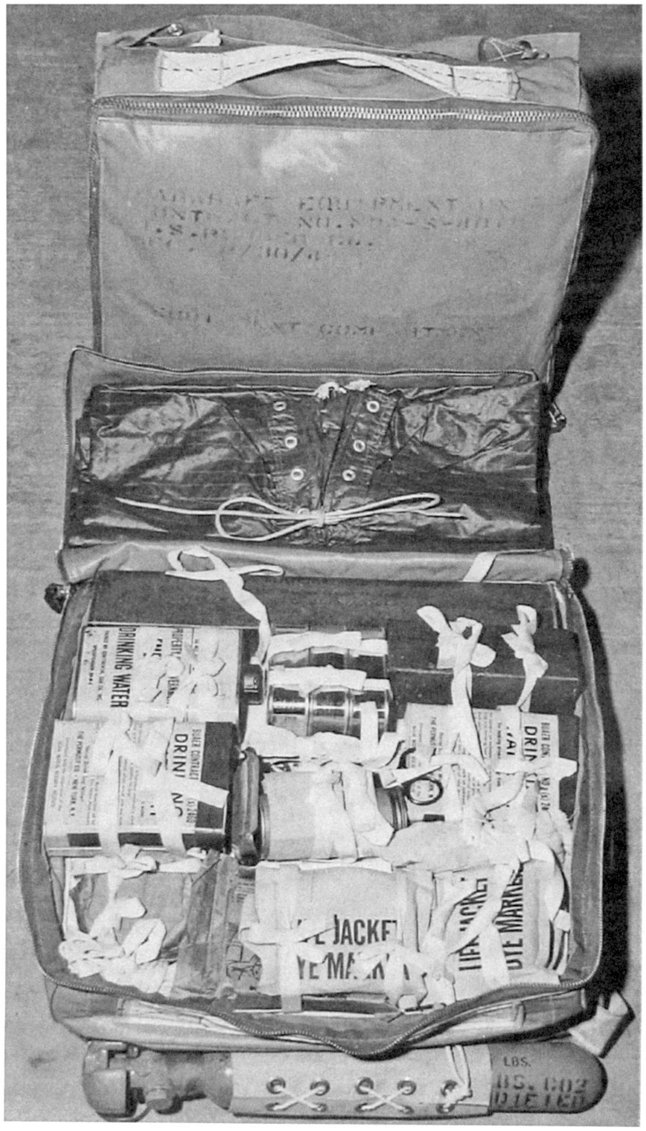


*(Right) Working in coordination with the New England Bedding Company, a suitable case was designed under BuAer Dwg. No. 9216 15" x 15" x 4" divided into two horizontal compartments. One half would contain the signaling and survival items, while the other half contained the life raft. In the original concept, a plastic container was recommended for containing the equipment, but was changed to tie tapes for securing items in place. The opposite side of the raft would be provided with snaps for quick opening, like that of the AN-R-2. 200 service test samples were procured from the New England Bedding Co. with all packing provided by them for service tests. Illustrated is one of the 200 seat packs designated as the Pararaft Model No.1. (National Archives)*

*(Left) The Pararaft Model No.1 is being demonstrated exhibiting its complete compliment of components and supplementary items at NAS Lakehurst, September 1944, where it was being service tested. At the bow of the raft is a packed assembly, open on both halves, illustrating the horizontal compartments. At the stern is an empty case illustrating the equipment side with tie tapes. In the inflated raft are the signal and survival items with the poncho being worn by a volunteer. To the left is the supplementary complement of Pilot's Personal Equipment, from bottom to top: life preserver with two sea markers, mosquito headnet, life preserver light, two hand held smoke signals, individual first aid kit, fishing kit and whistle with a nylon flying suit under them. New flying suits, including the nylon type, were designed to provide additional pockets for adequate stowage of the new items of personal equipment. (National Archives)*

For service testing it was recommended that the seat pack undergo simulated prolonged air combat patrol of at least four hours duration, in order to evaluate its comfort. Simulations were also recommended in abandon plane procedures. Unlike the AN-R-2 that was pressed into service before it was properly tested, the new seat pack was placed under controlled experiments to iron out any possible flaws that may be encountered.

After positive results, it was decided to adopt the new seat pack as the Model PK-1; Pararaft Kit NavAer Specification M-641, 1 November 1944. This specification officially superseded both the M-592 and AN-R-2, with both being cancelled for further procurement. M-592 would had been automatic upon adoption of the PK-1, but AN-6520-1 had to receive Army-Navy approval. With the USAAF in production of the Type C-2, authority to cancel it was given 20 November 1944. Existing contracts for the BuAer were allowed to be completed with manufacture extending into 1945. Before release of procurement for the PK-1, further refinement of details was required.

To avoid confusion and other manufacturing delays, organization between prime contractors and sub-contractors was organized. With all parties in agreement, four contracts were awarded 10 January 1945, for a combined manufacture of 50,000 PK-1 pararaft kits. Deliveries were expected to commence in April with commitments to the PTO. Air Cruisers Inc. was not expected to start production until June, as they were running at capacity, manufacturing rafts for the USAAF. The other three contractors would maintain the initial deliveries.

Upon award of contracts, the PK-1 was technically still under development, negotiating changes in the case and packaging of the components.

First aid provisions were no longer considered for inclusion with procurement of the individual aviator's first aid kit, which was added to the list of personal equipment. A magnifying lens was included in the fishing kit M-627, which was also added to the list.

The one item that was not carried over from the

M-592 was the machete, due to limited space availability. In lieu of the machete, an all-purpose giant jackknife was procured to be carried as personal equipment, replacing the sheath and scout knife.

By the close of January, all details orientated towards the pararaft and its supplemental compliment of personal equipment were determined. Limited quantities began reaching supply points in April, available under ASO stock number R83-K-709965.

The PK-1 with the Pilot's Personal Equipment was authorized for all crew positions in crews of three persons or less and VPB squadrons, replacing the M-592 and AN-R-2B.

Effective immediately, as of 7 May 1945, a directive was issued to discontinue installation of the back-pad kit and AN-R-2 in all carrier type airplanes only, and all squadrons commissioned as of 1 May were to receive the PK-1.

Technical Order 24-45, titled Pararaft Kit, Model PK-1, and Supplementary Air Sea Rescue Items of Personal Equipment Description, Inspection and Maintenance, was distributed 2 March 1945.

*Detailed are the known contracts for the Pararaft Model PK-1. Air Cruisers Inc. was not expected to start production until June as they were running at capacity manufacturing life rafts for the USAAF, the other three contractors would maintain the initial deliveries. The PK-1 would be superseded by the PK-2 BuAer Specification M-762 4 April 1946.*

| CONTRACT | AWARD DATE | CONTRACTOR | QUANTITY |
|---|---|---|---|
| NOA-5293 | January 1945 | Air Cruisers Inc. | 15,000 |
| NOA-5294 | January 1945 | New York Rubber Corp. | 5,000 |
| NOA-5296 | January 1945 | U.S. Rubber Co. | 10,000 |
| NOA-5319 | January 1945 | Rubbercraft Corp. | 20,000 |
| NOA-7056 | June 1945 | U.S. Rubber Co. | 3,500 |
| NOA-7151 | June 1945 | Rubbercraft Corp. | 1,500 |

*(Below) The overall construction of the raft remained unchanged from AN-R-2, but the bladder at the bow was now 10" in diameter from 12" of the AN-R-2. This reduction allowed the raft to be folded more compactly. Anti-chaffing carried over from AN-R-2B, with reinforcing around the CO2 inlet valve and a cotton bag around the bottle. The 16" oral inflation tube was to be screwed into the valve prior to packing, with a flap to cover the metal parts and snapped closed. Provisions for the radar reflector were now applied by the factory at the stern. The hand paddles and bailing cup were the same as outlined in AN-R-2 and were secured to the grab handles by 3' of cord prior to packing. The sea anchor was changed to a funnel shape and made of nylon, allowing it to be more compact when folded. A lanyard for attachment to the life preserver was provided and secured to the neck of the CO2 bottle.*

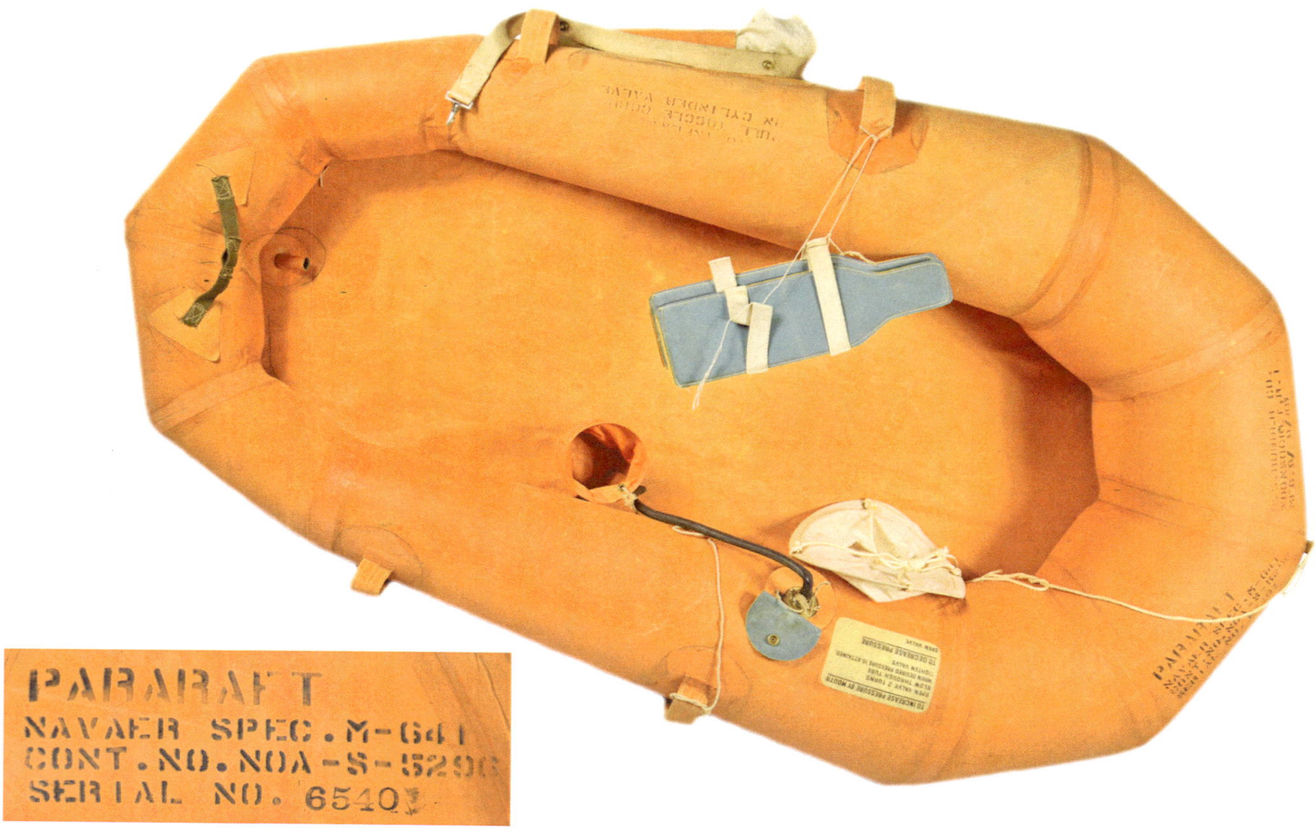

*(Above) The case as detailed in BuAer Dwg. No. 9216 was to be orange-yellow, differing from the field test sample of aircraft grey. The case eliminated the slot for the leg straps as it was it was no longer necessary with the introduction of the new pararaft container, and discontinued use of the standard seat type parachute, being replaced by the QAS. For comfort purposes the raft side was to be placed up. The flap was provided with snap fasteners on three sides with a grommet and lanyard for attachment to the life raft. Fully packed the complete unit weighs 24 pounds. The New England Bedding Co. was contracted for the manufacture of the cases.*

*(Top right) The poncho and water storage bag were placed on top*

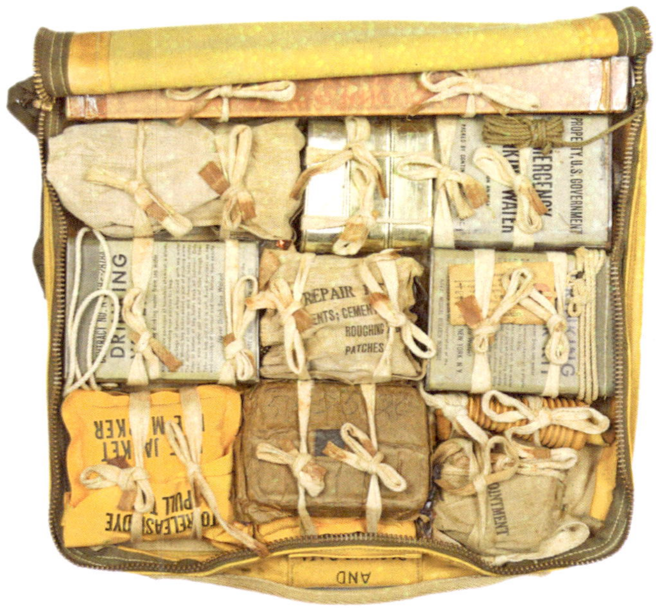

*(Right and bottom right) On the equipment side, identification and location of equipment was screen printed. This compartment was provided with a slide fastener. The supplemental Pilot's Personal Equipment included items that may be necessary prior to opening this compartment. Components were illustrated as screen printed on the case. Remaining components were secured in place by tie tapes.*

*(Below) In Amendment-2, dated 3 February 1945, the case was revised to include an inspection and data pocket with a card 5" x 3." This pocket was alternately used for the inclusion of the signal panel chart as detailed in Technical Note 91-44.*

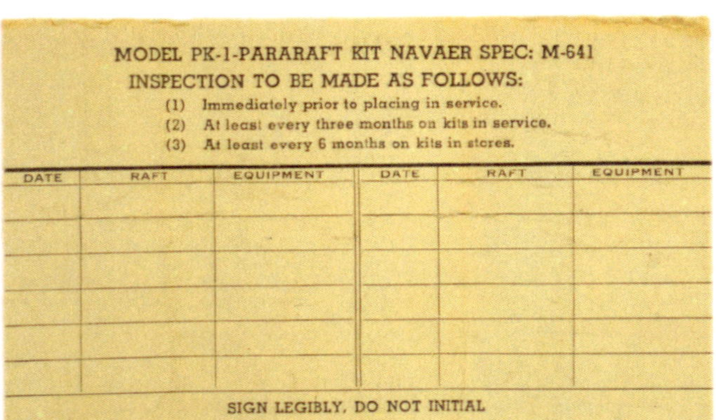

MODEL PK-1-PARARAFT KIT NAVAER SPEC: M-641
INSPECTION TO BE MADE AS FOLLOWS:
(1) Immediately prior to placing in service.
(2) At least every three months on kits in service.
(3) At least every 6 months on kits in stores.

| DATE | RAFT | EQUIPMENT | DATE | RAFT | EQUIPMENT |
|---|---|---|---|---|---|
| | | | | | |

SIGN LEGIBLY, DO NOT INITIAL

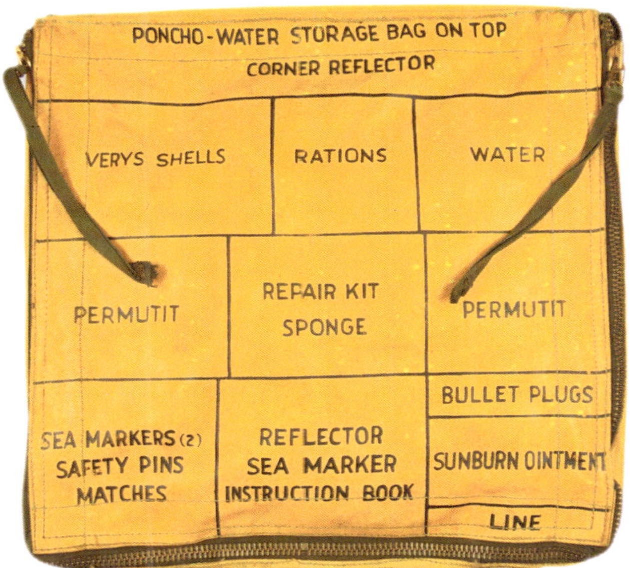

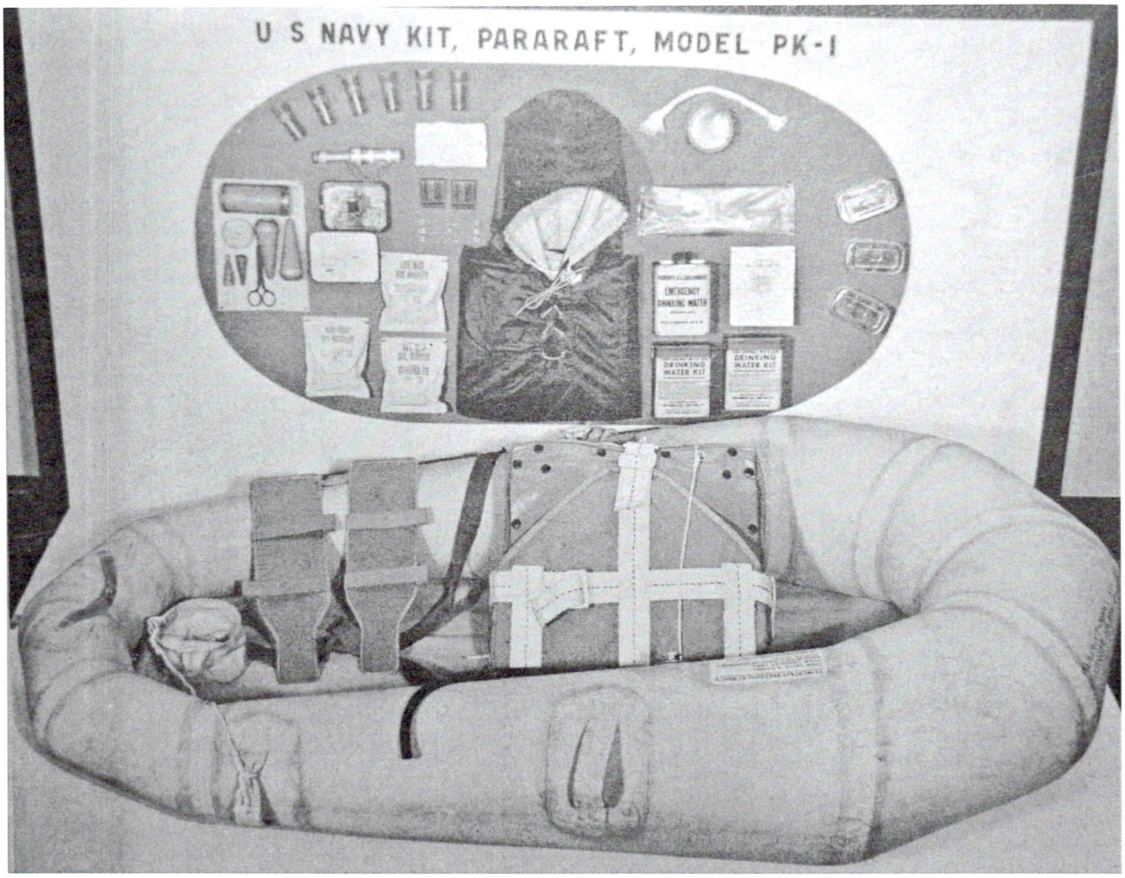


*Wartime image illustrating the complete complement of PK-1 pararaft accessories, March 1945. This unit and the supplementary personal equipment was a streamlined improvement over the M-592 and AN-R-2, retaining many of the same components by description.*

*(Below) The primary component was the raft. Accessories included hand paddles that remained unchanged from the AN-R-2, as with the bailing cup and bullet hole plugs. Improved items included a nylon sea anchor, provided as in specification M-3R for the MK-II life raft, stamped either PK-1 or MK-II, either were correct. The paddles, sea anchor and bailing cup would be secured and packed with the raft. Stored in the equipment compartment were the bullet hole plugs, sponge specification L-S-626 Size-4, 150' of nylon cord and a repair kit. The repair kit included 2 ounces of Type-B AN-C-54 rubber cement with three 2" and two 4" patches, packaged in a scrim covered waterproof wrapper.*

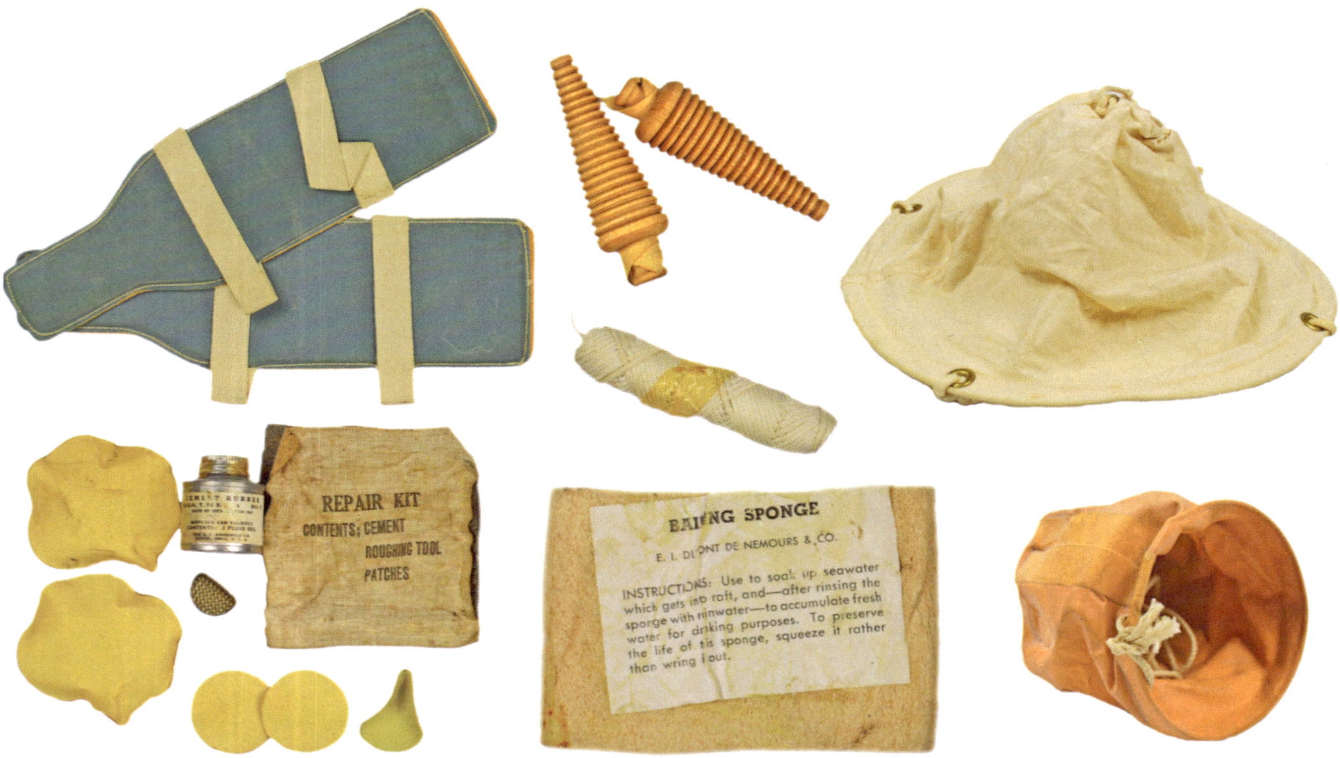

*(Below and right) The principle orientation for raft components was towards signaling. Provided was the MX-137/A radar reflector with attachments for installation on the stern of the raft as detailed in Technical Order 132-44.*

*(Below) Included were three M-566 sea markers, standard for emergency rescue equipment, as well as the M-580A signaling mirror and the MK-IV hand projector with six red Very signals. The hand projector and signals were packaged in a scrim covered waterproof wrapper. Since some supplementary items of personal equipment may have been needed prior to opening the seat pack, several were carried or worn on the person, including two MK-I MOD-0 smoke signals, .38 tracer ammunition, life preserver light and whistle. As discussed in the chapter on distress signals, T-49/M-75 two-star distress signals were optional as personal equipment. Multiple resources for signaling were at the immediate disposal of the individual, for day and night use.*

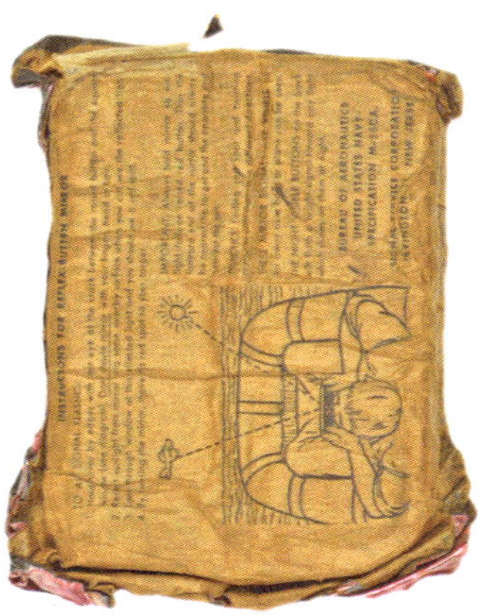

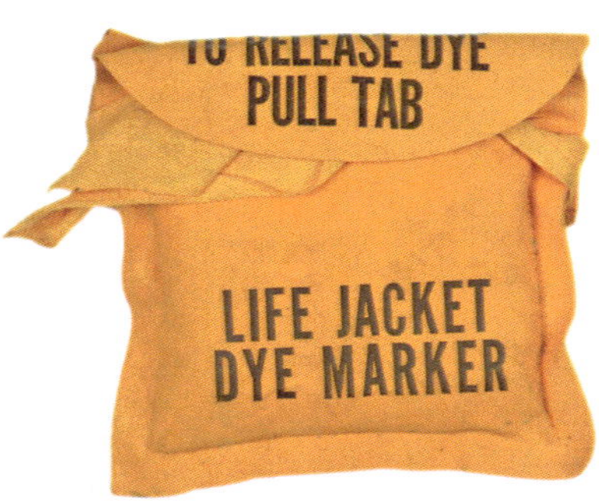

*(Above) Primary survival items included two M-613 chemical drinking water kits, one can AN-W-5B drinking water, three M-539B tablet rations and a five quart water storage bag BuAer Dwg. No. 8985.*

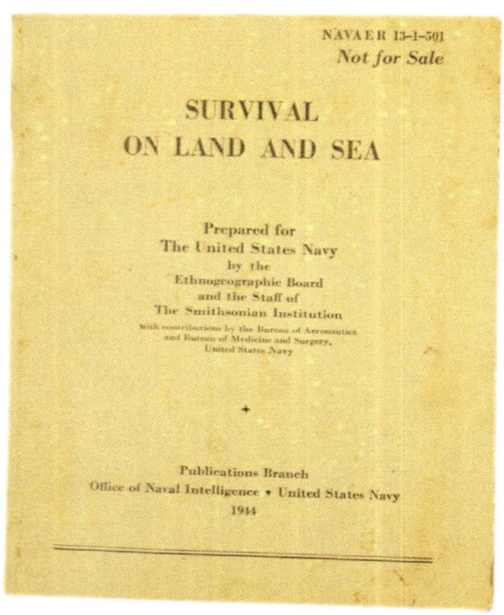


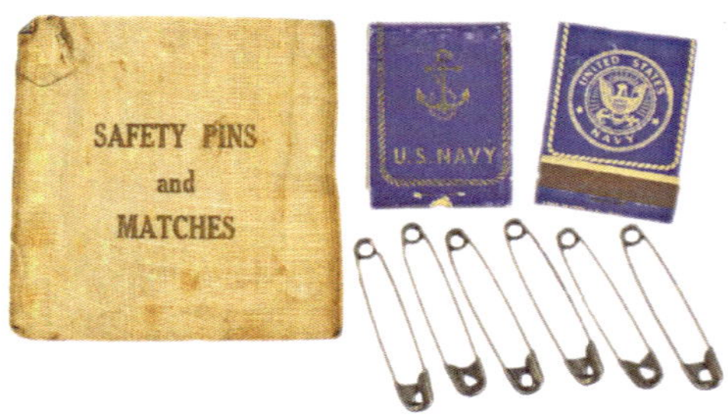

*Secondary survival items included a 4 ounce can of sunburn ointment and two books of safety matches with six two-inch safety pins, packaged in a scrim covered waterproof wrapper. A survival booklet titled Survival on Land and Sea – NavAer publication 13-1-501, was also included. The poncho was a revised form of the poncho raft cover, P-61 (from the M-592). The P-61 was too bulky for inclusion, under BuAer Dwg. No. 9217 the poncho was trimmed to a rectangular shape, eliminating the elastic edges. The hood was retained, made from a rip-stop nylon, colored non-specular blue on one side and orange-yellow on the other, with snap fasteners on two sides.*

*(Below) To better streamline the pararaft, a new cylinder was developed under NavAer Specification F-19, NAF Dwg. No. 49277, dated 1 December 1944. The new bottle measured 11-1/2" in length with a 2" diameter (compared to the AN-C-73 of 10" in length and a 2-1/2" diameter). It also utilized a new lever type valve, manufactured by Walter Kidde & Co. Technical Order 37-45, of April 1945, stated that all cylinders used on AN-R-2 rafts were to be updated with this new valve, available under stock number R58-WK-81336 and that they were already provided on pararafts specification M-641. As per guidelines established in September 1943, an instruction decal was applied. For anti-chaffing, a cotton sleeve was provided before installation on the life raft. A 3' cotton tape was provided for securing around the CO2 valve.*

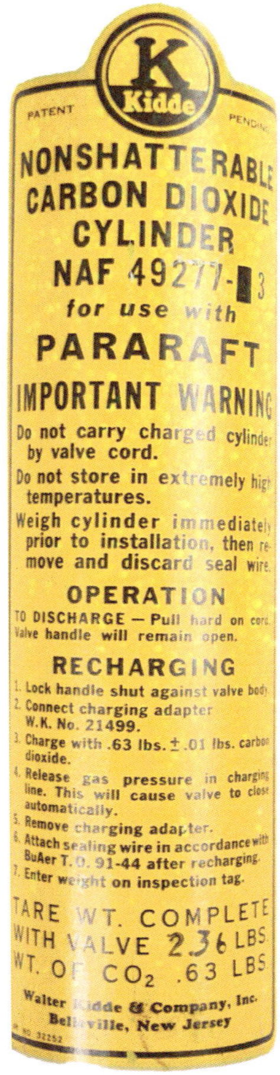


*(Below) The raft was packed with the cylinder towards the front. Two extra snaps were provided on two corners of the case, to permit the retaining line to emerge from either the right or the left. Male studs were provided on the lanyard to secure it in place. The purpose for choice of placement was relative to whether the seat or chest type parachute was being used. For use with the QAC, the clip should emerge on the left (as seen below. The raft side would theoretically be on the user's chest, with the front pointed down and the clip on the same side as the D ring of the life preserver.*

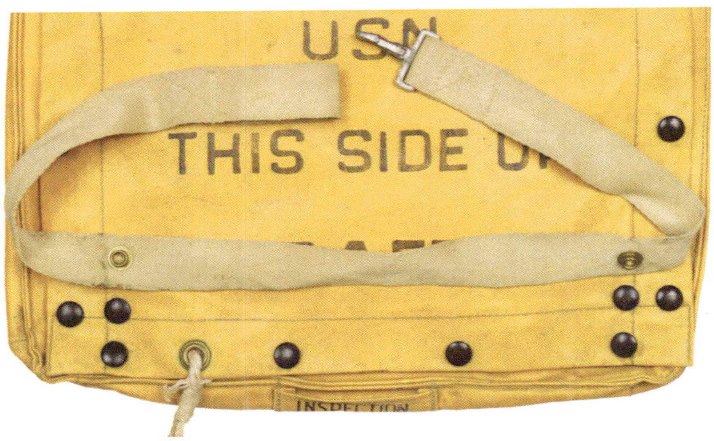

*(Below) When being used with the QAS and thus forming a seat cushion, the clip from the raft lanyard would need to emerge from the right (as shown) in order to line up with the D ring on the life preserver.*

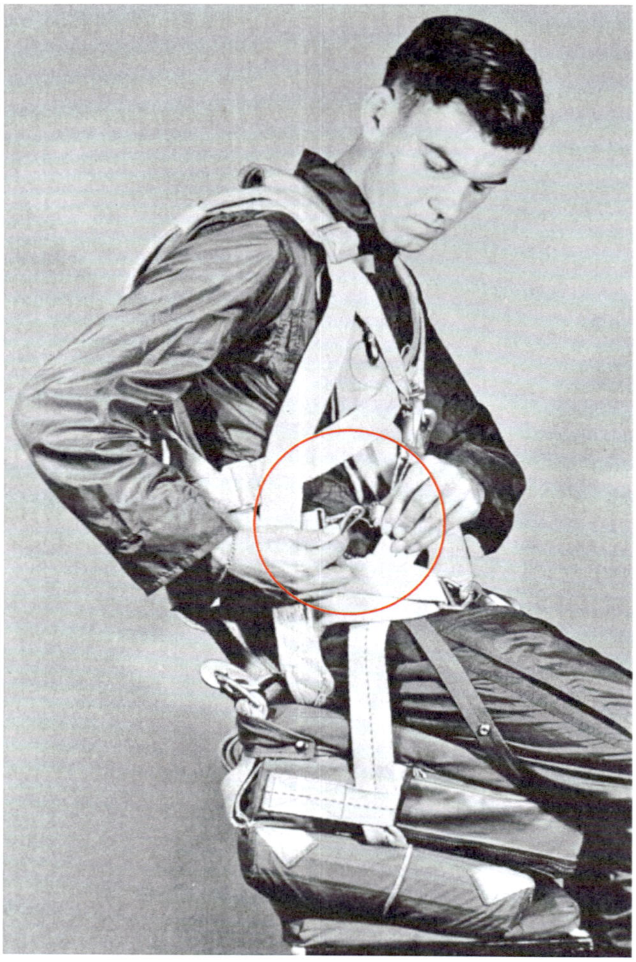

As part of the design emphasis of the PK-1, basic equipment was included in the seat pack, with other articles issued as supplementary personal gear. This personal gear was formally referenced as Pilot's Personal Equipment; items were systematically added to this list throughout the duration of the war. Aside from personal mandatory flying clothing, all other articles were squadron property and supplied as per allowance lists. Additionally, many articles that were available would be at the discretion of the Aviation Equipment Survival Officer (AESO).

Using carrier based aircraft as a base of reference, parachutes and pararafts were issued per aircraft, life preservers were provided on allowance per squadron, according to number of operational aircraft and their crew capacity, plus a percentage for spares. These would be the mandatory items utilized by pilots. The life preserver would be the first article issued as Pilot's Personal Equipment. As the war progressed, it became mandatory to issue one sea marker with it, then two. In January 1943, a side arm and sheath knife were added, followed by the issue of tracer ammunition, then a life preserver light etc. progressing with service requirements.

The bulk of emergency equipment, prior to the adoption of the PK-1, was developed at various times, as were items added to the list of Pilot's Personal Equipment, which created issues of integration and unnecessary bulk which compromised practicality. The development of the M-641 took conscious steps to combine the desirable features of the M-592, AN-R-2 and complement of personal equipment.

*(Far left) Ens. Gordon Marlow, a pilot and member of Air Group-10, wears typical flying clothing and personal equipment for Naval and Marine aviators from 1942 and into early 1943.*

*(Left) As the war progressed, new articles of clothing and personal equipment became available as can be seen in the picture of VB-9 pilot LT. jg. John Purio on the flight deck of the USS Lexington, February 1945. As part of his personal complement, he wears the AN-V-18 life preserver with three sea markers, life preserver light, whistle and a jackknife. He is wearing the original back pad kit as part of the complement of Standard Aircraft Equipment. (National Archives)*

*(Below) As detailed in Technical Order 24-45 titled Pararaft Kit, Model PK-1 and Supplementary Air-Sea Rescue Items of Personal Equipment Description, Inspection and Maintenance, the following items were to be issued in conjunction with the pararaft: One life preserver with two sea markers, one .38 caliber pistol, one shoulder holster, one wrist compass, one life preserver light, 17-L-16, one hat headnet M-565A, one fishing kit M-627, one giant jackknife, one whistle 75W1, one individual aviator's first aid kit, two distress signals Mk-I MOD-0. As part of description, life vests were to be updated as per Technical Note 15-45 with the addition of the smoke signal pockets. By this time, waterproof covers and tracer ammunition would have been provided for the revolver as standard. Shown below are the items outlined for use in conjunction with the PK-1 M-641 pararaft available to pilots and aircrew.*

*(Below) The new seat pack was to be universal for all climatic regions, with discussion on providing additional items regionally as needed, but it was determined this would create supply problems. By the time of standardization and eventual deliveries and distribution complying with Technical Order 24-45, other optional supplementary items including a scout knife, shark chaser and T-49 distress signals had become available. To supplement the PK-1 for use in tropical regions, the MK-I emergency flask kit was available, with the eventual addition of the Asiatic Kit. Illustrated are examples of additional optional items.*

Further steps were taken to accommodate service requirements and concerns, and in order to continue eliminating the undesirable conditions presented by the back pad kit, the BuAer initiated procurement of a new case, detailed under BuAer Dwg. No. 9216. The new case was designed to enable components from the M-592 and AN-R-2 to be placed in one seat pack. Procurement and manufacture of 20,000 empty cases was awarded to the New England Bedding Co. in November 1944, under contract NOA-4932, as an interim measure until the PK-1 became available.

Limited deliveries commenced in January 1945, continuing into May to central supply points. Once delivered, they could be requisitioned under ASO stock number R83-C-8670 titled as Case, Pararaft and Equipment Container, Model-A. Technical Order 9-45, titled Pararaft and Equipment Container Case, Model-A, was distributed 13 January 1945, directing which items from the M-592 and AN-R-2 were to be removed and placed in the case. Distribution and use would be the same as that outlined for the PK-1 and by Technical Order 24-45.

This contract was for the case only, with no provisions included. Dismantling the M-592 and AN-R-2 from stock and transferring the components to the case, as per Technical Order 9-45, was the responsibility of the receiving station. The case was manufactured as per the original sample, specified in AN-R-2 to be aircraft grey. No inspection and data pocket was provided. The contractor provided a basic, single-sided instruction sheet in an envelope attached to the case.

*(Below) Marine parachute rigger Sgt. Ira Bowman (Center) with two other squadron members of VMTB-143 exhibiting the Model-A pararaft, c.1945. (Courtesy Adam Lewis)*

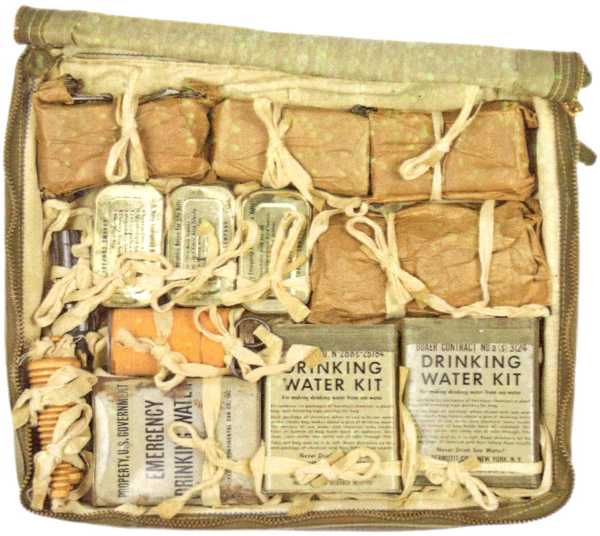


*(Top) Front and back of the new case.*

*(Above left) Components were to be secured in the equipment side with tie tapes, some being layered. Unit shown is packed as per technical order and the instruction sheet.*

*(Above right) The top layer of items removed exposing the components below. Added to the compliment from stock were two M-613 desalting kits and a bailing sponge with all other items organic to the M-592 and AN-R-2.*

*(Right) The single-sided instruction sheet provided by the contractor, detailing the packing of the Model-A. This sheet is a streamlined version of Technical Order 9-45. Items deleted from the AN-R-2 were the sail and first aid kit. Items deleted from the M-592 included the poncho, fishing kit, headnet, salt tablets, safety pins and tape, sharpening stone, mag lens, oil, jackknife, and machete. A radar reflector was to be added when available, projected in March 1945.*

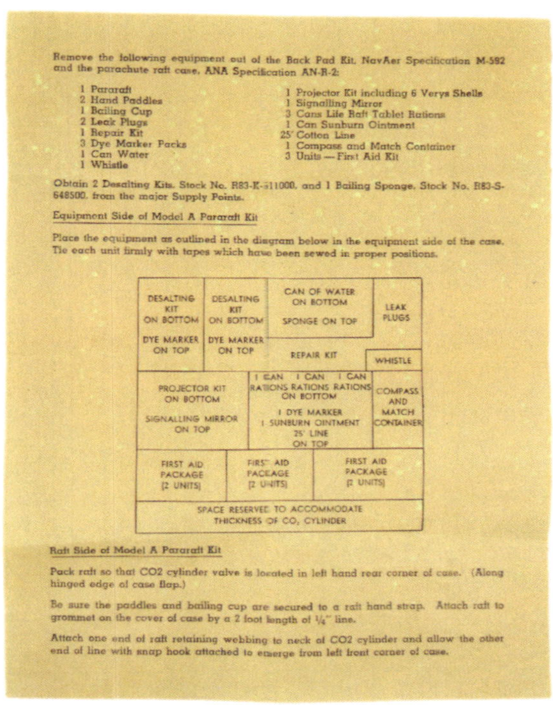

Remove the following equipment out of the Back Pad Kit, NavAer Specification M-592 and the parachute raft case, ANA Specification AN-R-2:

1 Pararaft
2 Hand Paddles
1 Bailing Cup
2 Leak Plugs
1 Repair Kit
3 Dye Marker Packs
1 Can Water
1 Whistle
1 Projector Kit including 6 Verys Shells
1 Signalling Mirror
3 Cans Life Raft Tablet Rations
1 Can Sunburn Ointment
25′ Cotton Line
1 Compass and Match Container
3 Units — First Aid Kit

Obtain 2 Desalting Kits, Stock No. R83-K-511000, and 1 Bailing Sponge, Stock No. R83-S-648500, from the major Supply Points.

Equipment Side of Model A Pararaft Kit

Place the equipment as outlined in the diagram below in the equipment side of the case. Tie each unit firmly with tapes which have been sewed in proper positions.



Raft Side of Model A Pararaft Kit

Pack raft so that CO2 cylinder valve is located in left hand rear corner of case. (Along hinged edge of case flap.)

Be sure the paddles and bailing cup are secured to a raft hand strap. Attach raft to grommet on the cover of case by a 2 foot length of ¼″ line.

Attach one end of raft retaining webbing to neck of CO2 cylinder and allow the other end of line with snap hook attached to emerge from left front corner of case.

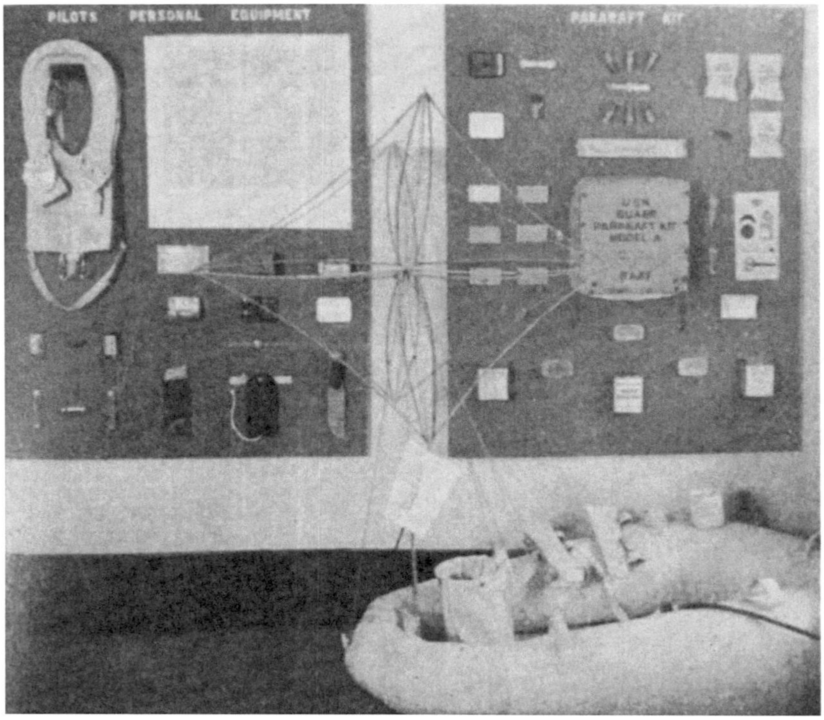


(Left) Display board at NAS Quonset Point R.I. exhibiting the Model-A and the supplement of Pilot's Personal Equipment, c.1945. The AN-R-2 life raft has been updated with the radar reflector. Equipment shown includes the life preserver with two sea markers and below it are two MK-I MOD-0 smoke signals and whistle, then two M-75 signals and a life preserver light. To the right, a cloth chart with the individual aviator's first aid kit below, plus the addition of the M-580A signaling mirror, which was optional for issue. A wrist compass is displayed along with the mosquito headnet, shark chaser and sheath knife. The use of the giant jackknife was optional and dependent on availability. Many pilots and aircrew retained the sheath or hunting knife.

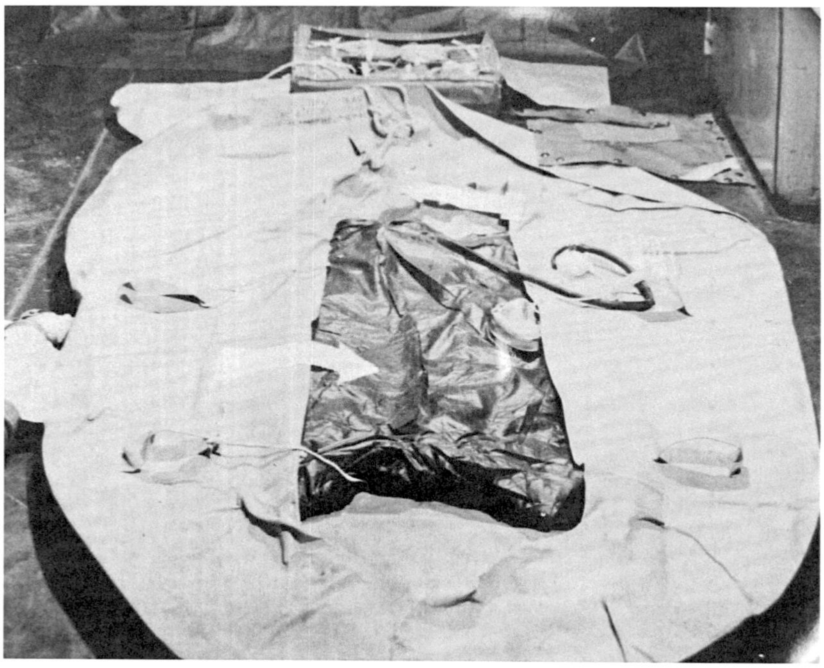

(Left and below) Conversion of the AN-R-2 pararaft to conform to the combination seat pack as detailed in ComAirPac Technical Bulletin 15TB45, dated 28 February 1945. The case was loosely converted to that of Dwg. No. 9216 by eliminating the slot and pockets. Tie tapes were provided on the bottom of the case for the components of the M-592, the poncho was laid out in the raft then folded and placed on top of the components. This conversion required no significant amount of additional material or fasteners. (National Archives)

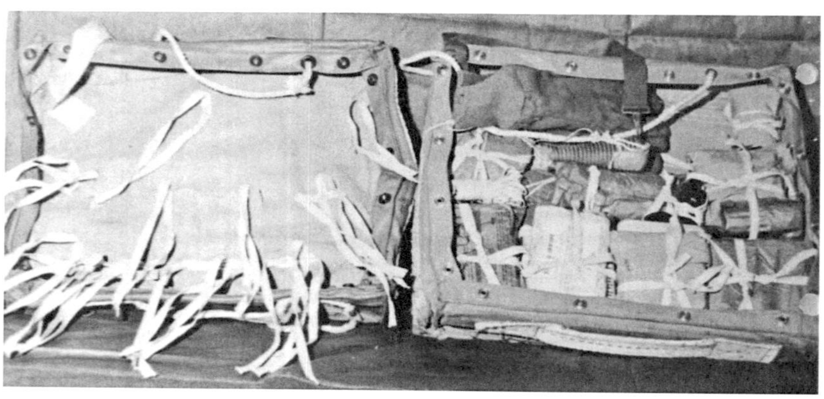

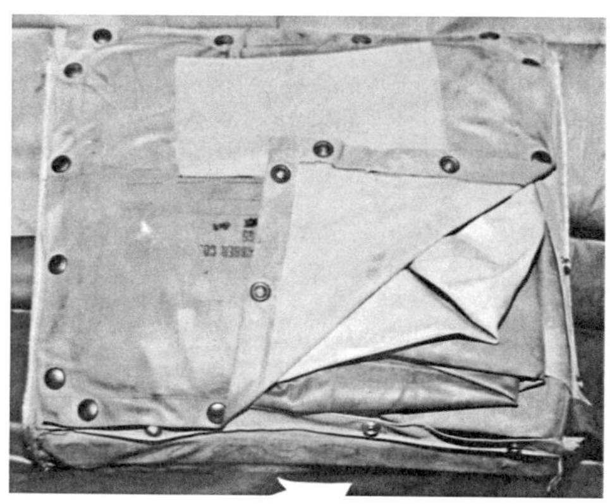

For operational requirements over water, the BuAer was certainly better equipped than they had been, having continually been improving their equipment since before the war. The USAAF was not, resulting in many deaths. USAAF Bombardment and transport aircraft were equipped with multi-person life rafts, but fighter or pursuit types carried no raft for the full year of 1942. In February 1943, the 8th Air Force reported that, in air-sea rescue activity to date, 62 out of 210 RAF aircrew had been saved. Out of 194 American aircrew, only 14 had been saved.

This statistic was the direct result of not having personal pararafts when bailing out. In November 1942, various representatives of commands within the 8th Air Force held a conference to discuss the status of flier's safety equipment. In the meeting it was agreed that there was a critical, deplorable, and confused condition that ultimately affected the lives of airmen, particularly when bailing out over water. By this juncture, enough combat experience had been gained to realize that American units were inadequately equipped in many respects. At this time there were no pararafts of U.S. manufacture available in that theater, and consequently fighter pilots had not been equipped with them. Only American pilots flying RAF Spitfires were so equipped.

At the conclusion of the meeting it was determined that flyers should be equipped with a suitable life raft to be worn on all combat operations. In further review, after the meeting Technical Order 04-15-2 was evaluated, showing the type of pararaft in development in the U.S. It was ascertained that this raft pack would be entirely unsuitable for the use for USAAF fighter pilots, because the seat pack pararaft and seat type parachute would be far too bulky for the limited space in cockpits.

By January 1943, arrangements had been made for the supply of the RAF Type-K Dinghy in the Type-B Back Pack, under reciprocal aid, in the quantity of 150 per month and continuing for 6 months. It was also during this time that the Bomber Command expressed the desire to distribute one-man rafts to aircrew, in addition to the provisions of the multi-person types. Bomber aircrew had been operating under the theory that the rafts would be available to them when dropped or detached from the aircraft by automatic switches. Crew members who had bailed out would swim back to the ditched aircraft and board the rafts. It was this theory that lead to many casualties.

In reality, crew members were often at a considerable distance from the aircraft and unable to reach its location, a situation compounded after dark and in high seas. Most critical was the water temperature. Survivors would need to reach the raft in a short time or risk hypothermia.

*Unknown P-38 pilot of the 94th FS is utilizing a Type-B back pack dinghy, Italy c.1944. (National Archives)*

*Unknown P-51 pilot lends another view at the Type-B dinghy pack with the lanyard and quick release snap, England, c.1943. (National Archives)*

(Left and below) Type-B RAF Dinghy Pack, Stores Reference Stores No. 27C/1900. It was constructed from a heavy canvas and designed to replace the standard parachute back pad cushion on seat and Observer type harnesses. The unit would incorporate the same components as the seat type packs. The external pocket was for a bail-out oxygen bottle. (Geoff Pringle Collection)

(Below) The Type-B pack was the initial model provided to USAAF fighter squadrons, but British resources were limited. Also issued was the Type-A Seat Pack, seen here worn by 8th AF fighter ace Lt. Colonel Francis Gabreski, England, January 1944.

(Below right) Type-A pack being utilized by the USAAF 27th Fighter-Bomber Group in North Africa, July 1943. There were multiple variations of the Type-A. This version included tabs with snap fasteners for securing to the parachute pack. (National Archives)

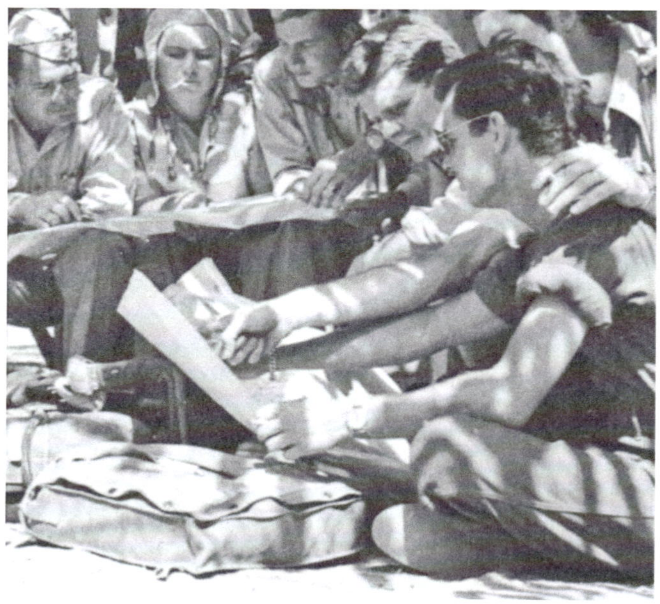

*(Right) During the meeting in November, it was determined that there was an urgent need for a parachute harness and dinghy pack for all flyers. The requirement was split into two primary groups: fighter pilots and crews of Medium and Heavy bombers. It was established that suitable equipment would be obtained from the British under reciprocal aid, for a limited time until supplies could be furnished from the U.S. A one-man raft was required for each member of bombardment crews to supplement the multi-person raft, in the event of a bail out. To satisfy the immediate requirements, dinghies were distributed upon availability to crew positions. Being an expedient, they were not always easily adaptable for attachment to the parachute harness. A Type-A Mark-III can be seen in the seat of the radio operator in this B-17, England, c.1943. (National Archives)*

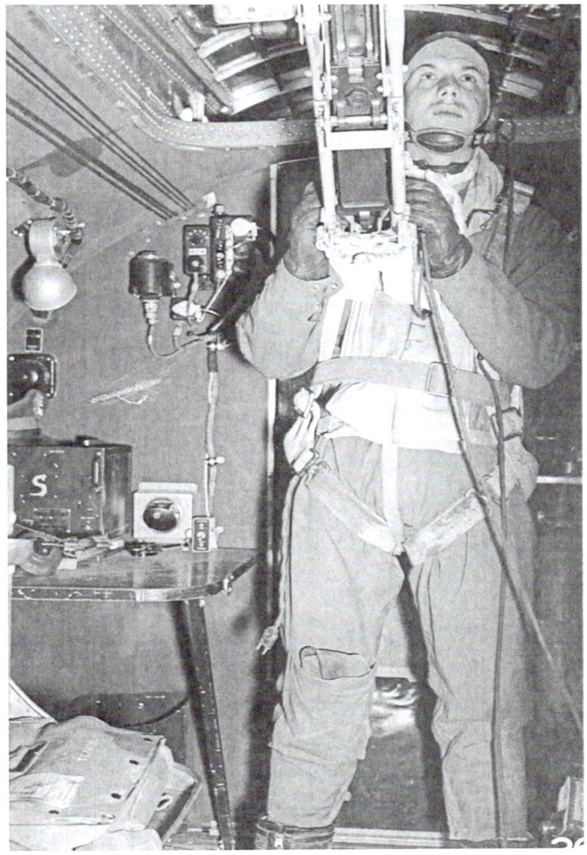

*(Below) Two examples of the RAF Seat Pack, Type-A. Top row is the Type-A MK-II, Stores Reference 27C/1899. Bottom row is the Type-A MK-III, Stores Reference 27C/2245. Both included a typical application of a four-language instruction label for Allied fliers in English, French, Polish and Czech. The principle difference was the attachment method. Four tabs were provided on the Type-A MK-II pack for attachment to the seat type parachute pack, with a lanyard and connector attaching to the life vest. The Type-A MK-III used special connectors for attaching to the parachute harness. It retained the lanyard and connector for attaching to the life vest. (Remi Bouche Collection)*

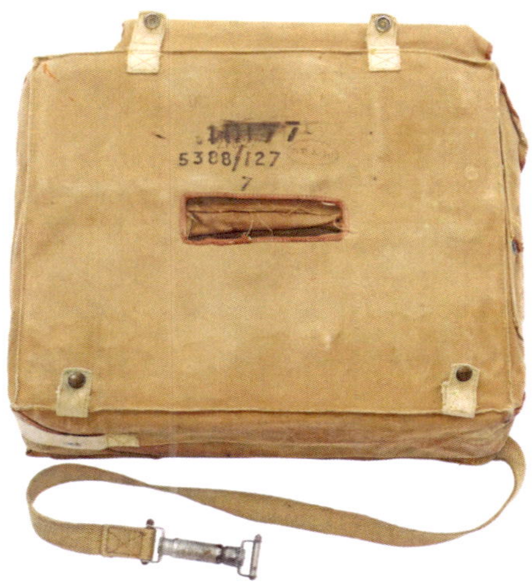

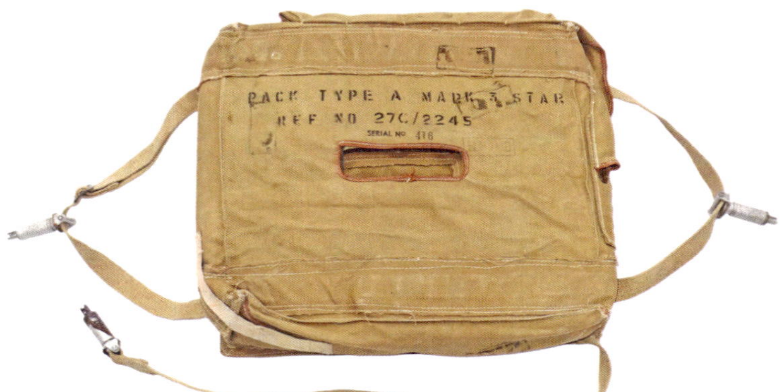

In January 1943, another meeting was held to further discuss and refine the requirements by fighter and bomber commands. The specific purpose was to determine what type of parachutes were desired and which type of dinghy was preferred, to find out if any changes were necessary and determine recommendations for U.S. units preparing for embarkation to the theater of operations.

During this time, five different parachutes were being utilized which created logistical issues. It was agreed that the best parachute-dinghy combination for bomber crews was that of the British Observer type harness (QAC) together with the quick attachable dinghy pack Type-C. Steps were taken to modify standard U.S. harnesses for conversation to British types.

At this time the Type-B dinghy pack with seat type parachute was preferred by fighter organizations. This would transition to the favor of the back type parachute, which required use with a seat pack dinghy with the Type-C pack the preferred model.

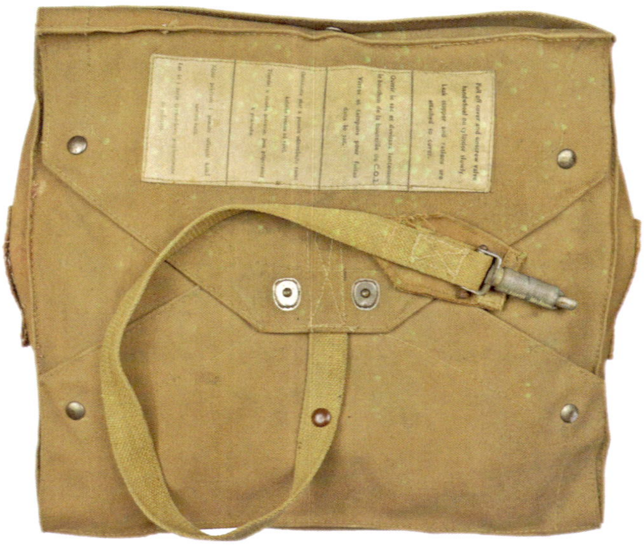

*The new Type-C dinghy seat packs were procured for use with the Observer parachute harnesses. The basis of issue was the pre-established guidelines, stating life rafts would be provided when engaged in operational flying over water. However, all operational flights in the ETO from England were performed over water, leaving only a limited supply to outfit squadrons until quantities manufactured in the United States could be obtained. The Type-C pack was another variation of pack for the Type-K Dinghy, improving upon the Type-A pack, which remained in service for use with seat type parachutes. The new pack was updated for quick attachment to the parachute harness and a new quick release of the lanyard. The case consisted of two separate pieces, referred to as an inner and an outer, together creating one unit with an envelope closure. Illustrated left is the inner case that contained the accessories. In the bottom left corner is a grommet with a line that connected to the raft. At the top is the lanyard that connected to the life vest.*

(Left) B-26 crew members load all their flight gear onto a jeep to be transported to the flight line, c.-1944. On the hood is a combination of back and chest type parachutes along with RAF Type-C dinghy packs. (National Archives)

(Below) USAAF personnel undergoing training with the RAF Type-K Dinghy. In 1942, the raft was updated with a spray shield rolled and stored in two sections at the bow and stern. The spray shield gave the occupant protection against the elements, alleviating exposure caused by weather conditions prevalent in the English Channel. A two-sectional wooden thwart, spanning the bladders and inserted into pockets, was provided for erecting a sail. The thwart supported the telescoping mast and sail, with guy lines connected at three points on the life raft for stability. (National Archives)

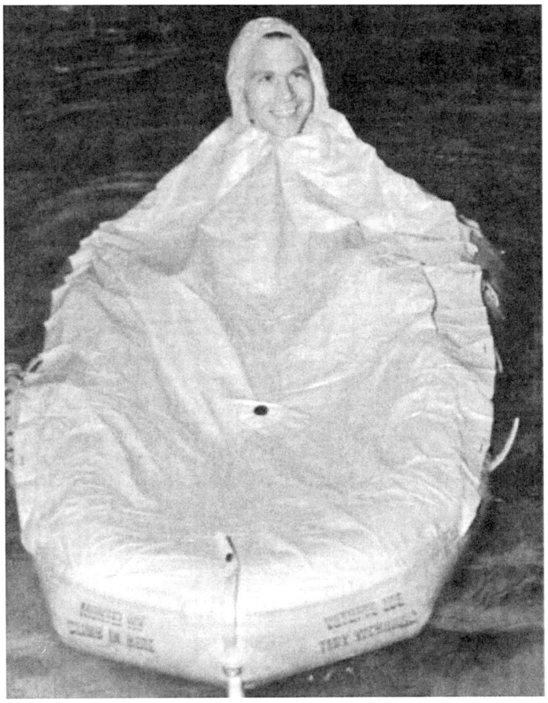

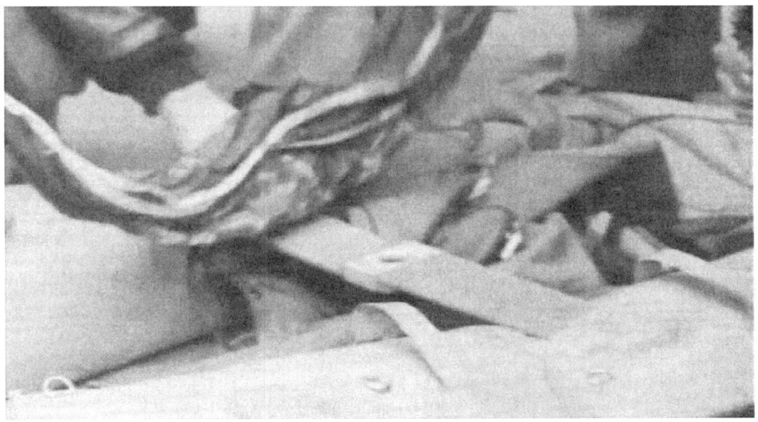

*(Below) As part of its compliment of accessories, the Type C Dinghy pack included the life raft, sea anchor, bailing cup, bellows pump, one pair hand paddles, two sets of nested bullet hole plugs, sail with mast, wood thwart, six MK-II distress signals stored three each in pouches, and one Mark-II flying ration. The exact dates of revisions and development of the Type-K accessories are not known, but it would also include a compass, heliograph mirror and bailing sponge. Shown are some of the accessories included with the Type-K dinghy.*

*(Below) The Type-A, Type-B and Type-C packs would be updated with a quick release snap. The male end would be integrated in the raft pack with the female adapter attached to a life preserver or another article.*

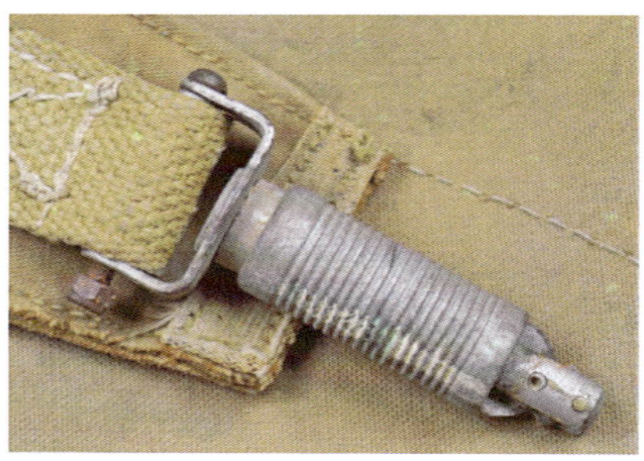

*(Left) A USAAF pilot has the female adapter on his Type B-4 life preserver at the waist strap connection, indicating he was utilizing one of the RAF dinghy packs.*

*(Right) Established from a meeting in December 1943, the 8th and 9th Air Forces would require 1,500 British type dinghies, which were found to be more satisfactory than the US type. This would supplement demand in 1944 until the new Type C-2 pararaft became available. Due to a shortage, these dinghies would be delivered without CO2 cylinders. The Director of Maintenance was ordered to fabricate adapters to facilitate the exchange and integration of US cylinders for use on the Type-K Dinghy. Coupling nuts were required for the transfer.*

*(Below) An example of the machined coupling nut for a US CO2 cylinder for use on the Type-K Dinghy.*

With preliminary guidelines established, Maj. Randolph Chaffee of the 8th Air Force flew back to the United States. His objective was to assist in the development and procurement of various items of individual flier's safety equipment. He conducted multiple meetings at Wright Field and Washington, expressing the recommended and requested features, and asking that they become standard for all fighter and bombardment crews in all theaters.

As part of his mission, the AN-R-2A was recommended to be revised to include quick attachable snaps and changes to its accessories. In the meetings from March and April 1943, two principle conclusions were established: first, that current pararaft manufacture was not to be delayed; and second, that modifications were to be minimal. Two modifications were outlined for the AN-R-2A: that the case should include quick attachable snaps and the life raft be provided with a spray shield. Deliberation of the addition of the spray shield extended until the end of 1943, and it is not evident this modification was ever applied to contracts of the AN-R-2A. It was concluded that the additional list of changes be field tested, which were projected to span a six-month period for engineering development.

The features presented by Chaffee included a collapsible mast with sail, a keel, a small compass for navigation, the use of floating wooden paddles in place of the present wire and fabric types, coating of the wooden paddles on one side with a reflective material, inclusion of rations as well as drinking water, inclusion of pyrotechnics or smoke signals and a bellows pump, among other minor adjustments in features. The recommended changes encompassed what was already included in the Type-K Dinghy. Provisions for the supply of pararafts to all crew positions and all aircraft operating over water were addressed in USAAF Military Requirement No.62, Supplement #2, 24 April 1943. This action would require a significant increase in the production of one-man parachute rafts.

*(Right) By September 1943, the new designs would progress to look more like that of the RAF Type-K Dinghy. A spray shield was applied to an AN-R-2 pararaft for field tests evaluating the concept and integration around May and June 1943. This was the preliminary model, used for all further development and each test including additional items and configurations.*

*(Right) One item included in the preliminary tests is item #6 in the photograph, an experimental flexible tube CO2 inflation system. (National Archives)*

*(Below) U.S. manufactured case, which is almost a direct copy of the RAF Type-C pack. (National Archives)*

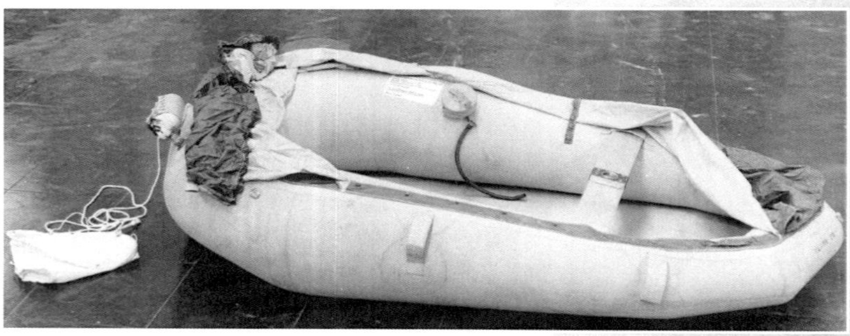

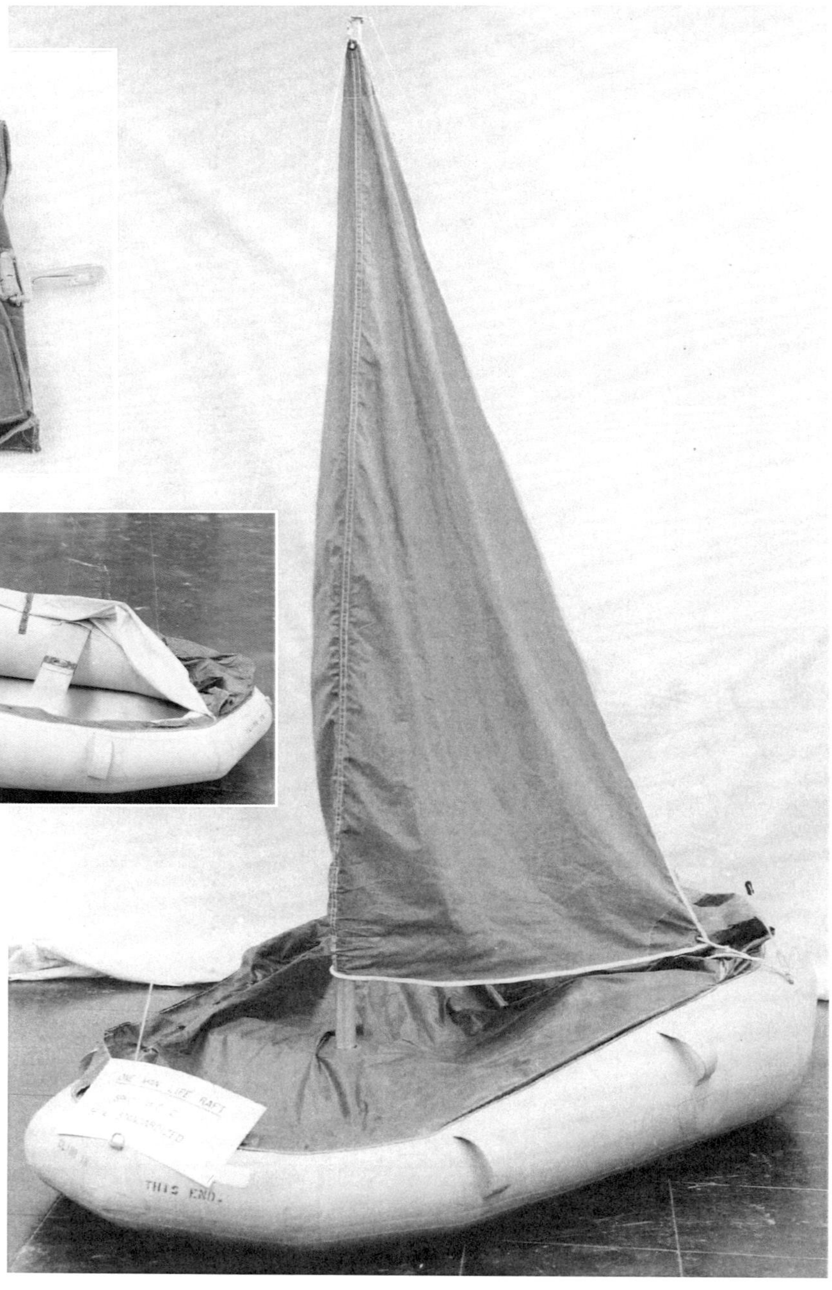


*(Above and right) By October 1943, the new pararaft began to take its form, building off the Type-K Dinghy. Preliminary specifications were drafted in December and designated Raft, Pneumatic, Type C-2, USAAF Specification No. 94-40725. These preliminary specifications were then distributed to potential contractors of all components for bidding. Final refinements were still being conducted into 1944. Illustrated are pre-production C-2 pararafts. (National Archives)*

*(Above and below) With all refinements calculated, specifications were released for standardization, 24 March, 1944. The Type C-2 was fabricated similarly to the AN-R-2, but with a reduction in bladder diameter, like that of the PK-1, to 10". Additional basic instructions were stenciled on the bladder of the life raft. The spray shield was vulcanized along 2/3 of the raft bladder, and the remaining circumference fitted with snap fasteners. The oral inflation tube was decreased to 6", and the raft was also provided with a bellows, hand pump and bailing cup. The CO2 bottle was moved towards the back as to not interfere with paddling action on the sides. A water pocket was added to the bottom for stabilizing, acting as a keel, and in turn contained the funnel shaped fabric sea anchor (Dwg. No. 44D18541). Upon release for standardization, change orders were also drafted and distributed to contractors on existing contracts for the AN-6520-1. Contracts 33038-AC-390 through 395 were authorized for change to the Type C-2 in April 1944 and tagged for expedited, priority shipment to the ETO. 5,000 were ready for overseas shipment by the middle of June. The total change order consisted of 24, 211 units, which was the remaining quantity outstanding for delivery of the AN-6520-1 on those contracts. Specification No. 94-40725 was approved 5 May 1944, with official standardization awarded 1 April 1944, to supersede specification AN-R-2A. Further quantity contracts were executed in August and into 1945. Once in service, they were to replace the AN-R-2 which, in turn, created surplus issues with discussion on converting the AN-R-2 to Type C-2 specifications – though this occurred on a limited scale. The life raft assembly was referenced under Dwg. No. 44J8538 and the complete assembly with case and accessories under Dwg. No.44G18535.*

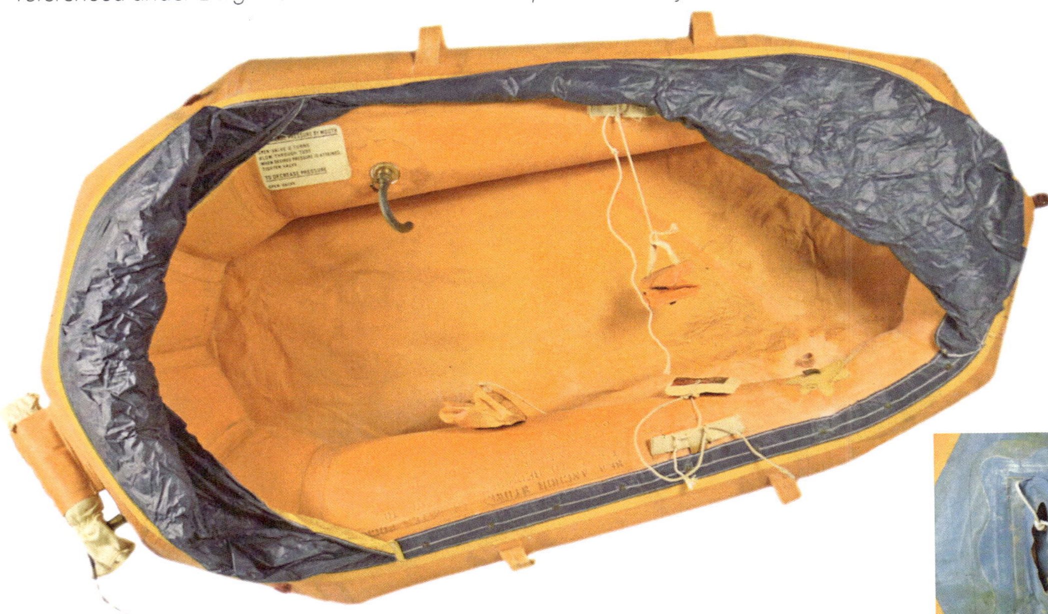

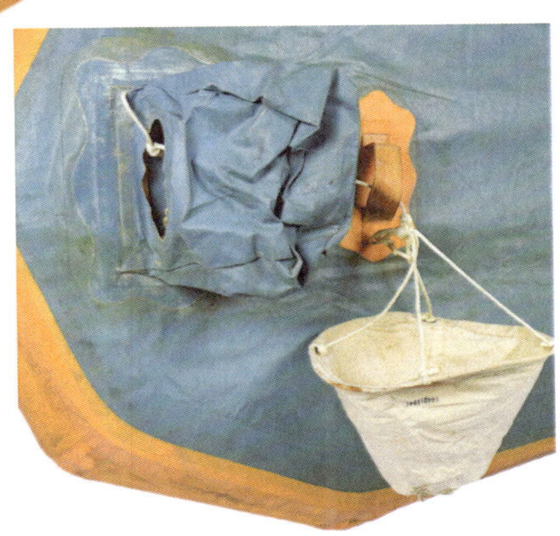

(Above) The original design of the case was almost identical to the RAF Type-C pack, but with a spring clip on the life preserver lanyard, and tabs provided at each corner for attachment of a seat cushion. This example was provided on the change order contract 33038-AC-393 from the Hood Rubber Co. This assembly was referenced under Dwg. No. 44B18565.

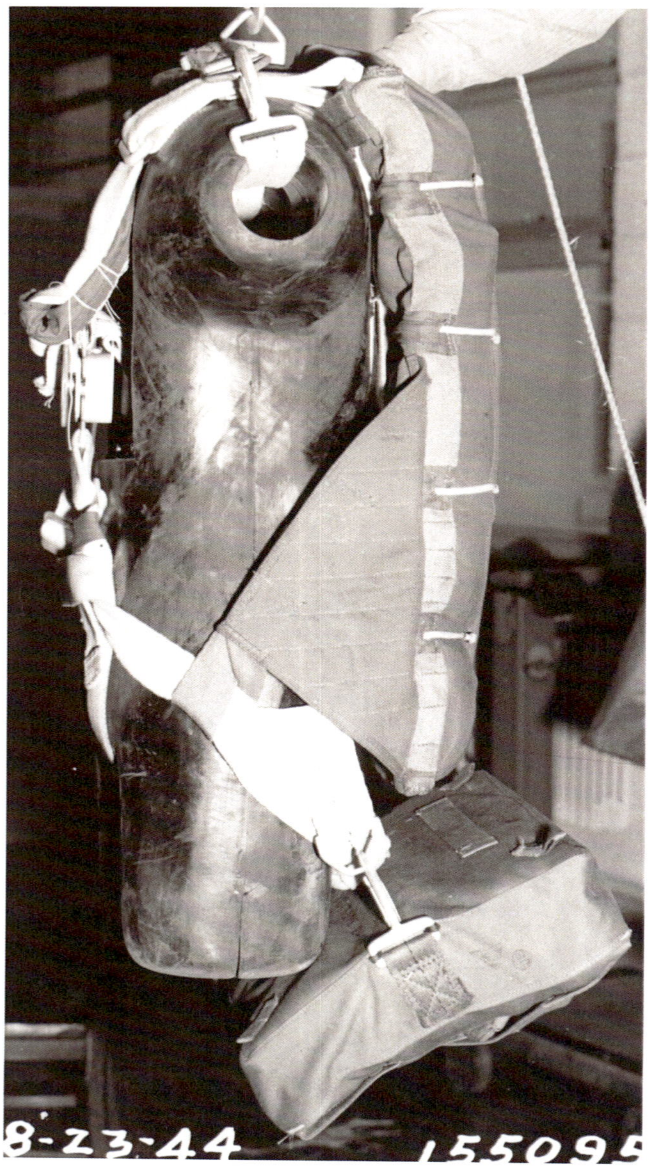


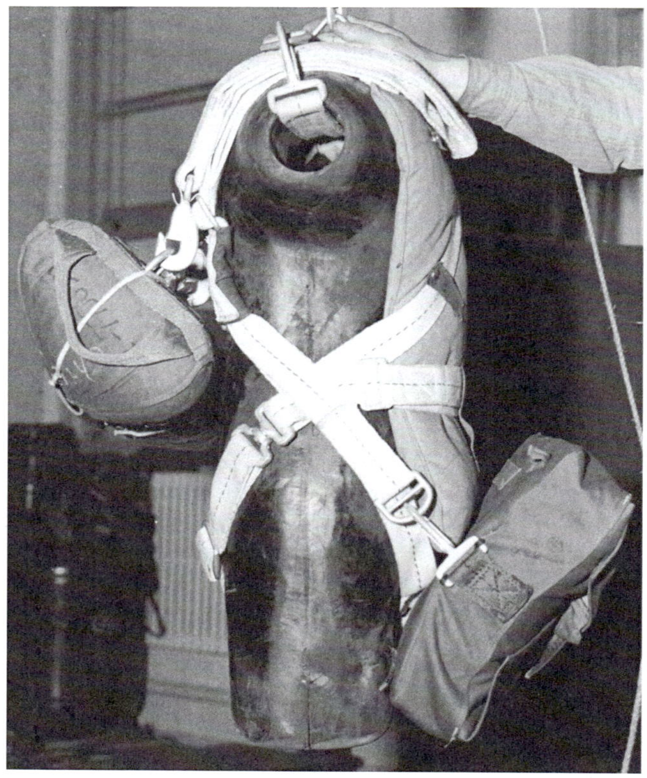

Left and above) Official images illustrating the correct attachment to the QAC and back type parachutes. (USAF)

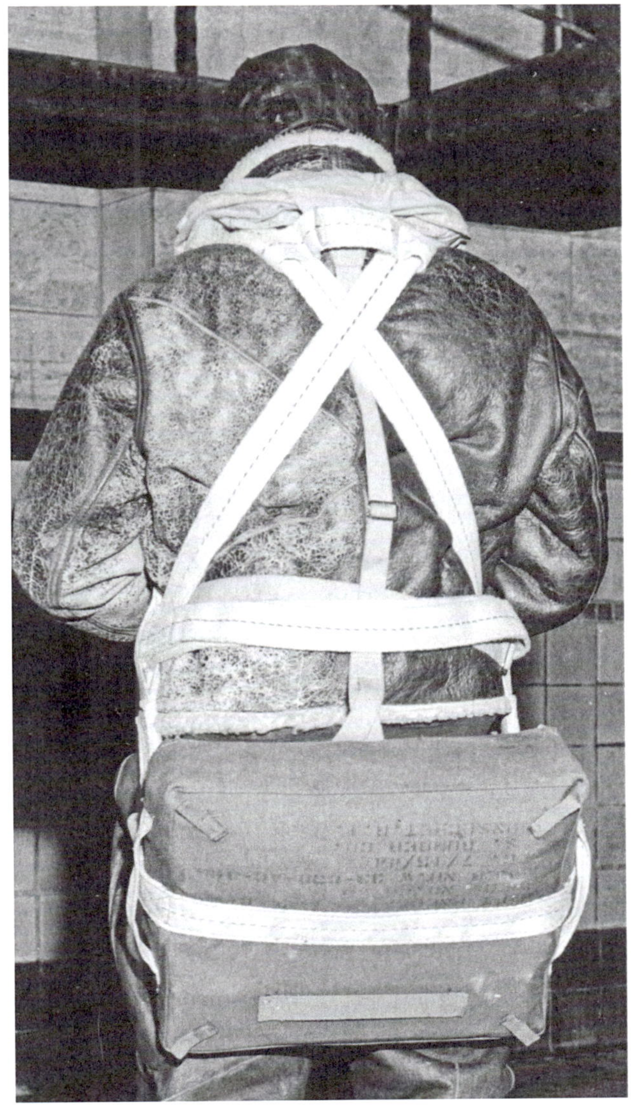

*(Above and below right) Service use and drop testing proved the configuration problematic in that the opening shock of the parachute would prematurely eject the raft assembly out of the case. To remedy the problem, Technical Order 04-15-10 was drafted and distributed 18 November 1944 to reverse the snap hooks. An example of a converted case is illustrated with the snap hooks reversed as applied by a service depot. The user would now sit on the envelope opening.*

*(Far right) The new revised case is demonstrated. (USAF)*

*(Below) Contractors were instructed to conform to the conversion under Technical Instruction TI-1713. This included converting to use of a commercial snap hook by National Automotive Fibres Inc. under their part No. N-1218 (National Automotive Fibres Inc. was also a contractor for the manufacture of cases, as was the Astrup Co.).*

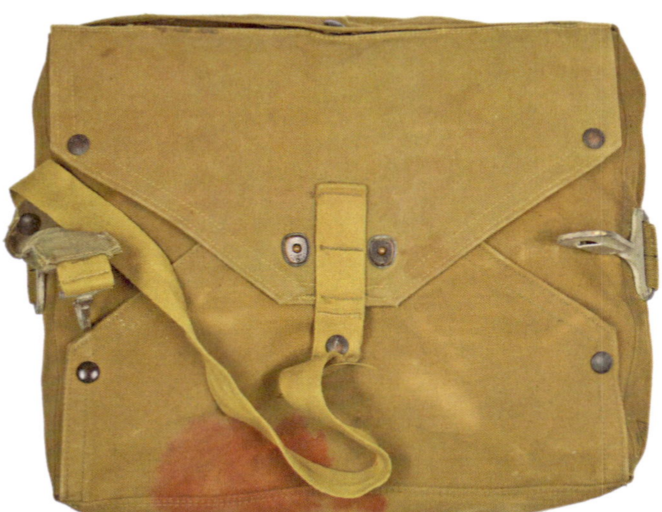

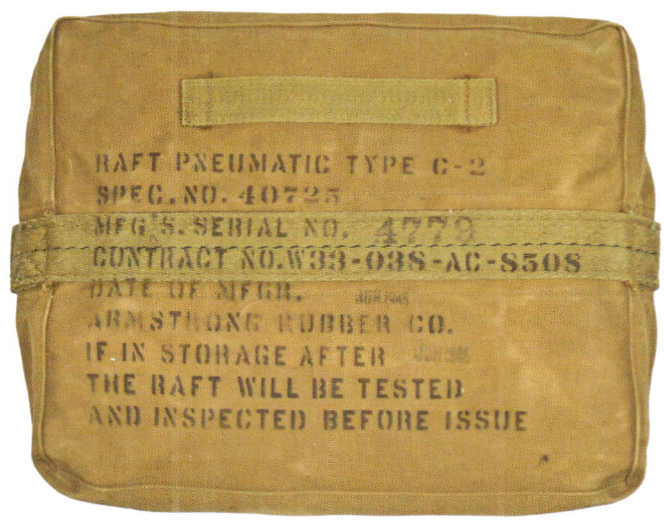

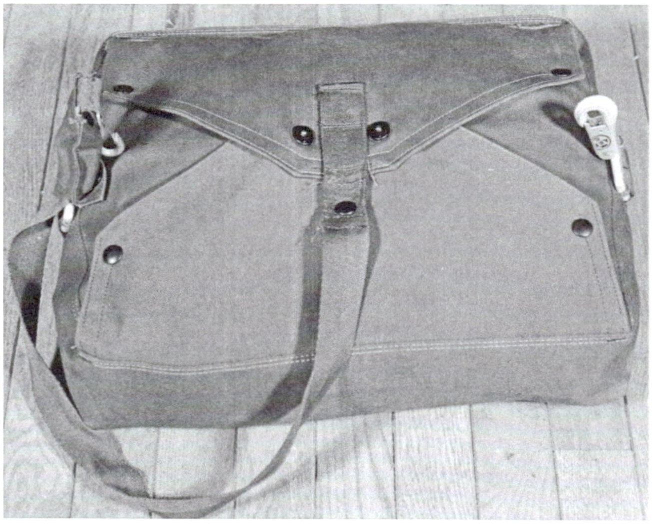

*(Above) Additional example of a factory converted pack with reversed snap hooks.*

*(Right) Factory applied reverse snap hooks. All cases manufactured after November 1944 would be in the reversed position. As part of final packing instructions of the complete assembly, the end of the life preserver lanyard was provided with a pocket and tab for stowage purposes and ease of locating. This was to be slipped over one of the snap hooks as shown.*

*(Below left) The inner case provided stowage for the accessories not attached to the life raft. Shown are which pockets the components were to be packed in as per specification.*

*(Below center) A pocket and strap were provided on the underside for the paddles/mast bridge.*

*(Below right) All accessories packed in inner case. Sail and mast in its cloth bag to be secured as illustrated. When complete and interlocked with the outer case, the complete assembly weighed 18 pounds.*

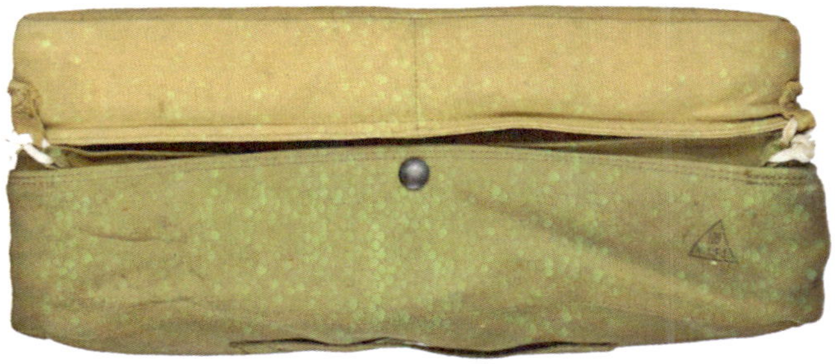

(Above left) In the inner case, a grommet was provided for a 3' length of nylon cord to be secured to the life raft CO2 bottle prior to packing.

(Above center) Integrated into the factory manufactured cases, after November, were two tabs on the back of the inner case. The inner case is shown locked to the outer by a single snap and to the left and right are tabs for attachment of the seat cushion. It is not secured at the four corners but with the two it does create a hinge for access to the envelope opening.

(Above right) Detail of corner tie.

(Below) Accessories for the Type C-2 pararaft. As per original specification, it was to contain the drinking water kit JJ-1, but due to the unavailability, principally through 1944, rafts were supplied a compliment of the water can AN-W-5. The need for distress signals was stressed before development, with the recommended inclusion of six, but this was finalized with the addition of four. Production of the signals had not commenced upon the initial deliveries, and was to be supplemented with the British Mark-II upon arrival in the ETO. As per specification, a cotton drawstring bag was provided for the signals. During field tests conducted in 1943, the repair kit and bullet plugs as provided in the AN-R-2 were found unsatisfactory, and it was decided to adopt repair plugs by A. Schrader & Sons under their part No. 2635-AK, assigned Dwg. No. 44B18563. Three plugs were provided. Also included was the signaling mirror, Specification No. 94-40653, two Vinylite bags by S. Buchsbaum & Co., Part No. SB-208, for water stowage, and other articles such as a pistol, compass and knife. A container of grease was provided, stowed in one bag for preservation (Dwg. No. 44A18567, provided by Livingston Plastics). A sponge, L-S-626 was also included, along with two wooden hand paddles with an applied retroreflective material. When interlocked, the paddles acted as the bridge or thwart for the sailing mast.

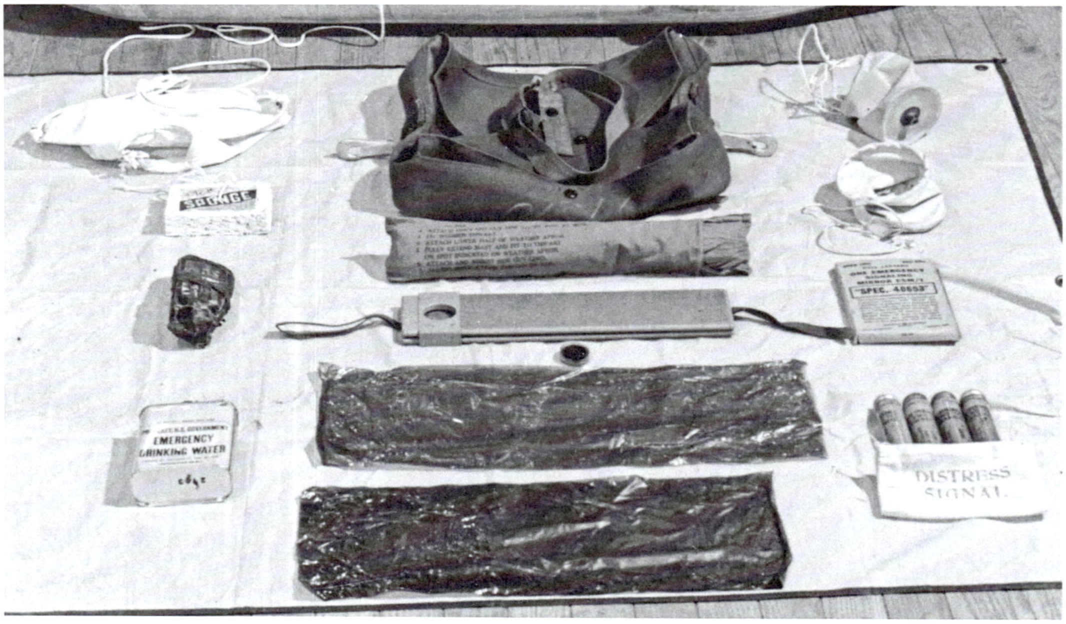


*(Left) Principle compliment of accessories for the Type C-2 life raft assembly. As mentioned, the drinking water kit JJ-1 would not become available until early 1945. The T-49/M-75 distress signals became available by January 1945. The signals included in this image are RAF type. (USAF)*

*(Left) For manual inflation, a 6" rubber oral tube was provided along with a rubberized fabric bellows pump, Dwg. No.44D18568.*

*(Right and below) The CO2 bottle provided would be as described in specification AN-C-73 but, as seen here, the bottle reflects specification No. 94-40725 and type designation of C-2. As part of original specification, they were to be provided with a lever valve as detailed in Dwg. No. 44D9888. Being a new item they were not available for the initial deliveries and early production rafts would be provided with the wheel type valve. Upon availability they were to be updated and marked as such on the case, as illustrated below. The Andrews coiled tube bottle (far right) included a lever type valve. A cotton bag was provided for the cylinders to prevent chaffing.*

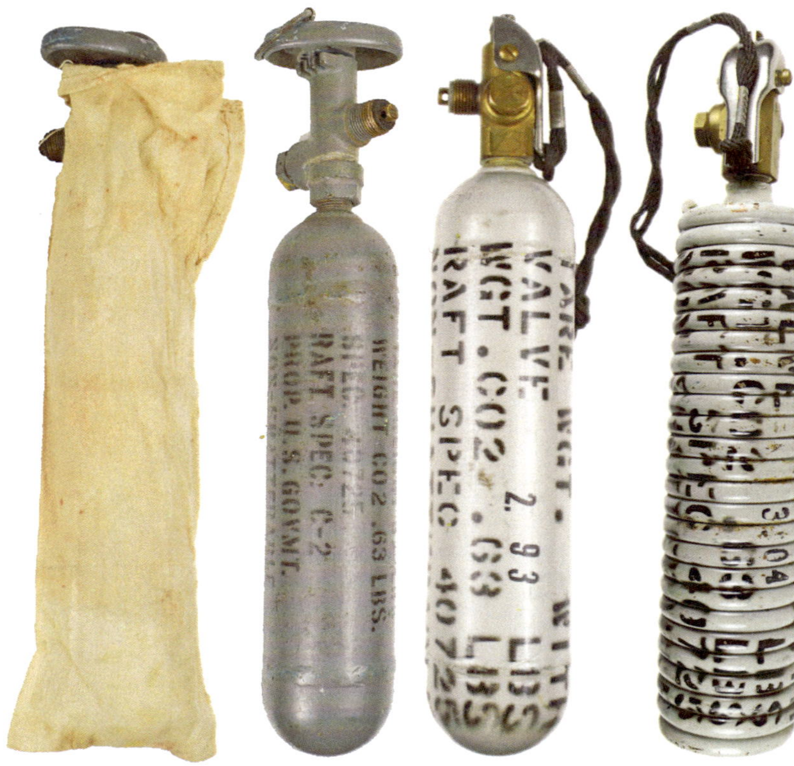

(Below) In 1944 two new signaling devices were being prepared for inclusion in life raft equipment: the MX-137/A radar reflector for one-man rafts; and the AN/CPT-2 radar beacon. In September 1944, modification kits were procured for service application in both the Type C-2 and AN-R-2. Technical Order 04-15-11 described the installation and its position in the raft as shown, additionally stating that Type C-2 life rafts manufactured after 1 March 1945 would have this application factory applied. Bottom right is an application on a C-2 raft. When the radar reflector or AN/CPT-2 beacon became available the sail and mast were to be removed, being stowed in its place in the inner case.

(Below) The two-piece spray shield was manufactured like the sails, being non-specular blue on the external surface and orange-yellow on the other. The volunteer to the right demonstrates the design principles, affording protection from the elements that may be encountered at sea. The two-piece design allows the user to extend his arms outside for paddling or sailing the raft. (USAF)

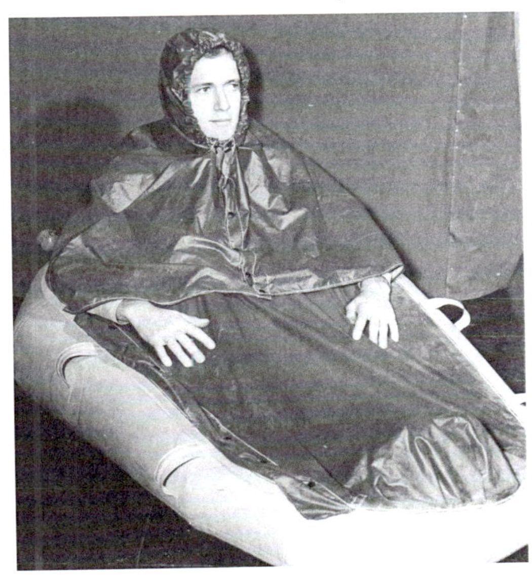

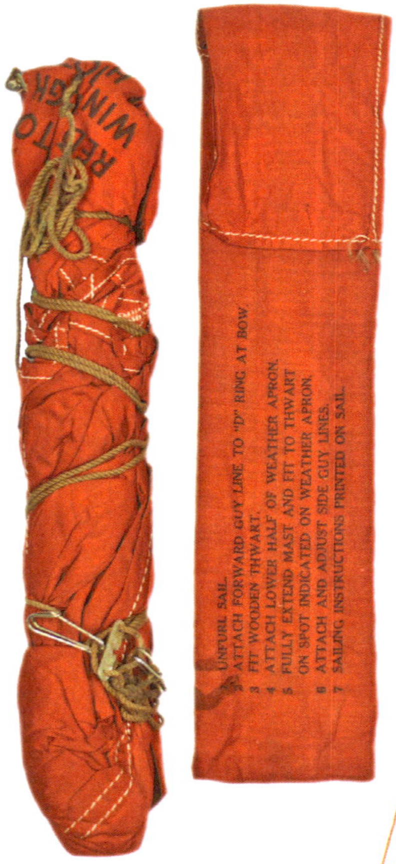

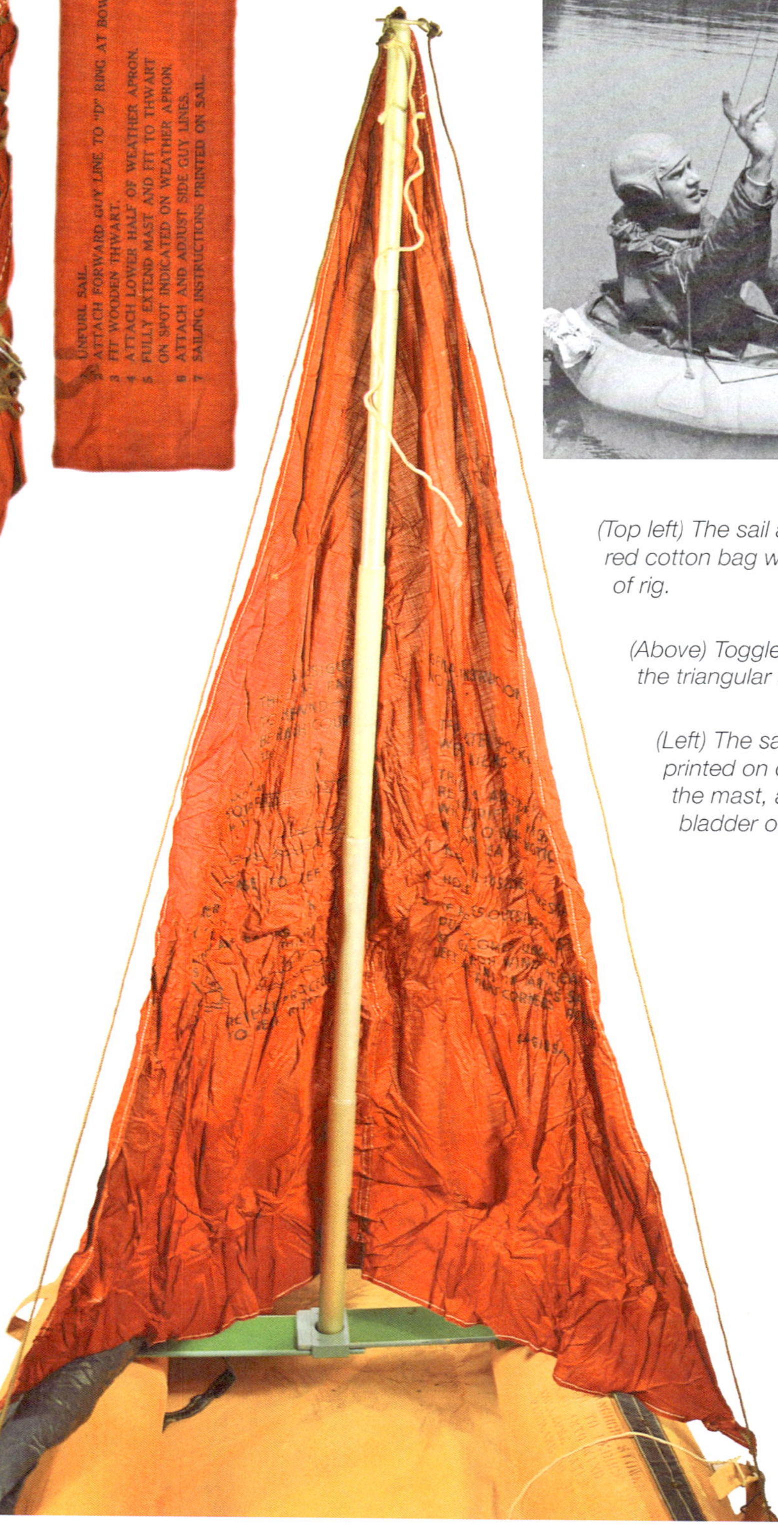

*(Top left) The sail and telescoping mast were stowed in a red cotton bag with printed instructions on the erection of rig.*

*(Above) Toggle lines were provided at two corners of the triangular sail for control while sailing. (USAF)*

*(Left) The sail was provided with sailing instructions printed on one side. Three guy lines supported the mast, attached to D-rings provided on the bladder of the raft.*

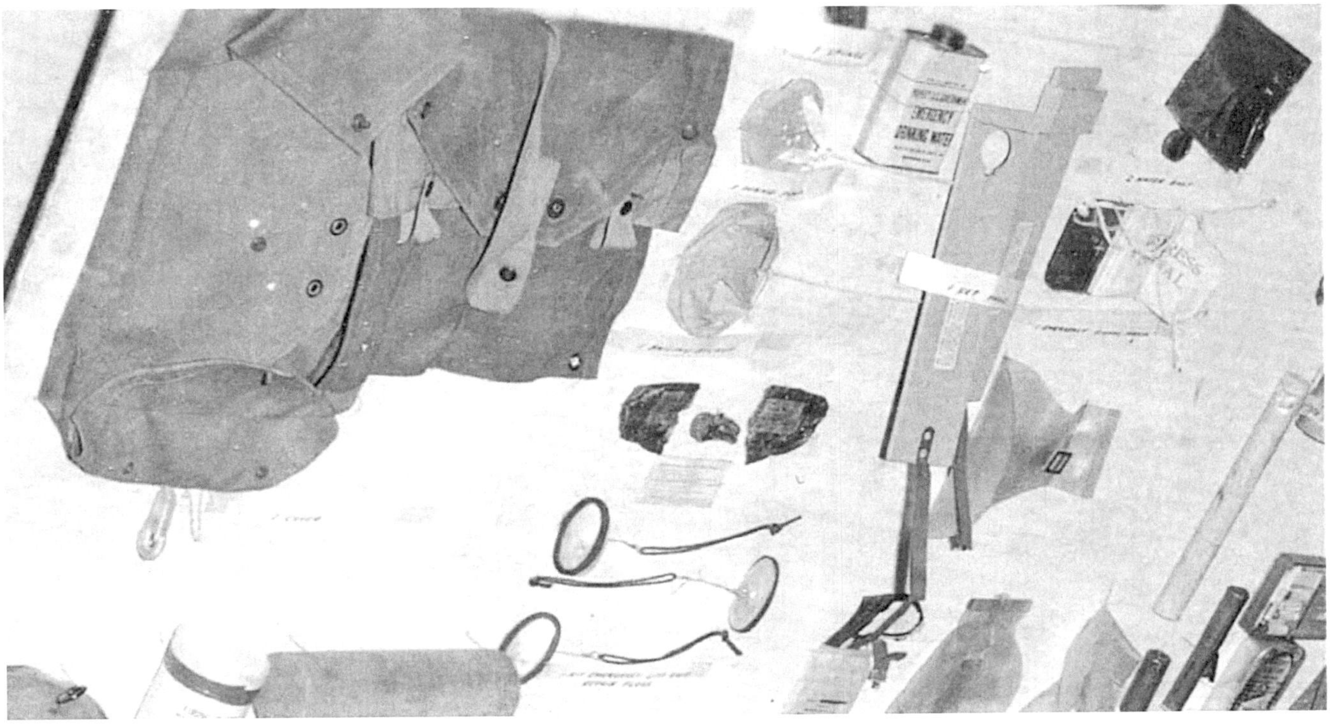

(Above) Accessories of a Type C-2 on a display board in the equipment room of the 500th BG on Saipan, July 1945. This was representative of the components that would have been delivered with the pararaft, minus the desalting kit and T-49/M75 distress signals due to their unavailability. The distress signal bag was repurposed for stowage of the signaling mirror.

(Below) Instruction on the use of the Type C-2 being conducted by two members of the 21st FG on Iwo Jima, April 1945. (National Archives)

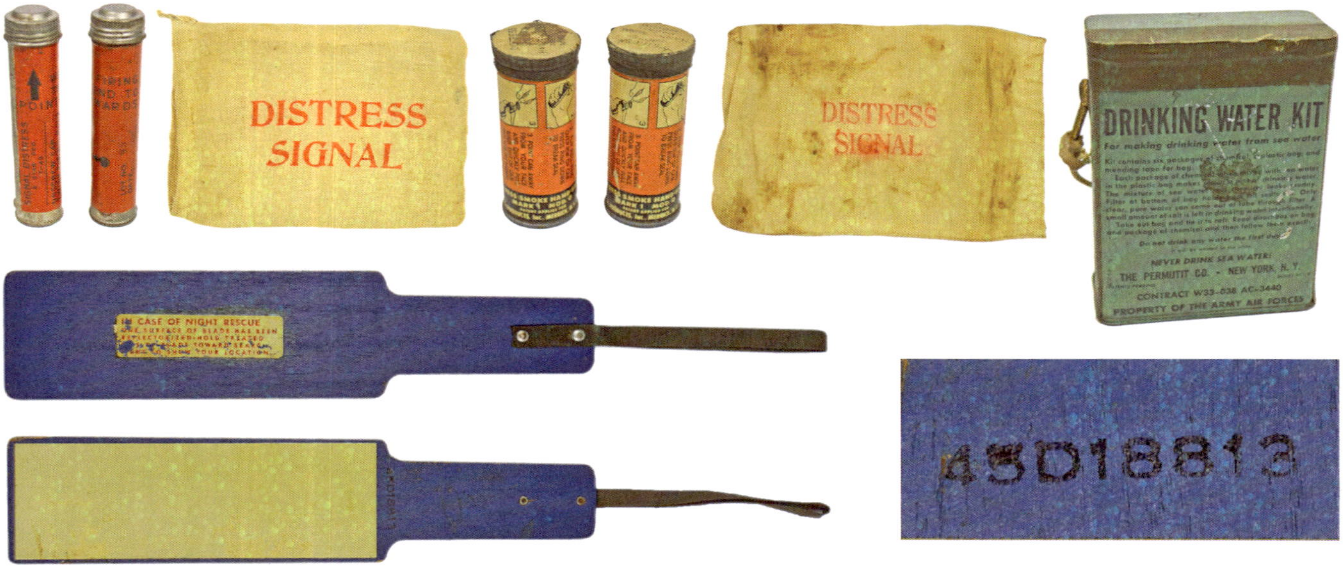


*(Above) Like the BuAer, the USAAF progressed towards provisions for signaling which led to a revision of the Type C-2. Approved was the deletion of two M-75 signals, being replaced by two MK-I MOD-1 smoke signals and adding an additional signal bag containing two of each signal per pouch. The sailing equipment was also deleted, providing room for the inclusion of the MX-137/A or AN/CPT-2. The wooden mast bridge paddles were replaced by two basic wood paddles as detailed under Dwg. No. 45D18813. These revisions were approved 21 May 1945, under Tentative Specification No. 94-40725A. Official approval of the specification was not made effective until 14 August 1945. Shown are examples of the updates of distress signals and paddles, by which date the JJ-1 would be available and included in the complement of accessories.*

*(Below) Raft accessories updated to C-2A specifications. On the floor of the raft are two distress signal bags with two MK-1 MOD-1 smoke signals and two M75 red distress signals, radar reflector, and JJ-1 desalting kit. An added item is the sea marker packet which would be included in the Type B-5 life preserver. (USAF)*

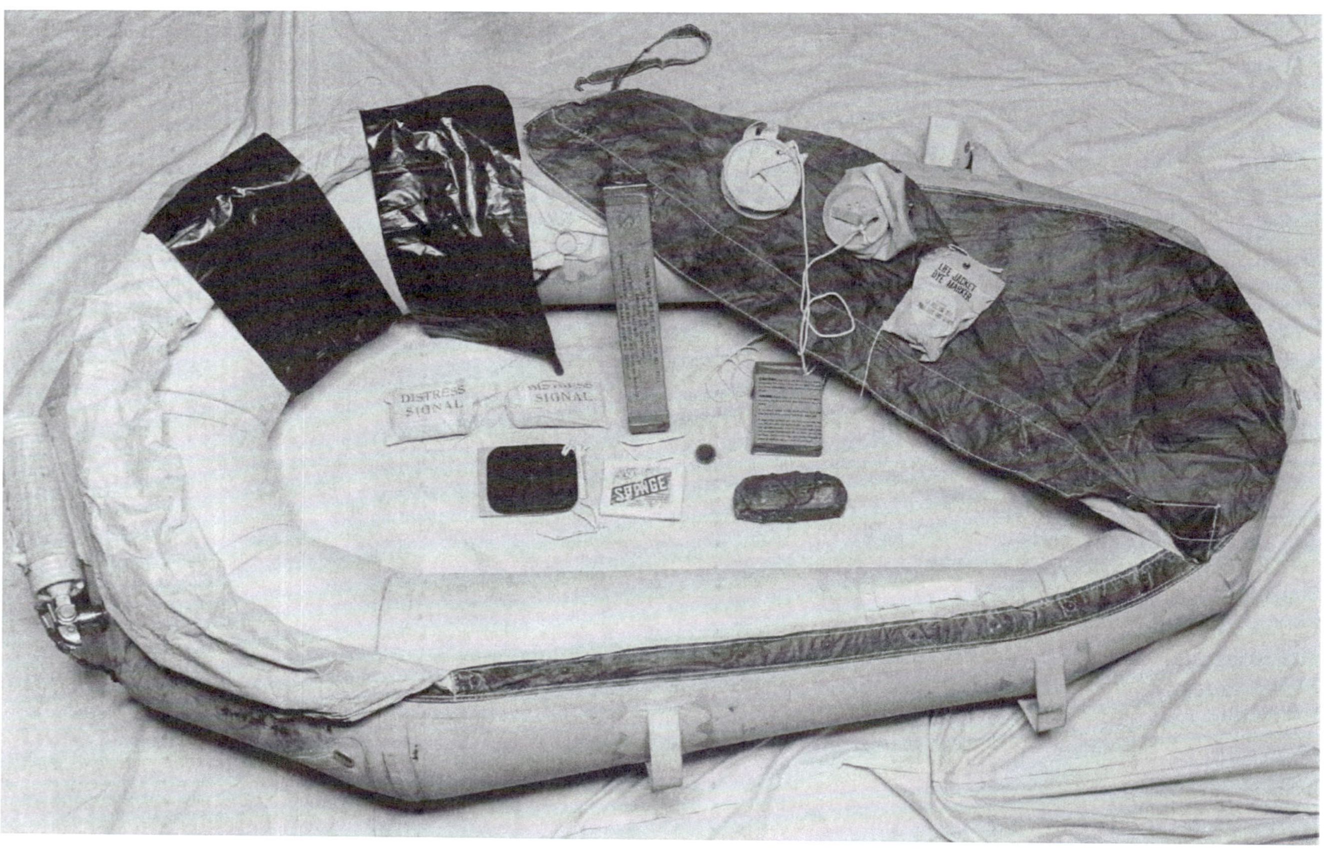

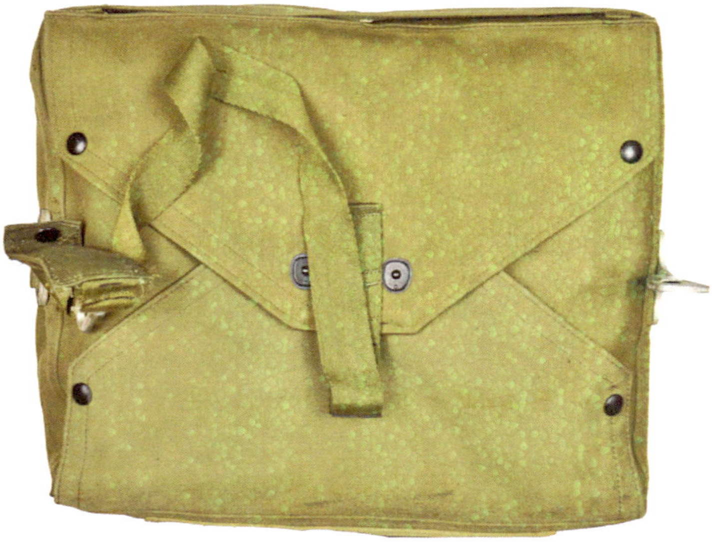

(Left and below left) Case for Type C-2A life raft. All contracts awarded after May 1945 would conform to Specification No. 94-40725A type designation C-2A. The complete assembly would be updated as per Dwg. No. 45G8189. The overall features of the raft remained unchanged, except for the factory applied adapter for the reflector and beacon units, and the elimination of the harness for securing the mast bridge.

(Below) Nomenclature on a Type C-2A life raft manufactured in August 1945.

(Below) The only change to the Type C-2A case was the positioning of the lanyard that secured the CO2 cylinder which was moved adjacent to the lanyard for the life preserver, utilizing a nylon tape. The two inner containers illustrated below were transitional. The lower unit retains the grommet on the left edge; the upper unit has eliminated the grommet.

(Below) Transitional markings on cases: the top photo reflects Type C-2. The middle photo shows the addition of a hand stamped "A." The lower photo shows the designation C-2A factory stenciled.

*(Below) Development of a new case began in April 1945, entering production in August, though it did not become available until 1946. This new design would carry over in the 1950s, with modifications. The opening was provided with a slide fastener and the top was padded for comfort. It had been revised with loops at the four corners for securing to seat type parachutes and retained provision for quick attachment. The accessories were now stowed in a bag with a zipper closure, having two nylon tapes, one connecting to the CO2 cylinder and the other to the life preserver. External tapes were provided for securing the radar reflector or beacon.*

*(Opposite) A list of known contracts and quantities for the Type C-2 pararaft. Not included are the change order quantities from contracts 33038-AC-390 through 395. Contracts awarded during May 1945 and after would conform to Specification No. 94-40725A Type C-2A.*

| CONTRACT | AWARD DATE | CONTRACTOR | QUANTITIES |
|---|---|---|---|
| 33038-AC-4375 | August 1944 | Armstrong Rubber Co. | 3,400 |
| 33038-AC-4376 | August 1944 | Air Cruisers Inc. | 3,000 |
| 33038-AC-4377 | August 1944 | Anderson Rubber Co. | 1,400 |
| 33038-AC-4378 | August 1944 | Kastar Inc. | 950 |
| 33038-AC-4379 | August 1944 | Rubbercraft Corp. | 1,000 |
| 33038-AC-4452 | September 1944 | Hood Rubber Co. | 5,000 |
| 33038-AC-4453 | September 1944 | U.S. Rubber Co. | 5,200 |
| 33038-AC-6209 | December 1944 | Armstrong Rubber Co. | 3,200 |
| 33038-AC-6210 | December 1944 | Rubbercraft Corp. | 3,200 |
| 33038-AC-6211 | December 1944 | Air Cruisers Inc. | 3,400 |
| 33038-AC-6212 | December 1944 | U.S. Rubber Co. | 3,300 |
| 33038-AC-6213 | December 1944 | General Tire & Rubber Co. | 3,500 |
| 33038-AC-8249 | February 1945 | General Tire & Rubber Co. | 3,000 |
| 33038-AC-8251 | February 1945 | Anderson Rubber Co. | 4,000 |
| 33038-AC-8508 | March 1945 | Armstrong Rubber Co. | 7,000 |
| 33038-AC-8509 | March 1945 | U.S. Rubber Co. | 6,900 |
| 33038-AC-10715 | May 1945 | General Tire & Rubber Co. | 3,000 |
| 33038-AC-10716 | May 1945 | Anderson Rubber Co. | 4,500 |
| 33038-AC-10717 | May 1945 | Air Cruisers Inc. | 3,600 |
| 33038-AC-11047 | May 1945 | Firestone Tire & Rubber Co. | 14,500 |
| 33038-AC-11048 | May 1945 | Armstrong Rubber Co. | 8,000 |
| 33038-AC-11049 | May 1945 | Rubbercraft Corp. | 4,700 |
| 33038-AC-11050 | May 1945 | U.S. Rubber Co. | 17,900 |
| 33038-AC-11051 | May 1945 | New York Rubber Corp. | 10,000 |
| P.O. 45-13088 | Unknown | Air Cruisers Inc. | Unknown |
| P.O. 45-647 | Unknown | Air Cruisers Inc. | Unknown |
| P.O. 45-4342 | Unknown | General Tire & Rubber Co. | Unknown |

*(Below) P-51 pilots stationed on Iwo Jima wearing Type C-2 pararafts attached to their back type parachutes, July 1945. The cases are pre-November 1944 and have been updated according to Technical Order 04-15-10. It was typical procedure for units preparing to ship overseas to be supplied with updated equipment, and a quantity of life rafts would have been expedited for shipment to the combat theater. Limited numbers were in use in both the PTO and ETO by the close of 1944. (National Archives)*

*(Right) P-51 pilot Lt. Clyde Decker using a modified Type C-2 pararaft with his seat type parachute, Iwo Jima, 1945. Right, an assembly of the Type C-2 pararaft being demonstrated. (National Archives)*

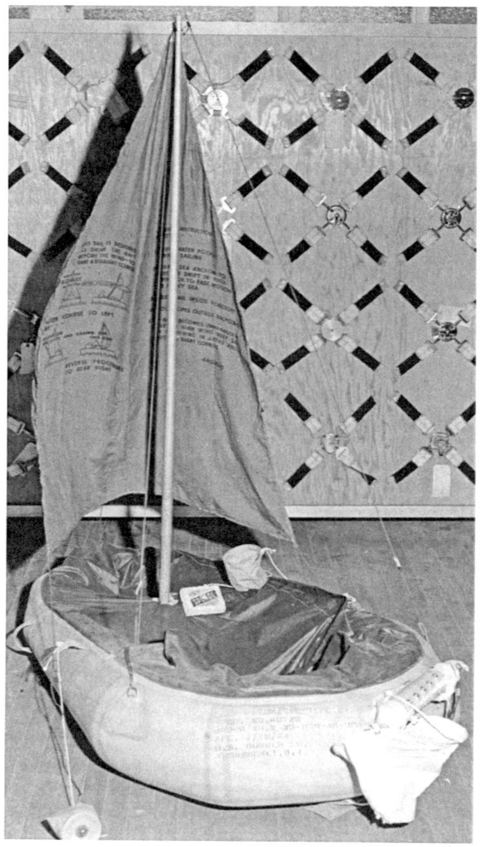

*(Above and above right) The BuAer developed a compact pararaft in 1945, not unlike the pre war USAAF concept of the 42B3447. Its intended purpose was to provide a personal life raft to aircrew in confined spaces in VTB, VSB and VPB squadrons and for use in VF class aircraft, stowed in the cockpit for air droppable purposes and other specialized applications. It was standardized as the Packet Raft, Model PR-1 NavAer Specification M-706. The basic complement included a raft, sea anchor and CO2 cylinder, packaged in a unit 12" x 6" x 2". It entered production before VJ-Day, but was not available in quantity until January 1946.*

*(Right) To better assist in air-sea rescue efforts, the BuAer developed a series of air droppable life raft assemblies in 1944. One of these was rated for one-man use, designated Rescue Assembly, Aircraft Droppable, AR-1, detailed under NavAer Specification M-699. The assembly consisted of two units: one with the raft, and the other with emergency provisions. The complete assembly, as pictured, includes one projector kit, eight M-566 sea markers, one signal mirror, one M-613 desalting kit, one corner reflector, one M-641 pararaft, one two cell flashlight and six MK-I MOD-0 smoke signals. The unit was eventually standardized 18 May 1945, with a procurement of 15,000 assemblies. Deliveries were due in August of that year. The production model was integrated into one complete unit eliminating the need for two separate bags. (National Archives)*

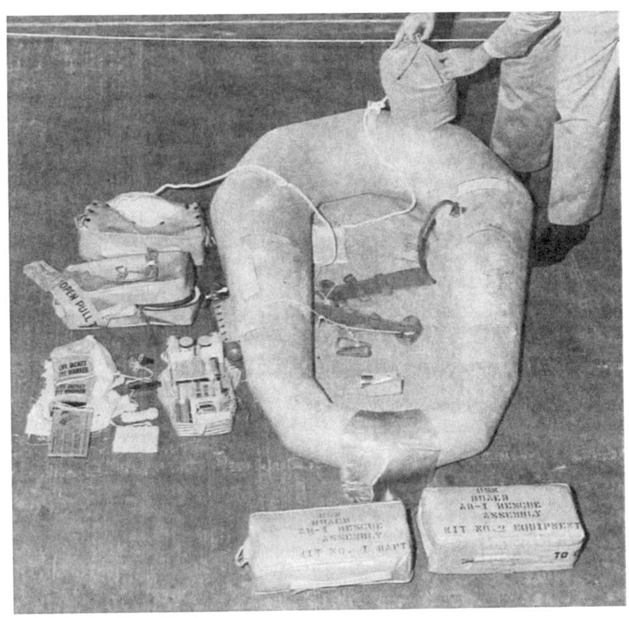

*(Right) Rescue Assembly, Aircraft Droppable, AR-1 undergoing field tests, summer 1944. (National Archives)*

*(Below and opposite page) Illustrated for comparison purposes is the German Luftwaffe one-man pararaft, (Einmannschlauchboot) Type A-2.*

*(Below) Back pack container for the Luftwaffe Einmannschlauchboot Type A-2. (National Archives)*

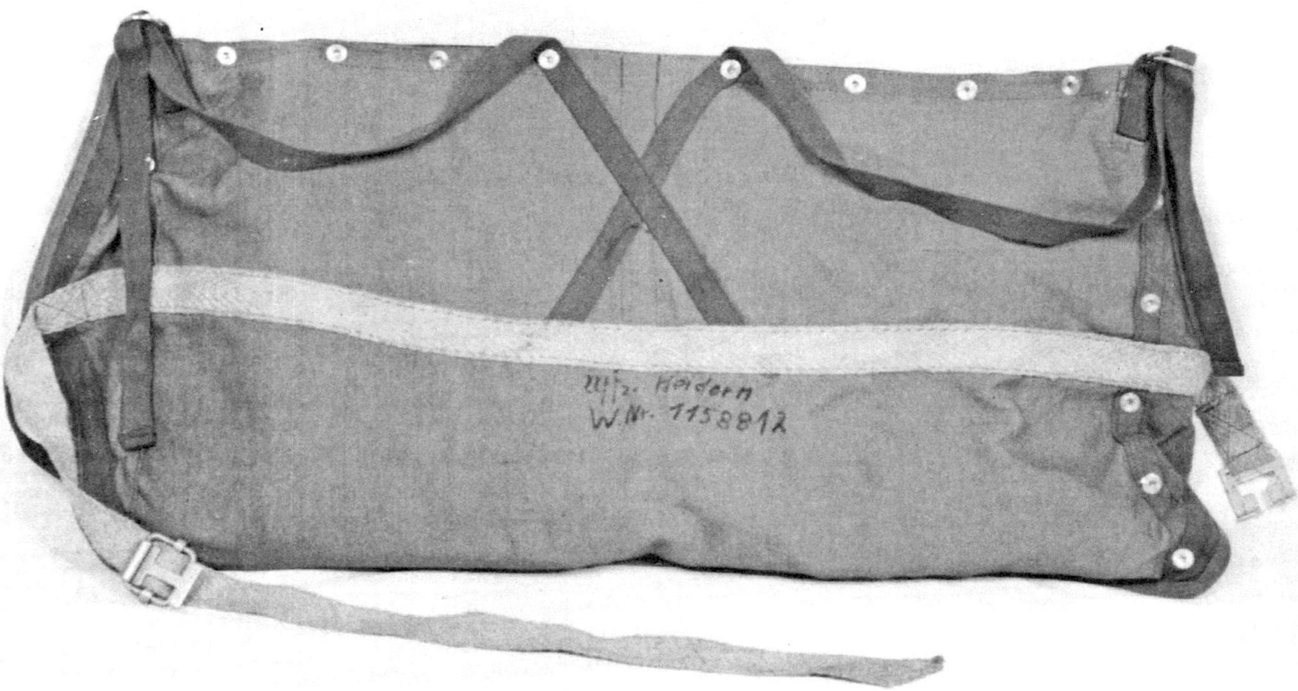

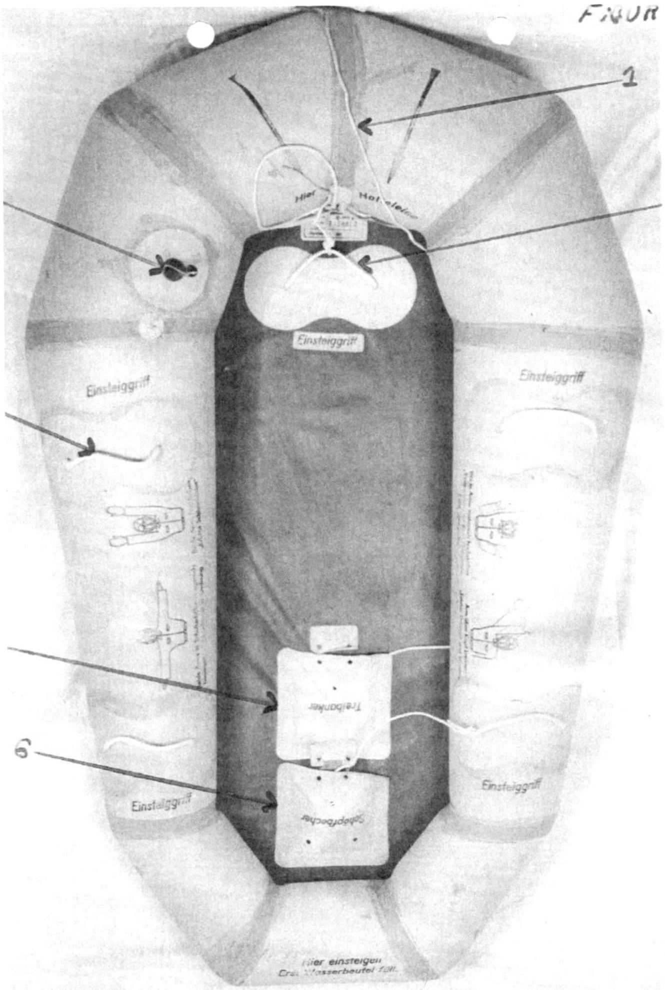


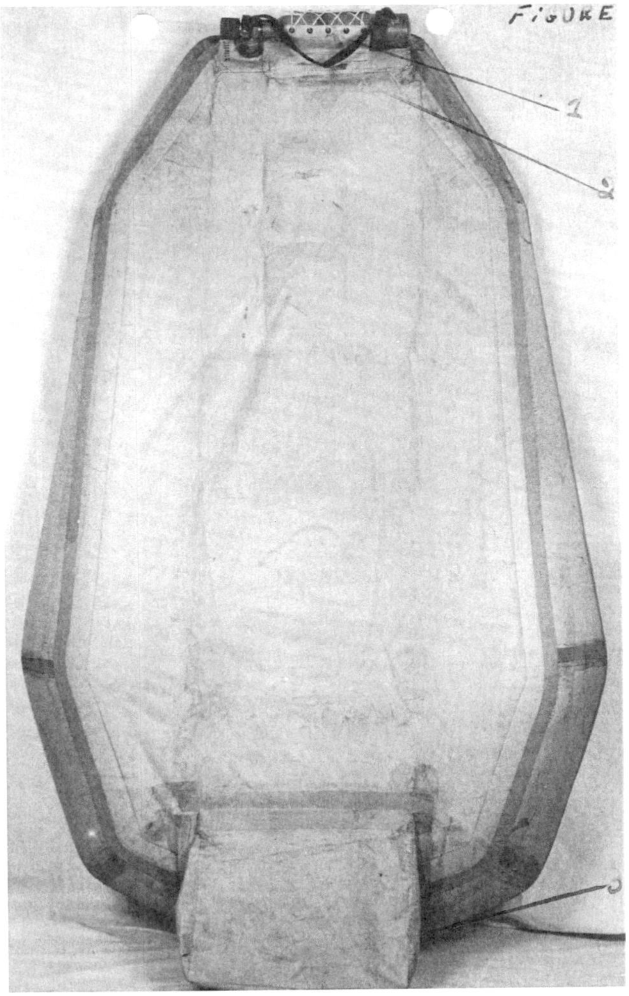


(Above and below) There is a distinct similarity between US, British and German rafts, and it is known that the Americans copied the British, but it is not clear if the Germans copied the British, vice versa, or coincidence. (National Archives)

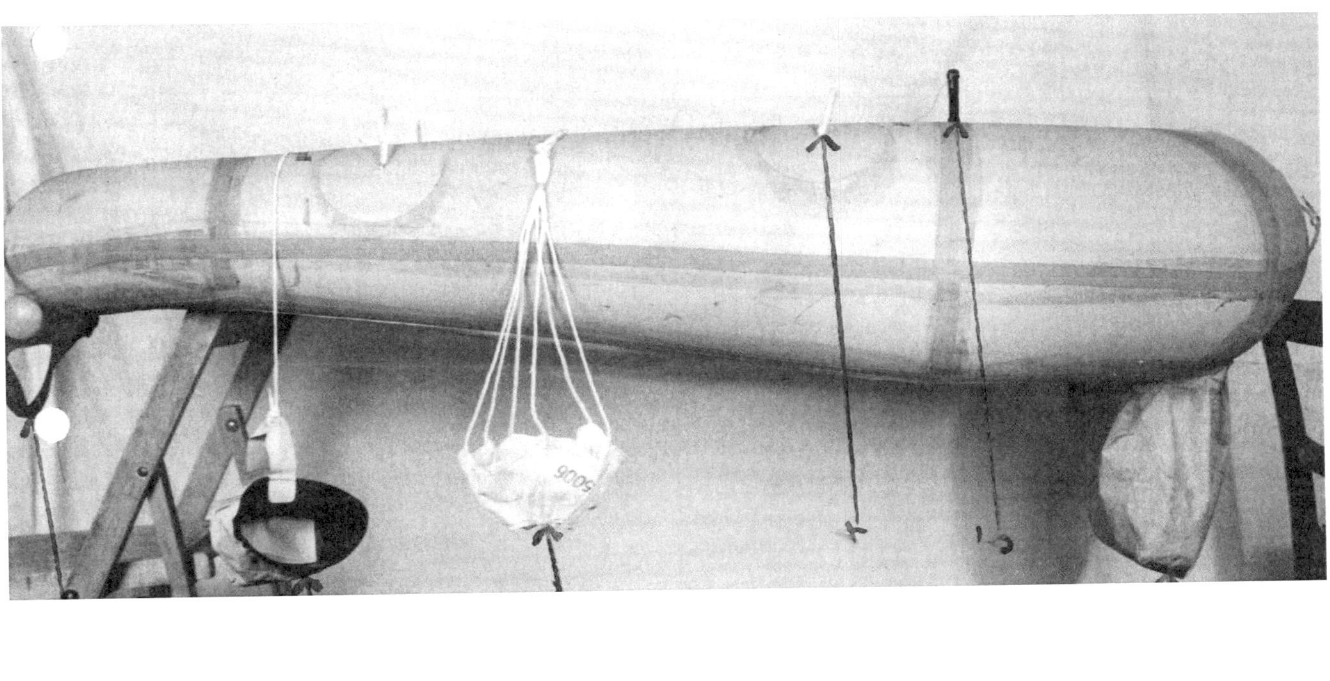

The AN-R-2 pararaft was originally intended to be used solely with the standard seat type parachute. As illustrated previously, the first production runs by both services differed from each other. The BuAer initiated production with the center slotted cases, then revised the contracts to a slot in a forward position by the close of 1942.

The USAAF eliminated the slot entirely, changing the method in which the unit was to be secured to the parachute harness.

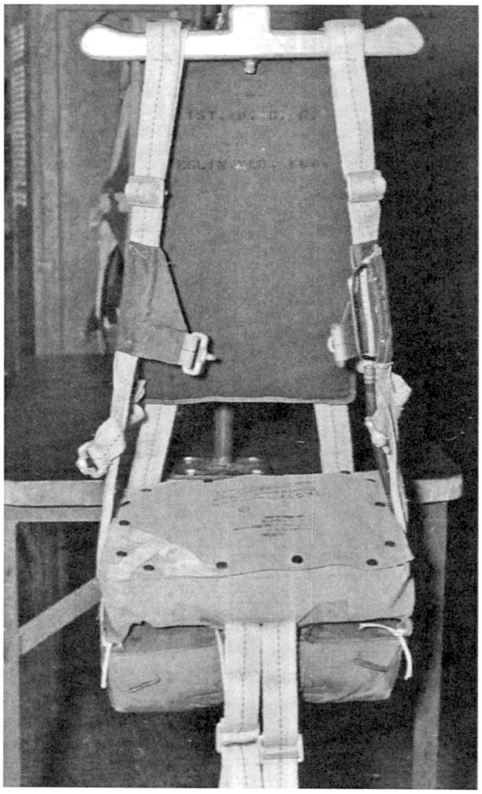

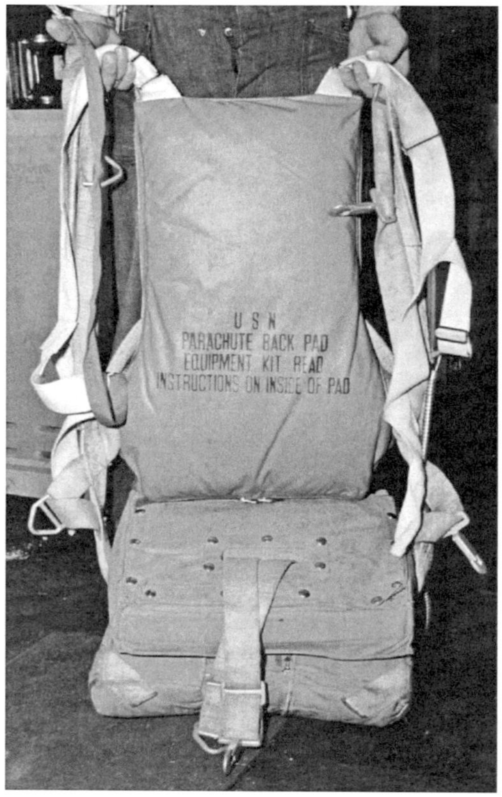


*(Left) The BuAer assembly, including the compliment of the original back pad kit used in conjunction with the pararaft and the standard seat type parachute. The pararaft was held into position by simply inserting the parachute leg straps through the slot. The same method applied for the center slotted case. NAS Norfolk, July 1943. (National Archives)*

*(Above) The first technical order, T.O. 04-15-2 18 September 1942, for the USAAF one-man parachute raft included the center slotted case and described attachment to the seat, back, and troop-type parachutes by means of the leg straps through the slot. This technical order was drafted in July 1942, prior to the acceptance of the non-slotted type and production pararaft. Technical Order 04-15-1, dated 26 May 1943, would correct the error and include a description of attachment for the non-slotted case: with the seat cushion removed, the pararaft case was tacked to the parachute pack at the four corners, the legs straps were then slipped through the strap provided on the front.*

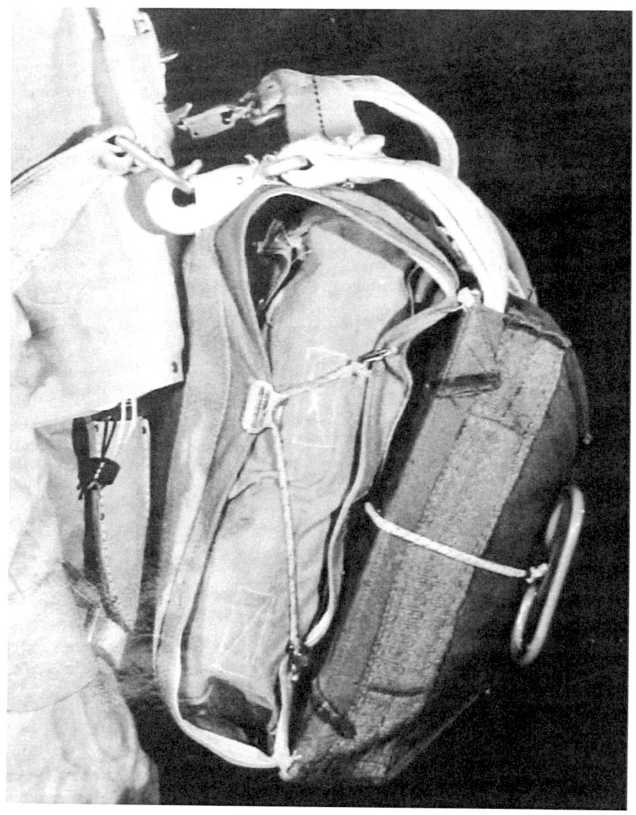

*(Right) To meet the extended requirements for the pararaft to be available to all personnel, the raft pack would need to be adaptable for the US Navy Quick Attachable Chest type parachute. The BuAer Engineering Division developed a separate canvas container which was tacked to the parachute pack at the four corners, with the pararaft being secured in the container by means of a bungee cord on one side. This container was developed in conjunction with the NAF Dwg. No. 69233 back pad kit between Dec. 1942 and Jan. 1943. Pictured is the complete QAC assembly with back pad kit and pararaft, with the container attached to the QAC parachute pack. (National Archives)*

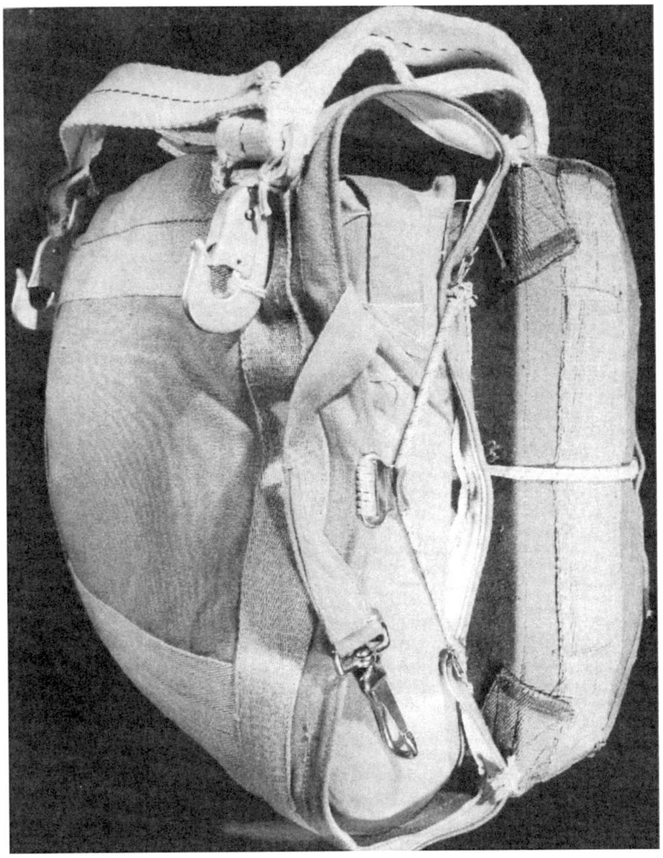

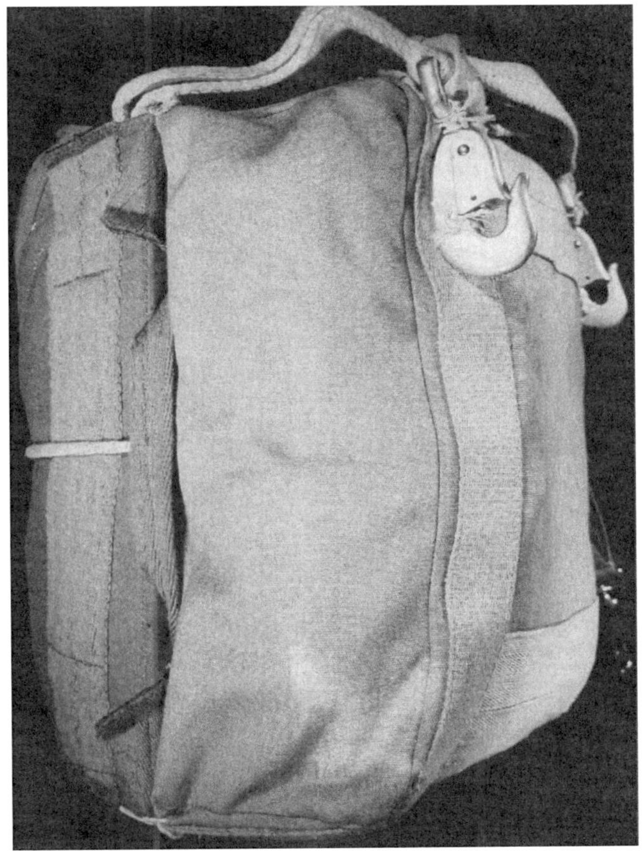

*(Above) Details of the experimental USN QAC pararaft container. (National Archives)*

*(Below) Following dummy and live test jumps, the pararaft container was approved for the standard QAC parachute. Technical Note 41-43 was drafted and released 11 June 1943, describing the attachment and fabrication of the container. This new assembly met service requirements to ensure that a life raft would be available to personnel equipped with the QAC parachute, in the event it became necessary to abandon the aircraft over water. Images and drawing are from Technical Note 41-43. (National Archives)*

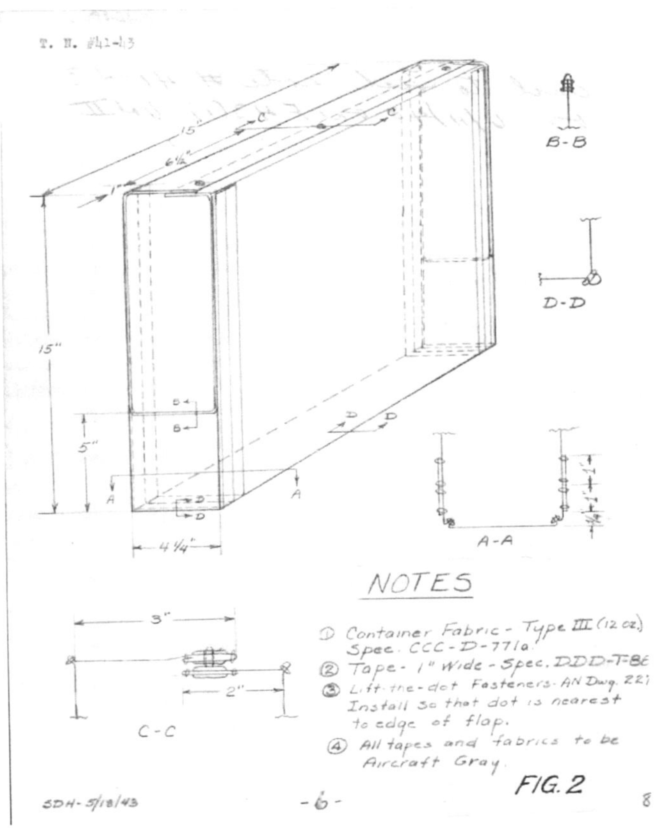


FIG. 2

*(Below and right) Details of the USN pararaft pack as approved for use with the QAC harness from Technical Note 41-43. (National Archives)*

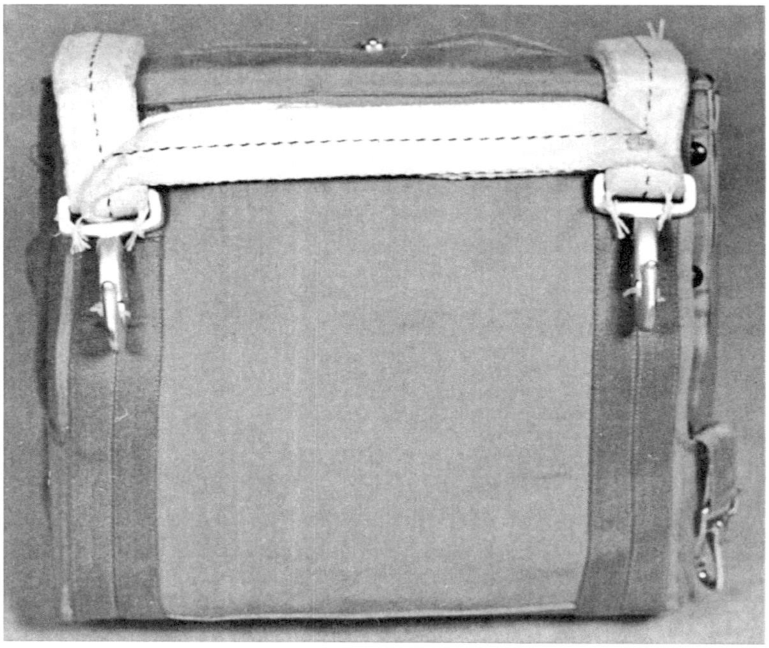

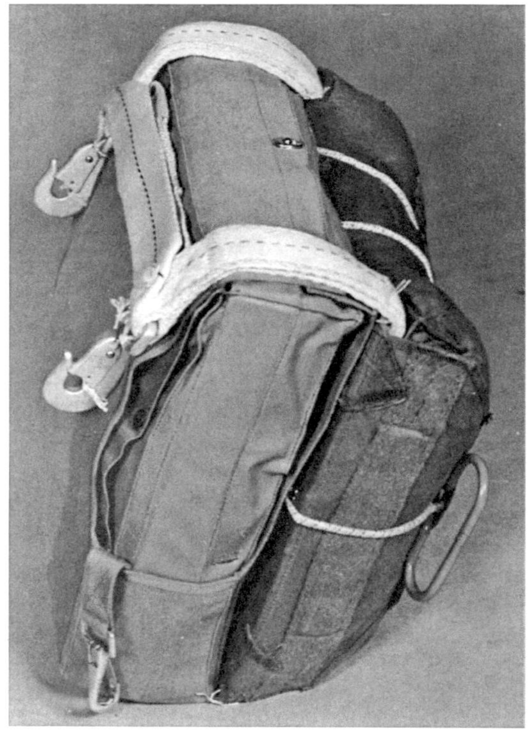

*(Left) In the spring of 1943, the newly developed USN Quick Attachable Seat Type Parachute was introduced. This new assembly detached the parachute, which remained in the aircraft, while the harness with emergency back pad kit could be properly fitted and remain with the individual. The leg straps on this unit were no longer integrated onto the parachute pack and unable to secure the pararaft. As prescribed in Technical Note 22-43 of 30 April 1943, and revised Technical Note 54-43 of 24, July 1943, for the QAS, the pararaft was to be tacked to the parachute pack at each of the four corners. BuAer understood that this was not the best method for attachment, and upon distribution of the technical notes requested designs and recommendations for an improved application. The flaw encountered was that the pararaft tended to break free from both the parachute and the life vest upon the opening shock of the canopy, losing the pack. A member of VF-1 is shown wearing a modified QAS parachute harness. Note inset detail of the standard attachment of the pararaft to the parachute pack, July 1943. (National Archives)*

(Above) An AN-R-2A case that was fabricated at NAS Quonset Point, R.I. Integrated into the case are tabs at the four corners for attachment to the QAS parachute.

(Right and below) The immediate principle concern with the QAS was to prevent the loss of the pararaft during parachute descents. To help restrain and secure the raft, two keeper straps were added to the case which the harness retaining belts would pass through. When the harness was worn, the pararaft was then secured in place by the tacking on the four corners and the parachute harness via the keeper straps assuring the unit would stay with the user. This concept was devised within the Naval Aircraft Factory and illustrated in the following images, September 1943. (National Archives)

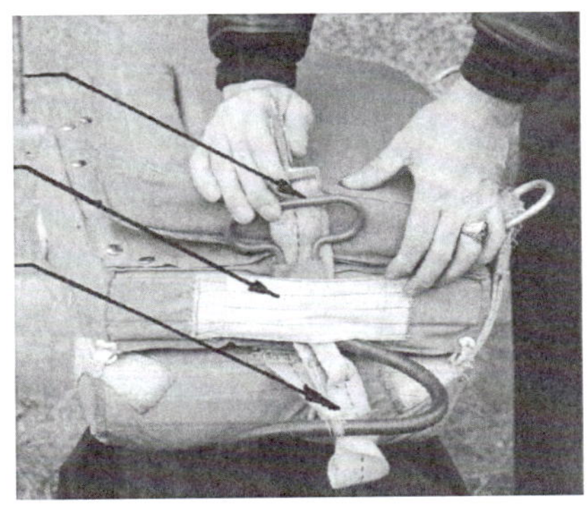

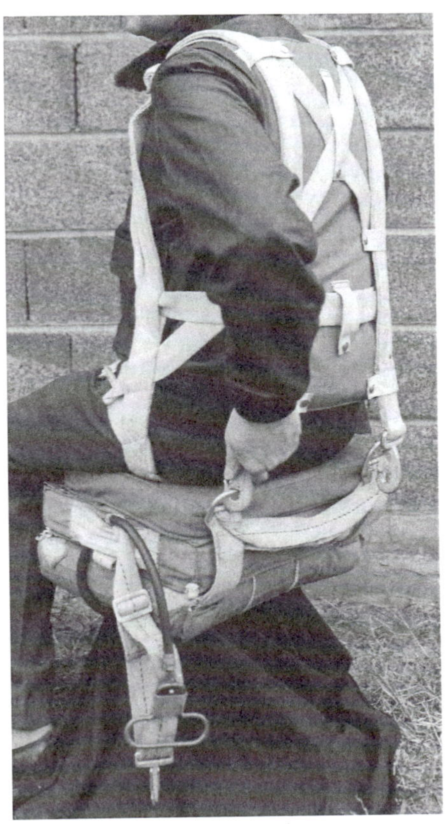

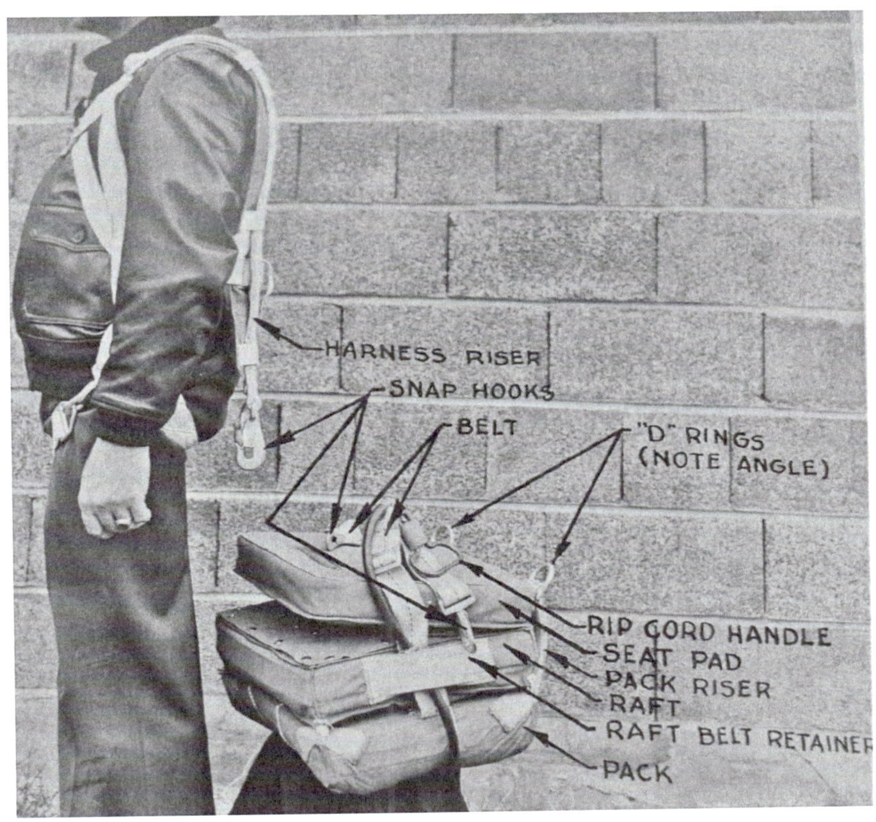

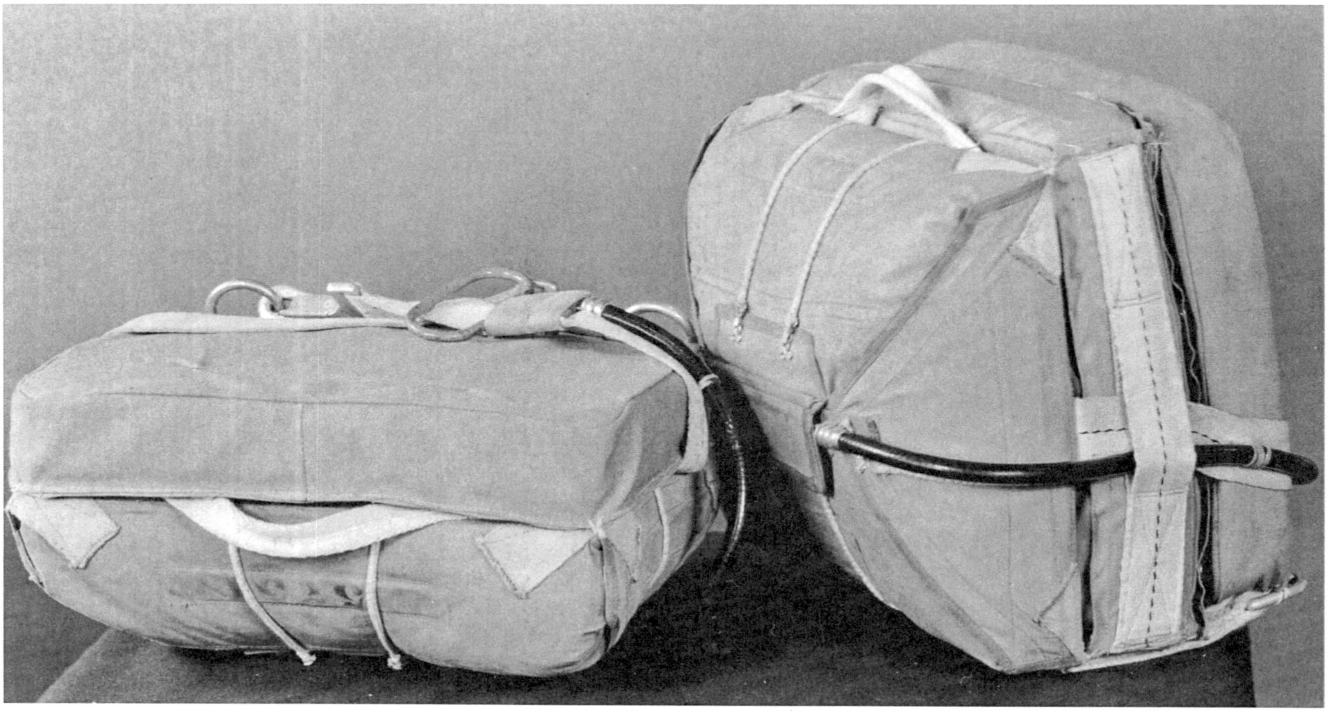

(Above) The concept of the keeper strap was introduced in revised Technical Note 74-44 of 30 August 1944 for the QAS, superseding Technical Note 54-43. By this date, numerous methods for securing the pararaft were being utilized and were becoming mainstream. Technical Note 74-44 was an expedient measure, formally addressing service concerns, and was intended to be a temporary measure. Pictured above are two separate QAS parachutes, on the left is the standard configuration with just the seat cushion tacked to the pack at each of the four corners. On the right, the parachute has the pararaft attached, and the pararaft case has the keeper straps, as per the drawing included in Technical Note 74-44. Additionally, it is tacked to the parachute, and the seat cushion is tacked to the removable top cover of the pararaft pack. (National Archives)

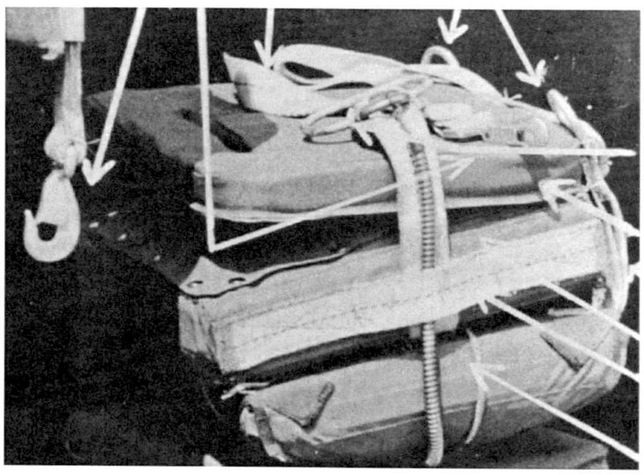

(Right) Picture from T.N. 74-44 offers another view of the parachute assembly. With the keeper straps, it was not necessary to tack the raft pack to the parachute, however it was stated to do so in the technical note.

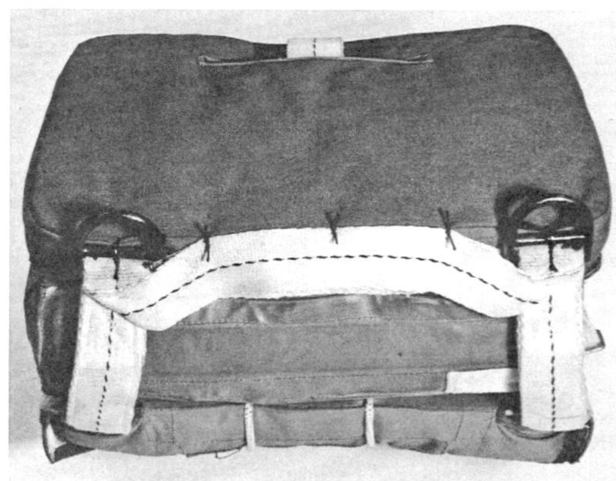

(Far right) An Alternative method, which allowed better access to the raft, was to tack the seat cushion to the D ring risers on the back of the parachute. (National Archives)

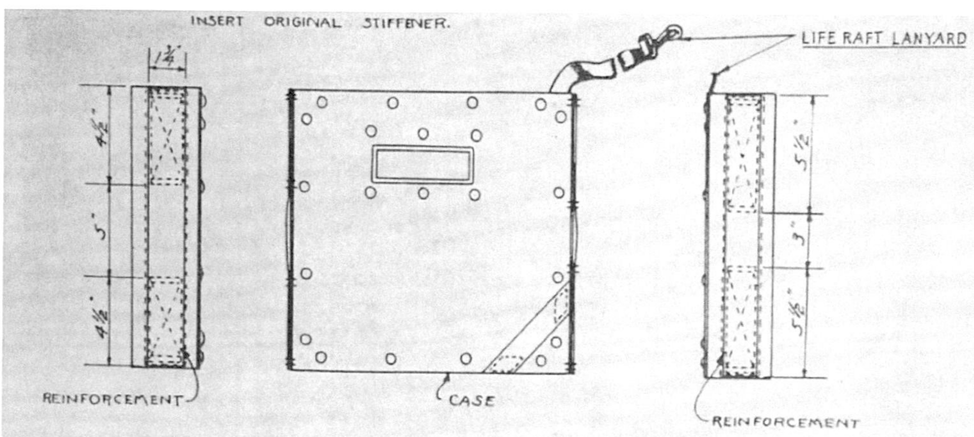


(Right) The drawing included in Technical Note 74-44 for the attachment of the keeper straps to the pararaft case.

(Above and right) Keeper strap modification on a US Navy contract AN-6520-1 pararaft case.

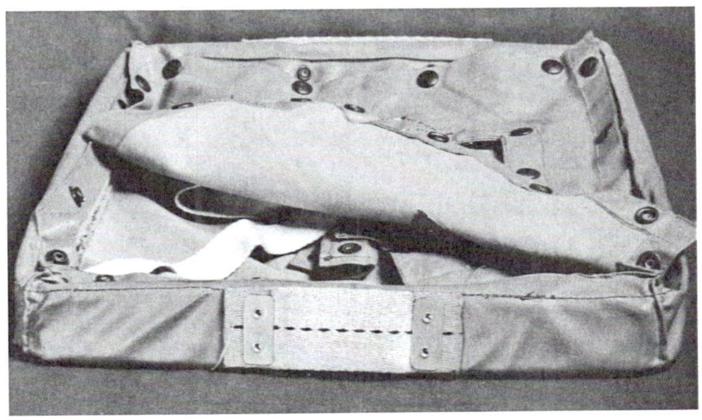

(Above and right) Due to the complicated process of attaching the web keepers, some were secured with rivets. It was stated that the process outlined in Technical Note 74-44 required 2 hours for the application, for each case, by having to remove several layers of material and re-sewing the assembly. By utilizing rivets, the process was cut to less than half the time. Two variations of riveted keeper straps are illustrated. Both modifications are documented to NAS Seattle, WA. (National Archives)

*(Below) Typical service modification for securing the pararaft. The exact origins are not known, but this method was the most common, service-wide, and documented as early as the summer of 1943. In this modification, a length of parachute webbing was added to the parachute, perpendicular to the retaining lap belt with a snap hook and V ring on each of the ends. This retaining system could be adjusted to secure just the pararaft, as pictured (top right), or lengthened to accommodate both the raft pack and seat cushion (bottom left). The securing methods presented thus far prevented loss during a parachute descent, but impaired the recovery of the pararaft once in the water. The user would slip from the parachute, while still descending, at the point his feet touched the water and if the lanyard from the life vest to the raft were to snap, the parachute was prone to drift away, with the loss of all equipment. By the beginning of 1944, it became the practice to add quick release methods, altering the procedure to release the raft from the parachute about 10-15' above the water, before slipping from the parachute. The picture bottom right illustrates one quick release method by modifying the snap hook with a pull pin for quick release. (National Archives)*

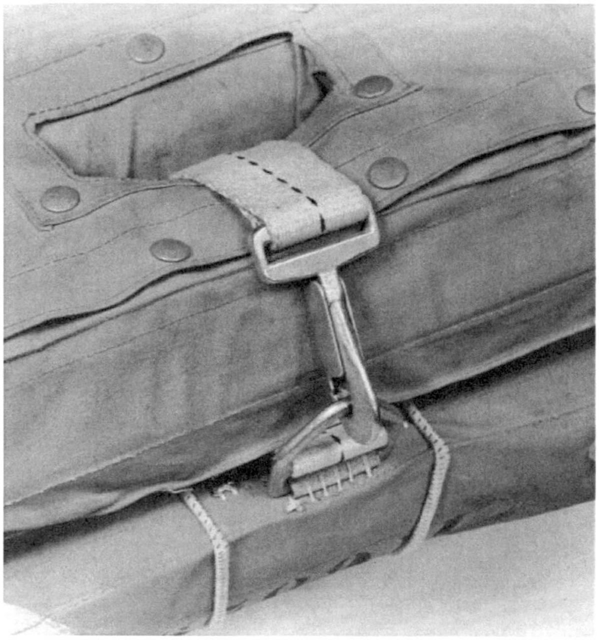

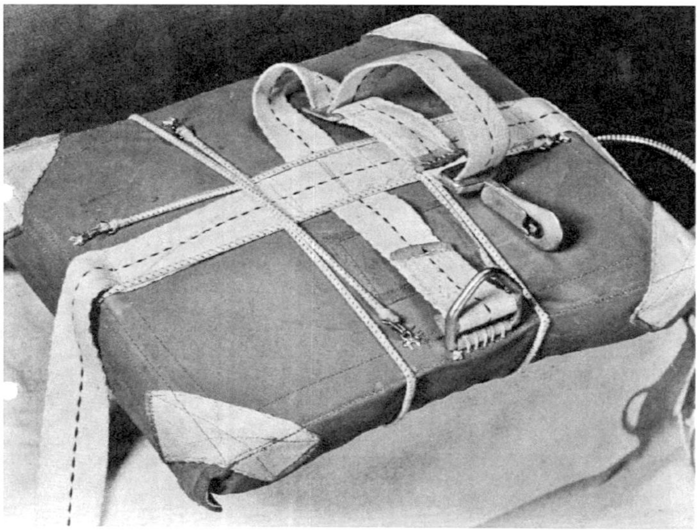

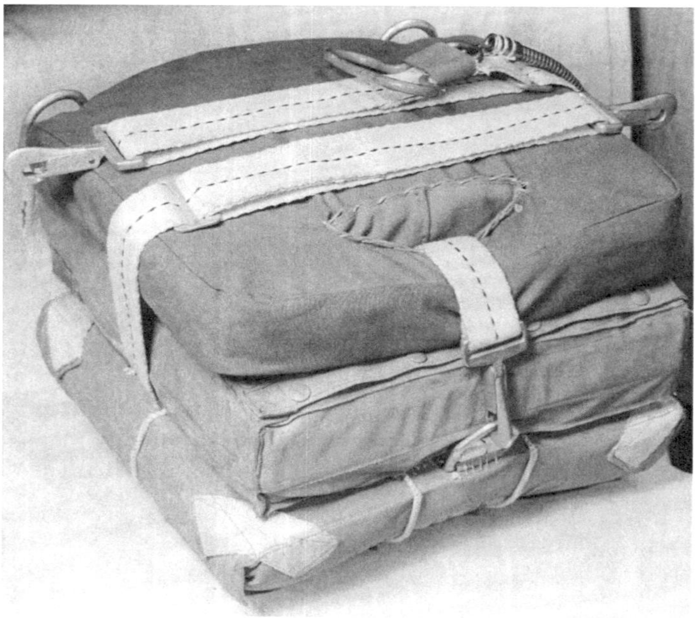

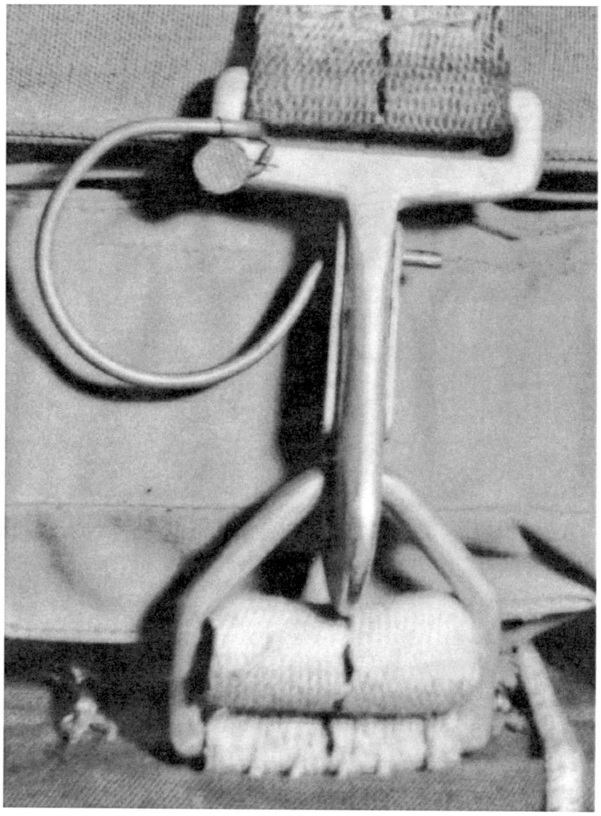

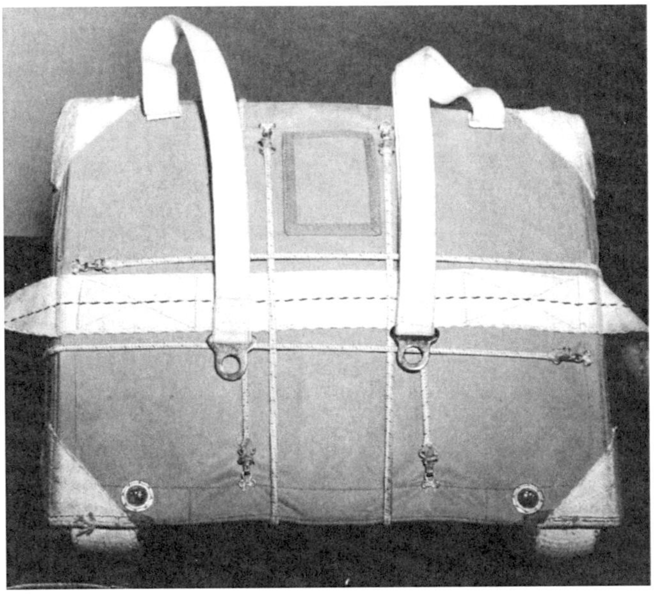

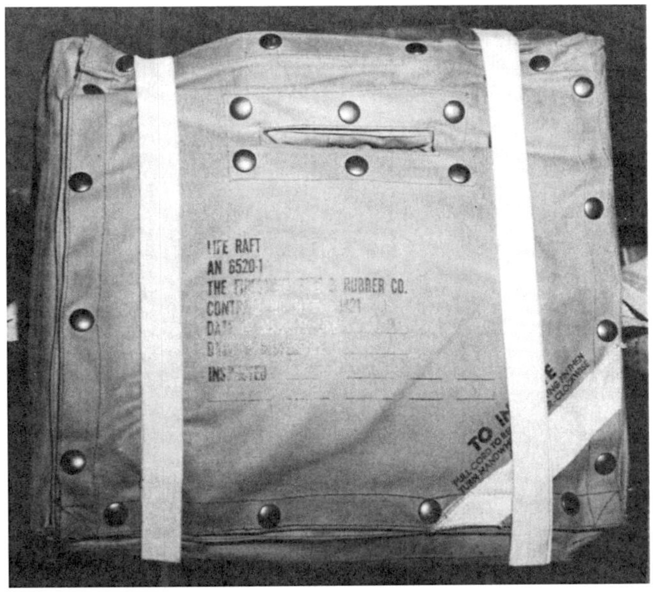


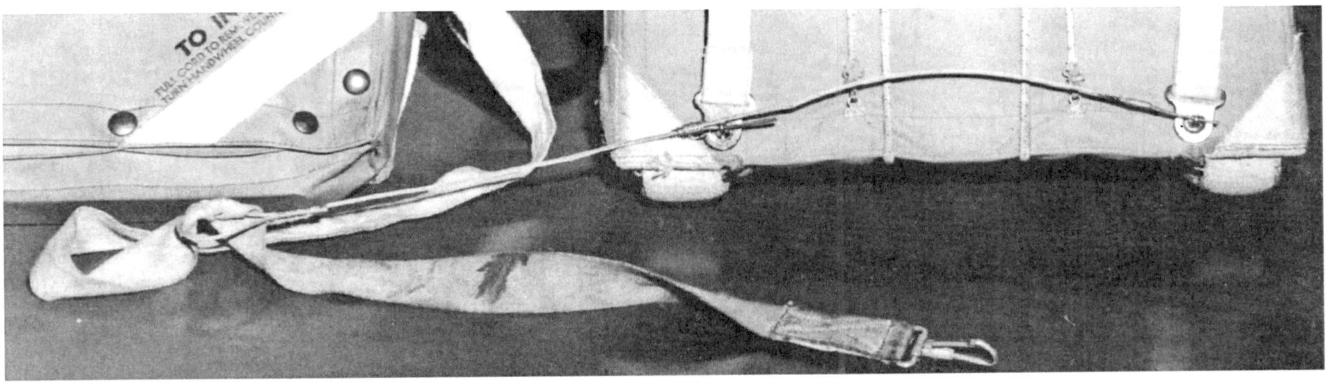

(Above) Quick release pararaft detainment system for the QAS, developed by CASU-25, February 1944. This application adapted the standard parachute rip cord method for quick release. The rip cord pins were secured half way down the life raft lanyard, which pulled free when the user slipped from their parachute harness, assuring the raft remained with him and eliminating the possibility of getting entangled with the harness and parachute. (National Archives)

(Below) A significant modification in the genealogy of the attachment of the pararaft to US Navy parachutes. The container utilized for the QAC was adapted for use on the QAS, with the principle modification for a quick release. Replacing the Lift-A-Dot fastener was a split connector link, keeping the container closed. A pull tab was provided, pulling the link free from the reinforcement webbing loops and allowing the raft pack to be removed from the container. The container was attached to the parachute in the same method prescribed in Technical Note 54-43. This method was submitted to the BuAer Engineering Division in March 1944 from NAS Jacksonville, FL. (National Archives)

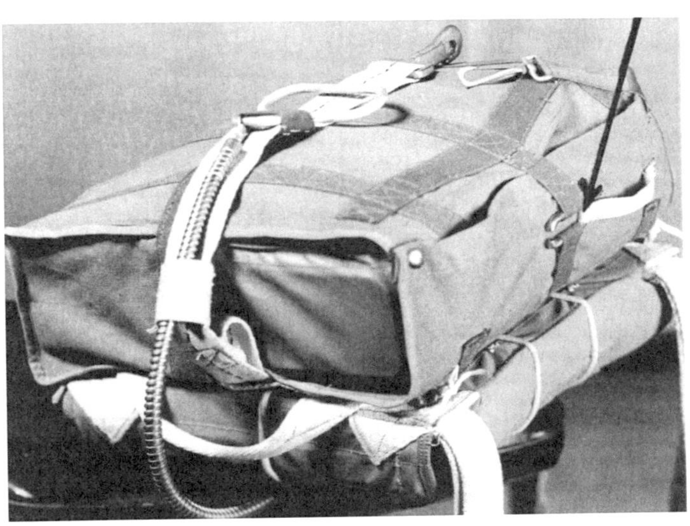

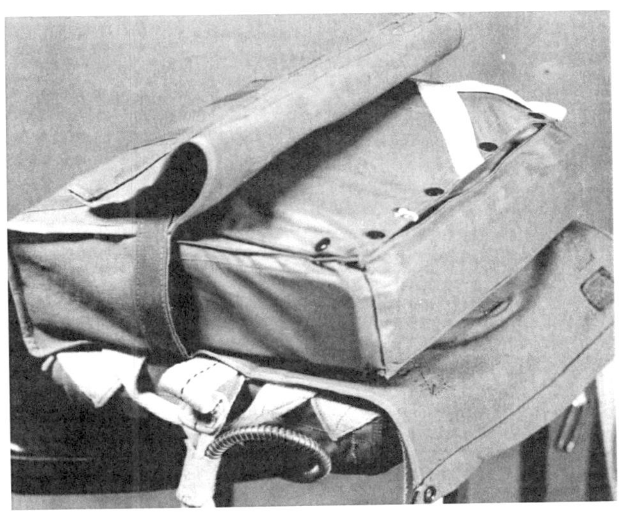

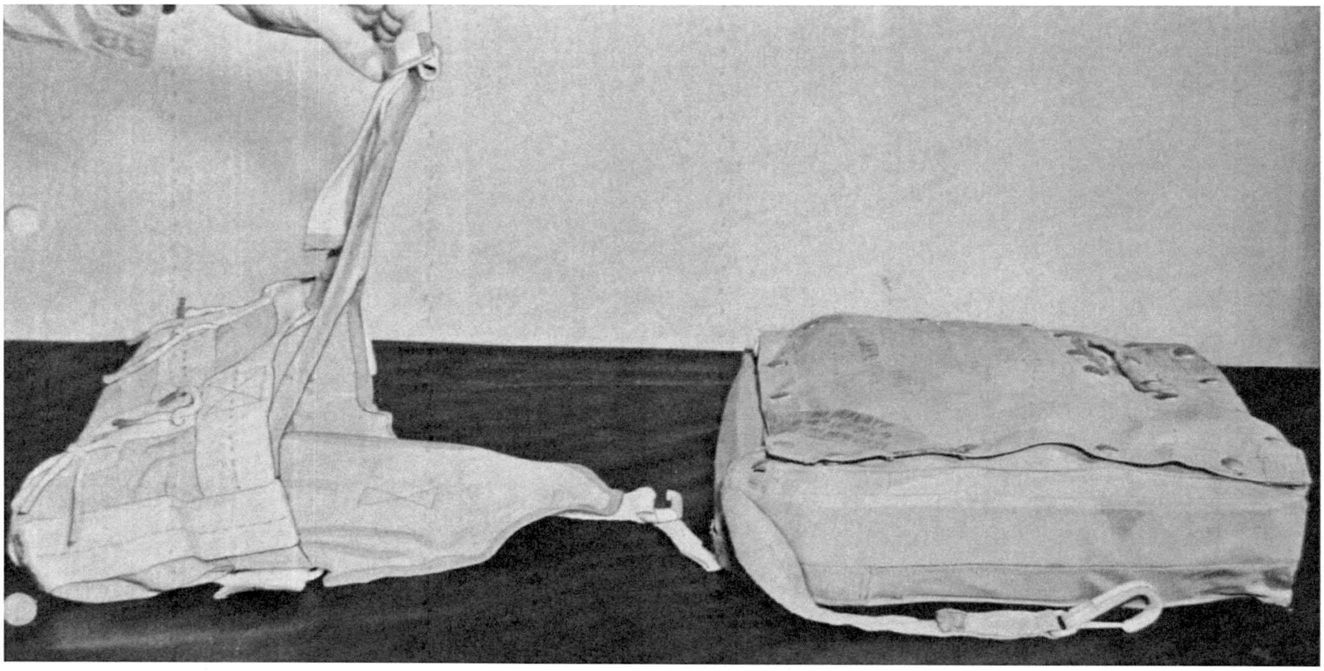

*(Above and right) By the spring of 1944, the BuAer was working on the details for a combination seat kit to replace the AN-R-2 pararaft and back pad kit to satisfy service requirements. In turn they were testing and evaluating additional requirements to meet the need for a standardized universal method to secure the pararaft to all types of parachutes. The prescribed guidelines included that the pararaft was to be secure from being accidently torn loose, but to permit easy removal upon a water landing or a parachute decent. The Engineering Division developed a pararaft container utilizing the methods contained in Technical Note 41-43 from NAS Jacksonville. This new container was first incorporated into Technical Order 114-44, of 1 September 1944, for the attachment of the pararaft to the QAC parachute. The pictures on this page are from Technical Order 114-44, illustrating the details of this new container and its attachment method at three points on the QAC parachute pack. The closure of the container utilized the split connector link for quick release purposes. The containers were to be manufactured locally, as per detailed drawings provided in the technical order. (National Archives)*

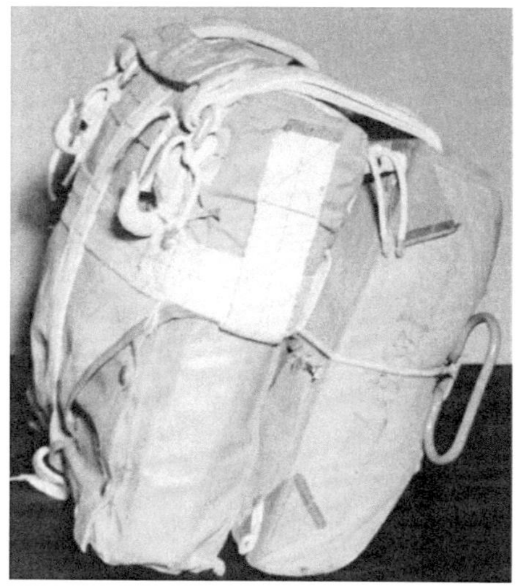

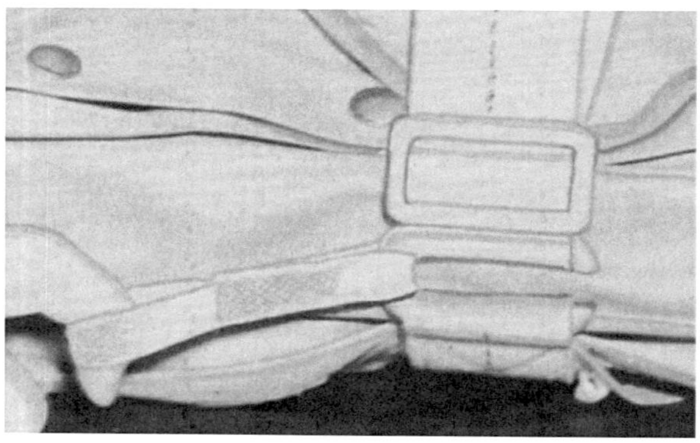

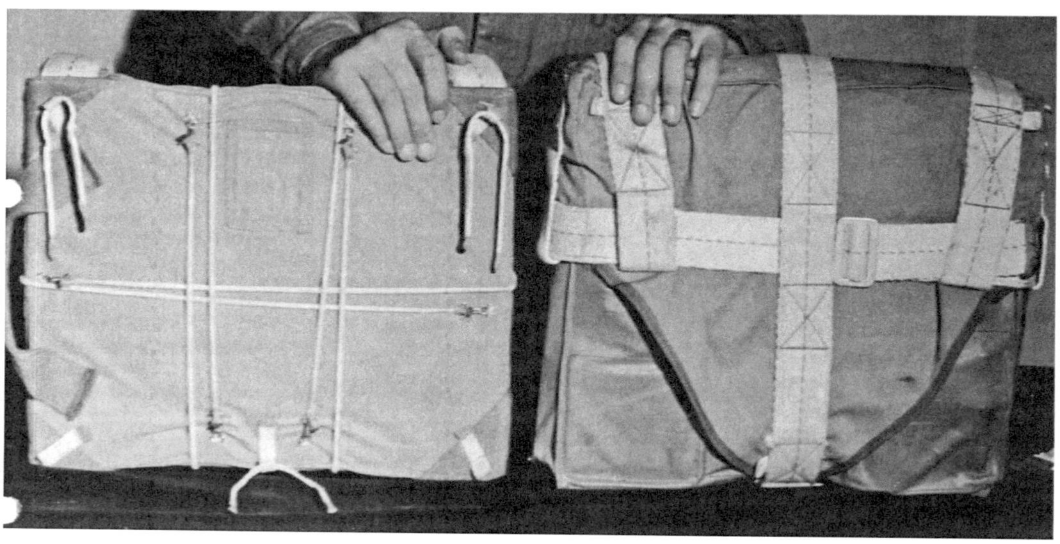

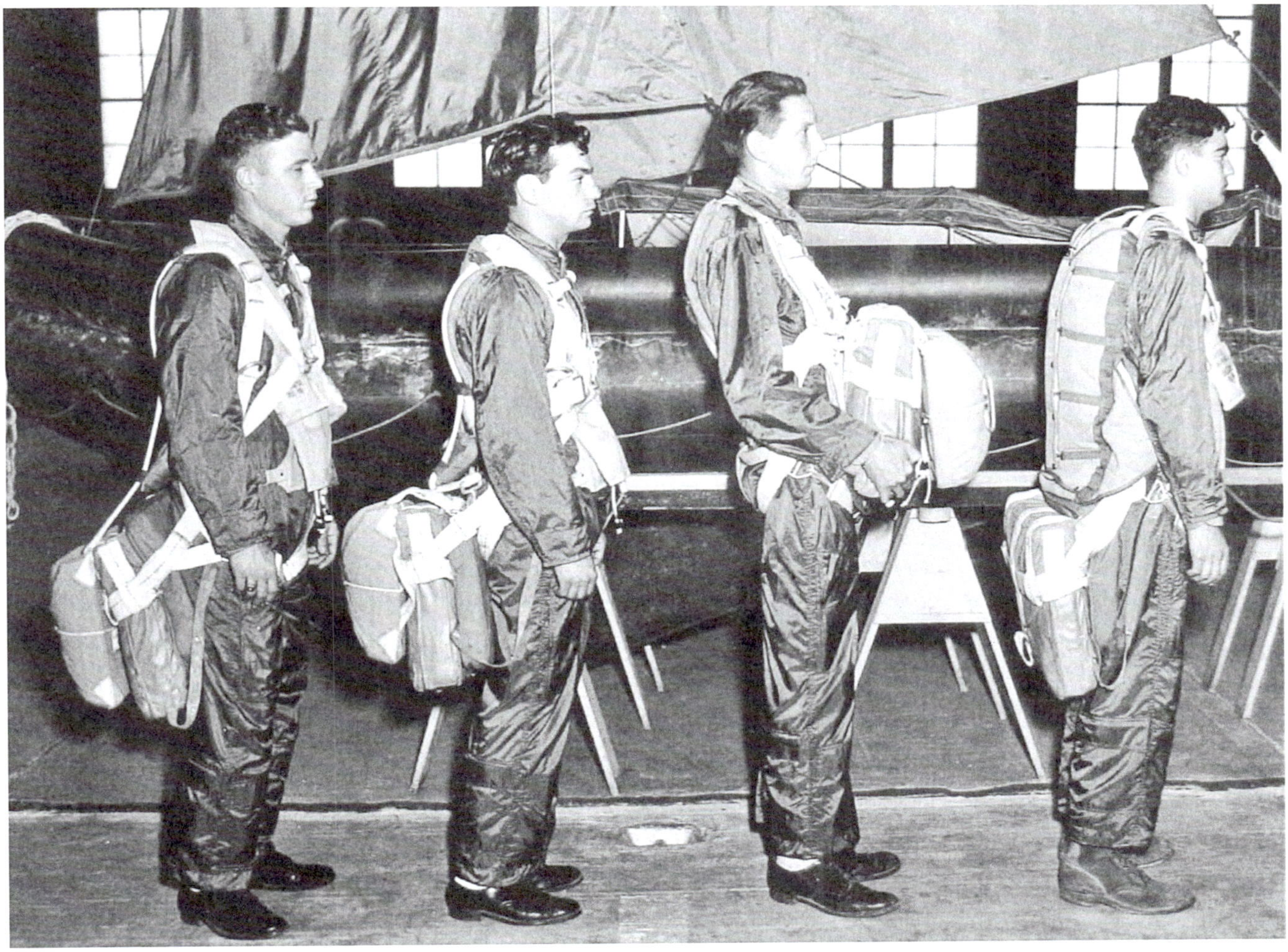

(Above) Technical Order 114-44 was another expedient measure, like that of Technical Note 74-44. The BuAer was still investigating a means of quick release and the adaptability to other parachute types. Following positive results, the BuAer standardized the new container under Dwg. No. 9218, 5 October 1944, titled Container, Pararaft Kit. In addition, a suitable quick release snap was adopted, titled Hendershall Type Quick Release Snap. In November, contracts were awarded for the Model-A pararaft case, and in January 1945, contracts were released for the PK-1 pararaft, prompting the procurement of the new container. An initial order for 50,000 containers was awarded to Sears & Roebuck Co. to equal the amount of PK-1 kits on order, which were to be supplied to the contractor and delivered complete. An additional 40,000 containers were added to the order to supplement service requirements for the Model-A and AN-R-2 kits, available under ASO stock number R83-C-8690. Only the contract containers included the Hendershall snap due to availability issues, leaving service fabricated containers to utilize the expedient split link. Upon release of the contracts mentioned above, new technical orders for the QAS (22-45, 26 February 1945), Standard Back (18-45, 31 January 1945), and Standard Seat (17-45, 6 February 1945), type parachutes were distributed for the attachment and fabrication of the pararaft container Dwg. No. 9218. Pictured from left to right are the Standard Seat, QAS, QAC, and Standard Back type parachutes utilizing the new pararaft container. (National Archives)

(Right) The Hendershall Type Quick Release snap.

*(Above and right) The 50,000 containers fabricated by the Sears & Roebuck Co. for the PK-1 had the nomenclature printed on one side. As part of the standard feature for the PK-1 case, a hand-hold was sewn to the front to help with the removal from the container, as was the Model-A.*

*(Left and below) An example of one of the 40,000 containers fabricated for use as a replacement for the PK-1 and use for the Model-A and AN-R-2.*

*(Left) AN-6520-1, modified locally by adding the hand-hold, indicating it was used in conjunction with a pararaft container.*

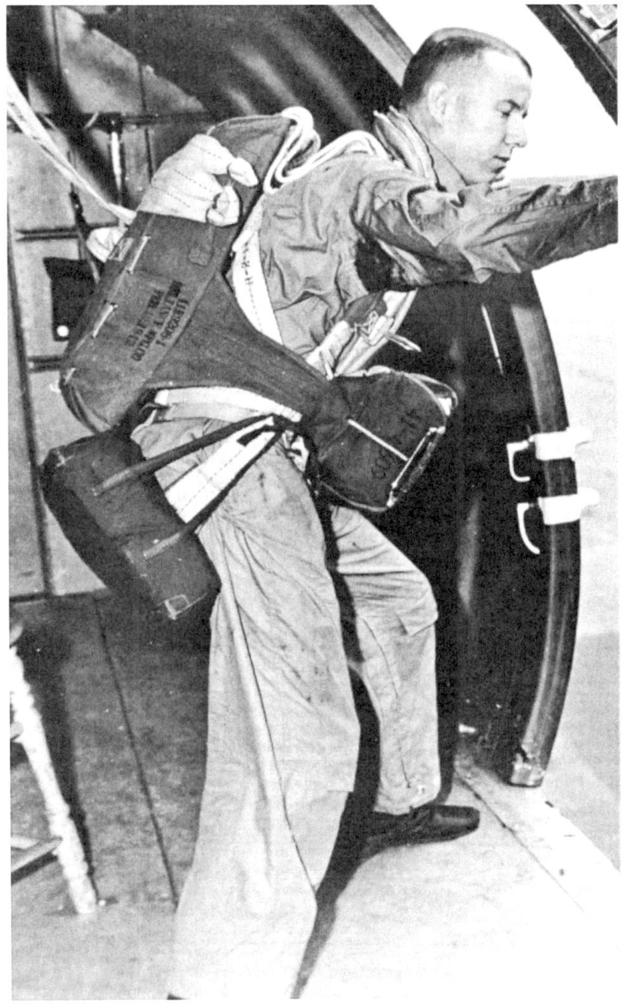

*(Above) 1st Lt. David Allen of the 502nd Parachute Battalion demonstrating the new USAAF one-man parachute life raft, tied to a troop type parachute harness, at Indian Lake, Ohio, August 1942. Four cotton straps were integrated into the design of the non-slotted pararaft case for attachment to parachutes other than the standard seat type. The same attachment method would apply to the QAC and Back types. (National Archives)*

Like that of the US Navy, new service requirements were established for the issuance of pararafts to all USAAF bombardment aircraft. Prior to USAAF units operating overseas, life rafts of any size were only provided if the flight path was over water. This policy created a deficiency in safety equipment for units operating in the ETO, especially Great Britain, as each mission, or sortie, required the flight path to be over water.

To supplement this deficiency, life rafts and emergency equipment were requisitioned from the RAF. The two principle parachutes utilized by the US Navy in 1942 and into 1943 were the standard seat type and the QAC, and to meet their service requirements, an application was devised for the attachment of the pararaft to the QAC.

The back-type parachute became the preference for USAAF fighter pilots operating in the ETO and the QAC for USAAF bombardment crews. As part of policy No.33, the positions of pilot, co-pilot, bombardier and navigator were to be issued the back-type parachute with all other crew members receiving the QAC in bombardment aircraft.

To satisfy the service requirement to issue a pararaft to all crew positions, the method of attachment of the RAF Type C pack of the Type K Dinghy was adopted and in turn requisitioned in quantity. The RAF method incorporated snap hooks on the raft case and D rings on the parachute leg straps for quick attachment and release. To convey the extreme importance of the issues encountered from combat experiences in the ETO, Major R.W. Chaffee flew back to the continental United States and scheduled a series of meetings with various commands.

In a conference held at the Material Command Wright Field in April of 1943, it was agreed to develop and standardize a universal pararaft and parachute harness attachment method.

*(Left and above) The method of attachment presented by Major Chaffee was that used by the RAF, with the D ring on the harness and the snap hook on the case. For reasons unknown, the AAF decided to do the opposite and placed the D ring on the pararaft case and the snap hooks on the harness. Samples were fabricated and tested at the AAF Proving Ground Command Eglin Field, FL in June 1943. (National Archives)*

Following the successful drop and test jumps, Technical Instructions No. CTI-1370 was drafted, 18 June 1943, for the modification of the pararaft. CTI-1370 was distributed to contractors of the QAC and back type parachutes in July 1943. Dwg. No. 43G26486 detailed the pararaft modification with a length of parachute webbing across the back of the case, terminating on the sides with parachute snap hooks.

Prior to change orders in production it was

decided to reverse the D rings and snaps to reduce complications and maintain a universal method between RAF and USAAF equipment.

By July 1943, the bulk of the 1942 contracted pararafts had been delivered. Limited quantities of the March and April contracts were delivered as per specification AN-R-2A, which were of the slotted cases. These units would not include the quick attachable parachute modifications.

Only cases manufactured after July 1943 would have the snap provisions factory applied, Technical Instructions CTI-1370 and Dwg. No. 43G26486 were additionally distributed to service activities to modify those in stock.

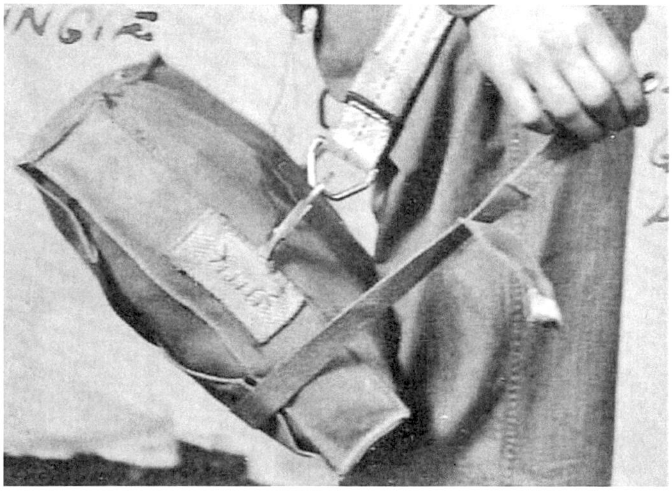

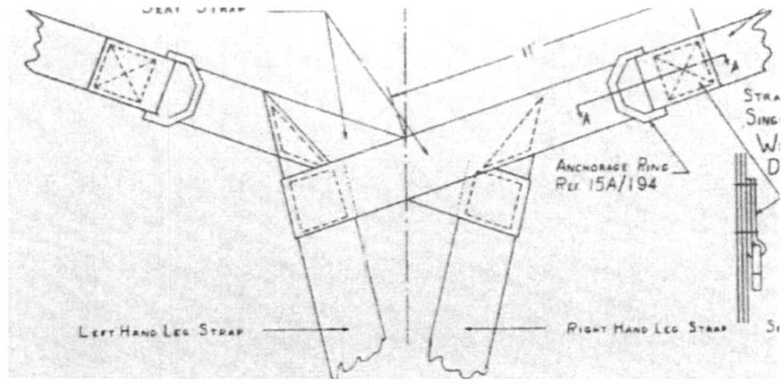


(Left) USAAF QAC parachute modified to match the RAF Observer harness with D rings and a Type-C dinghy pack.

(Above) Part of the drawing in CTI-1370 for the modification of parachute harnesses.

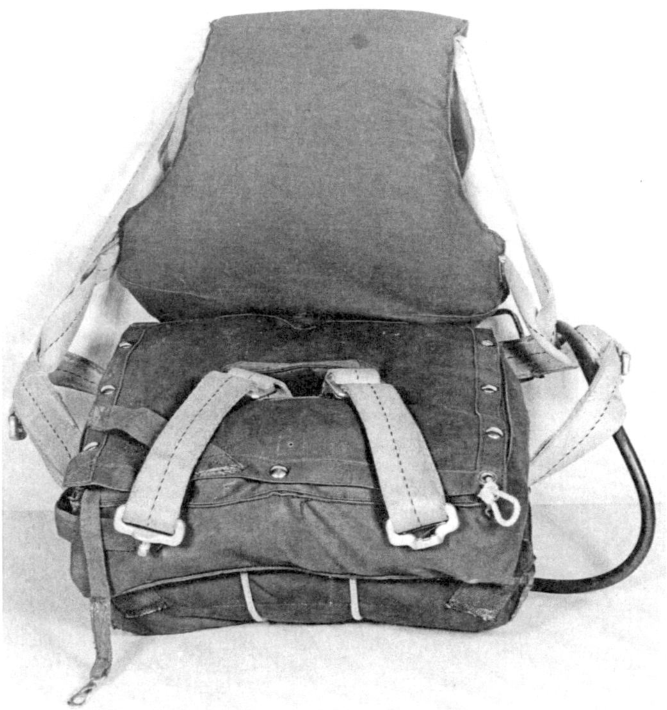

(Left and below) The BuAer center-slotted M-524 pararaft fitted to a standard USAAF seat type parachute for evaluation. After field testing, the USAAF determined it was not optimal for their needs. (USAF)

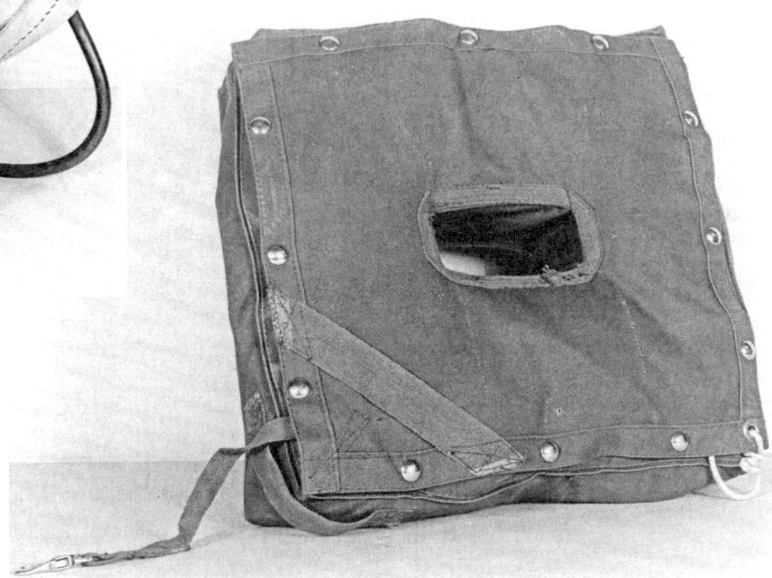

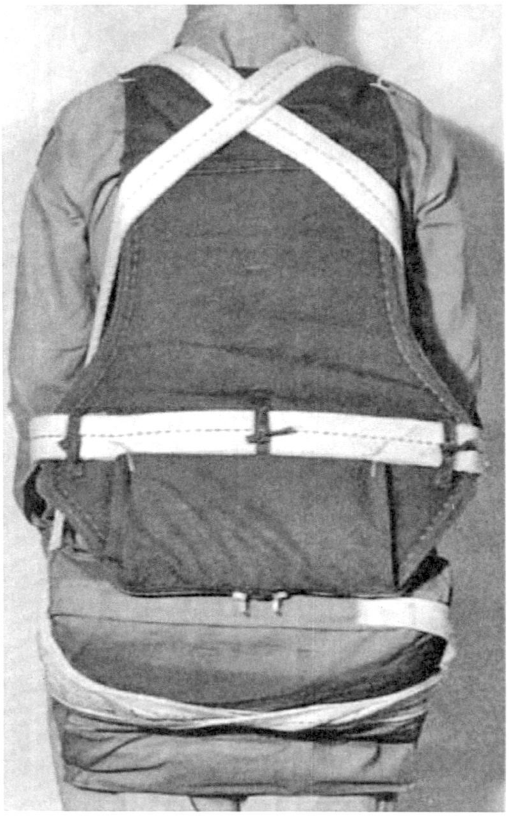

(Above) For back and QAC parachutes, the pararaft was to be secured by the leg straps through the slot, as detailed in Technical Order 04-15-2. At the time of the draft for this order, the center slotted type was the standard by specification, the case being used in the image is of the forward slotted type. This method presented a long list of deficiencies, yet remained the standard application until being addressed by Dwg. No. 43G26486. (National Archives)

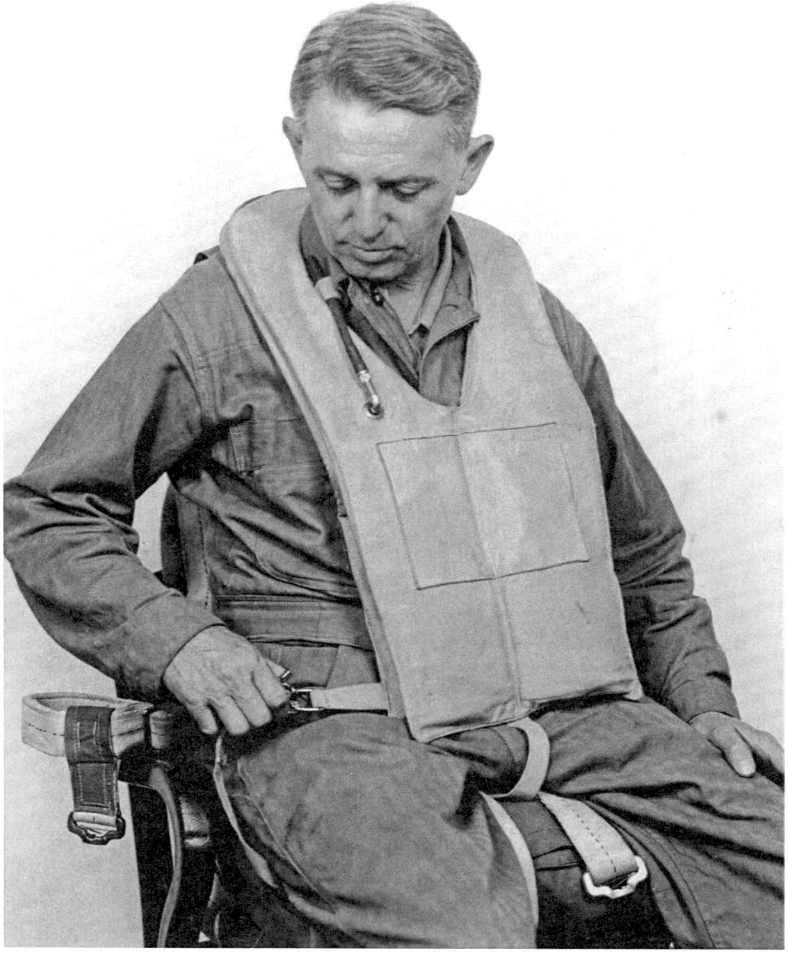

(Above) A volunteer demonstrates the attachment of the pararaft lanyard to the life preserver D-ring. The life preserver was intended to be worn with the oral inflation tubes on the right-hand side, and the waist strap connection was also on the right. The pararaft lanyard placement was designed to integrate with the prescribed wearing of the life preserver. (USAF)

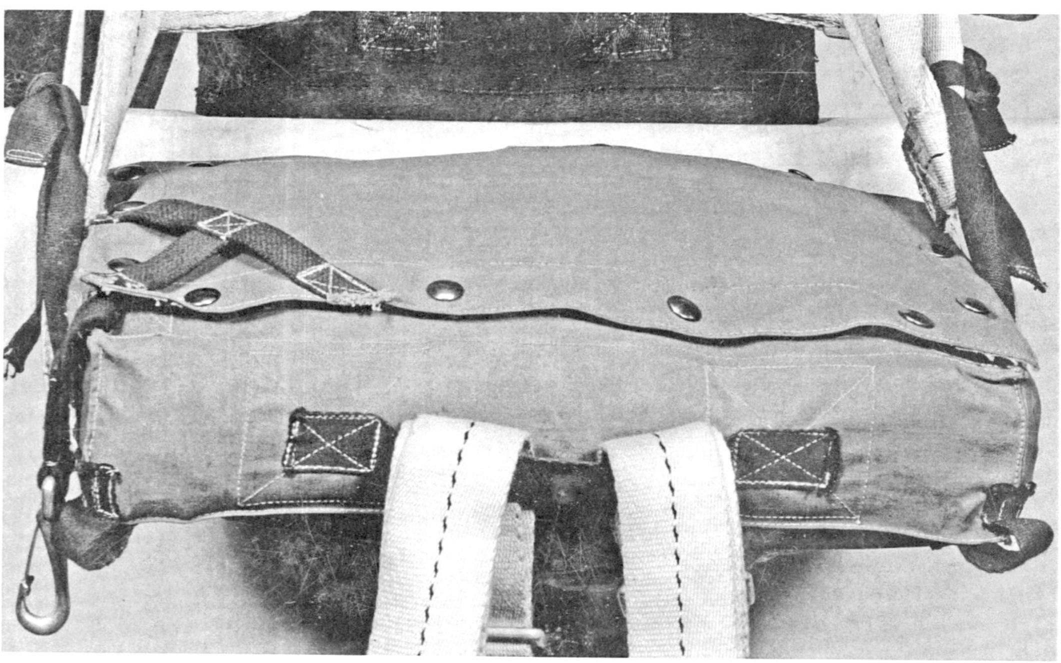

(Left) This image illustrates the purpose of the four tie tapes on the USAAF first generation non-slotted AN-R-2 pararaft. As can be seen, the tapes are tied to the vertical webbing on the harness, a system adaptable to both the back type and QAC parachutes. The leg straps could go over the pararaft or under it, either method securing it to close to the body of the user. The method shown is under the pararaft unit. (USAF)

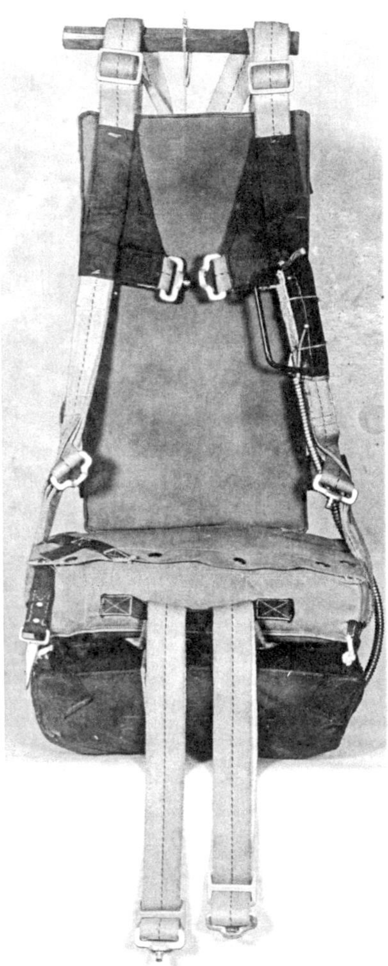

*(Left and above) Attachment of the non-slotted case to the standard seat type parachute. Tabs are provided at the four corners of the case then secured to parachute pack replacing the seat cushion. (USAF)*

*(Right) Revised Technical Order 04-15-1 of 15 August 1944 included these illustrations of the new AN-R-2A pararaft and attachment to QAC parachutes. Shown worn in conjunction are parachute emergency kits. (USAF)*

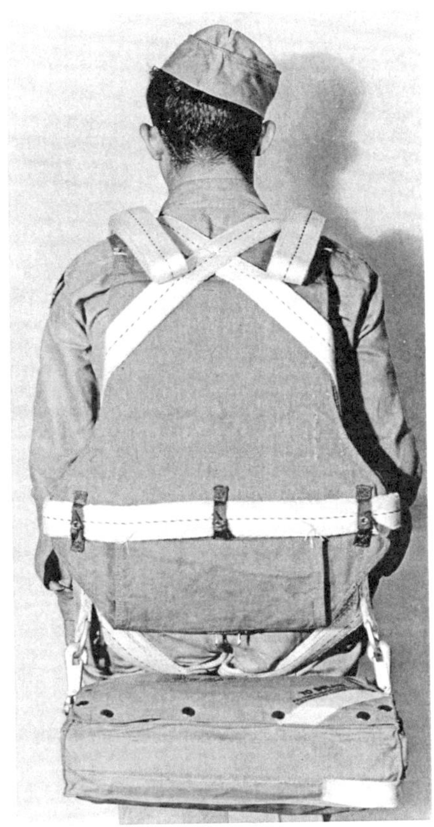

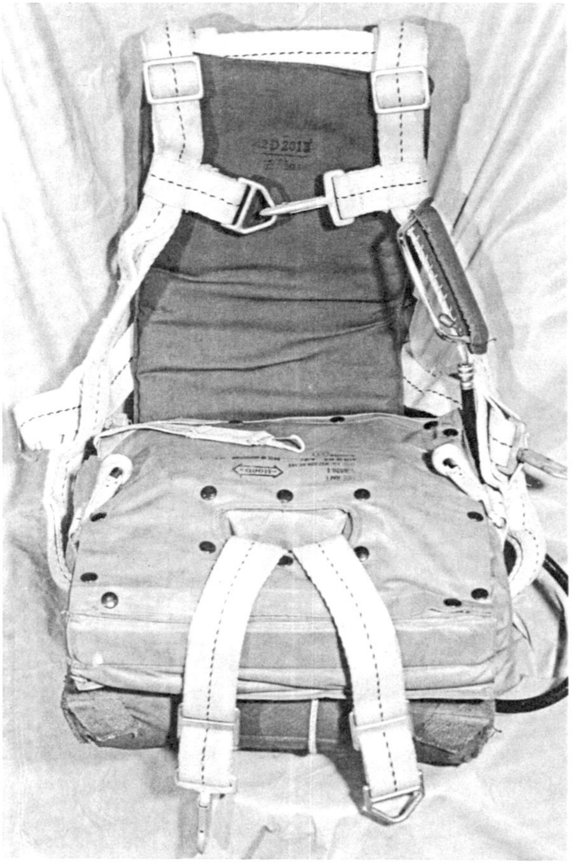


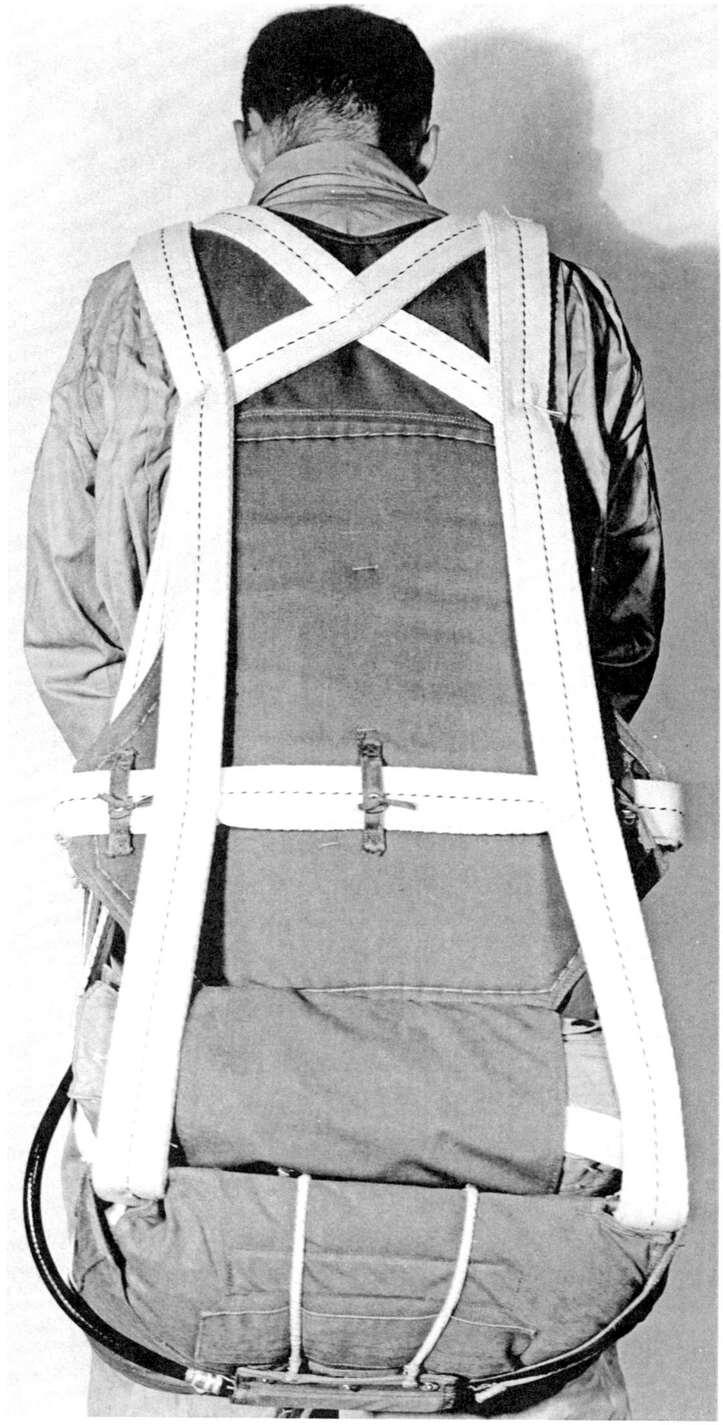

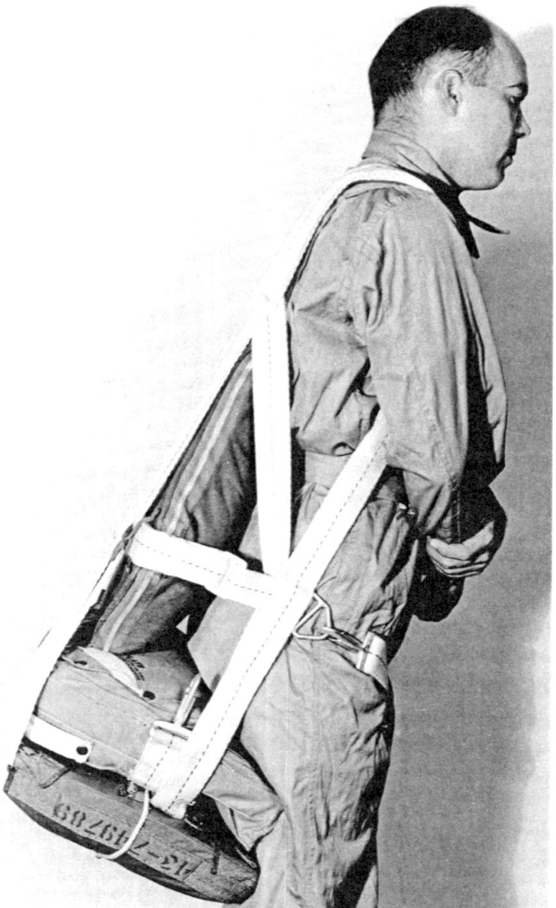


*Revised Technical Order 04-15-1 of 15 August 1944 also illustrated the AN-R-2A pararaft and attachment to the standard seat type parachutes. Additionally, it outlined that the snaps were to be cut off to increase comfort when used in conjunction with the standard seat type parachute. This was an optional recommendation. The attachment would be the same for the back-type parachute as it was with the QAC. (USAF)*

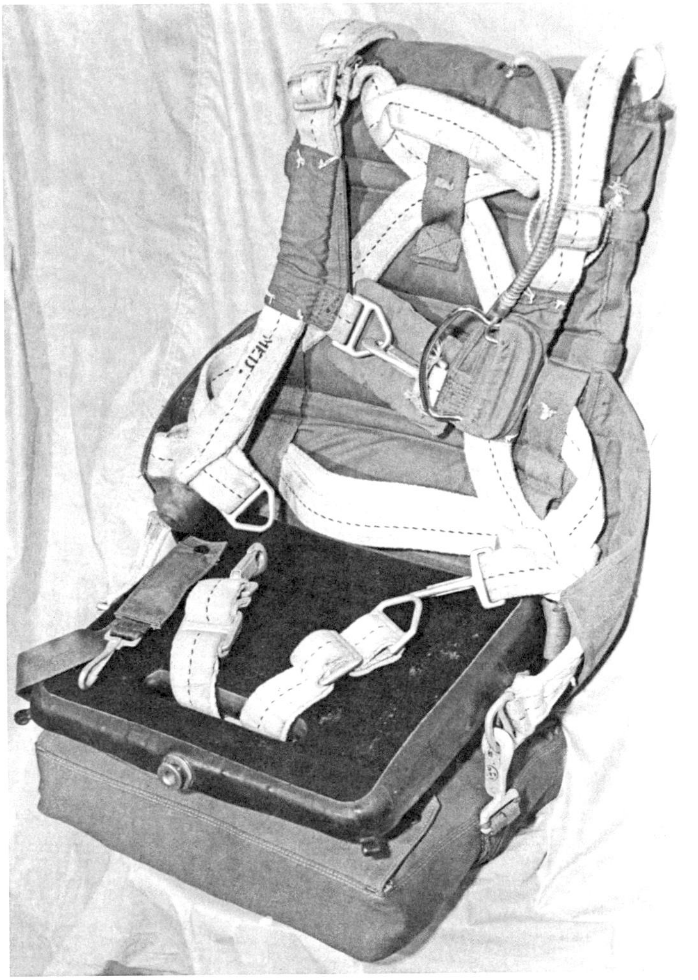

*(Above left and right) The attachment methods for the AN-R-2A applied to both the Type C-2 and C-2A cases, remaining standard into the post-war era. After the revision which reverse the snaps, consideration was not given for attachment to the seat type parachute, primarily due to it being phased from service in preference of the back type and QAC. All subsequent revisions to Technical Order 04-15-1 included only the attachment of the Type C-2 for the back and QAC parachute harnesses. Aboard bombardment aircraft, the pararaft pack was to be stowed in a convenient location and attached in preparation for bailing out or ditching. The same principle applied to the Type C-2 pararaft. (USAF)*

# LIFE PRESERVERS

There has been much deliberation and argument over the decades regarding who developed the life preserver that saved so many lives during WWII. Unfortunately, no single entity can lay claim as the sole source. Documentation discussing the early development is sparse, but enough can be found to lay the ground work leading up to the all-so-familiar models utilized during the war.

The real development and standardization for both aeronautical services began in the 1920s. Multiple developments were occurring simultaneously regarding flotation devices for aircraft and individuals. For aircraft, the services took steps to standardize life preserver cushions for seats.

On 19 February 1924, the USAAC standardized the Type A-1 life preserver seat type cushion, under Specification No. 40216, Part No. 074272, followed by the Type A-2 back type, under Part No. 076216 on 9 August 1929. These were intended to be standard for all aircraft, however they were not compatible with all models of airplanes. Aircraft manufacturer designs that could not incorporate the standard A-series cushions were instructed to supply their own life preserver cushions, prior to delivery.

Future development in aircraft would include more universal seat designs. On 25 November 1930, these cushions were revised to change the color from the original production of olive drab to yellow, for better visibility, referenced under Army/Navy Dwg. No. AN-7501, the Type A-2 back cushion was redesignated as the Type A-3.

The Type A-1 seat type remained unchanged except for the color, the Type A-2 back cushion was revised, eliminating the lift-a-dot fasteners to a slide fastener or hook. The Type A-4 was a back cushion, like that of the Type A-3, designed for cabin chairs referenced under Dwg. No. 32G2005, standardized 4 May 1932.

In 1934 and 1935, a Type A-5 life preserver was field tested. Intended to supersede the A-1 and A-3 for use in amphibious aircraft, it was described as a circular ring, constructed of cotton duck fabric filled with kapok, with an attached length of rope to throw to survivors in the water. Standardization was not approved, opting for the use of pneumatic type life preservers instead.

*(Below) Under the joint Army/Navy specifications of 1930, cushions were to be fabricated with a fire-resistant duck covering, colored yellow. As part of this specification they were to be property marked U.S. Property, but, as seen below, this was not entirely complied with. Seat and back cushions remained the standard, but their use was slowly discontinued by the late 1930s and through WWII, especially in pursuit or VF class aircraft. With the progressive standardization of pneumatic life preservers, and parachute designs incorporated with pararafts, they became unnecessary and their bulkiness created issues. They did however remain in use on bombardment and transport aircraft in various crew positions.*

*(Below left) Two B-17 gunners use the AN-7501 seat and back cushions on a practice bombing run in Hawaii, January 1942.*

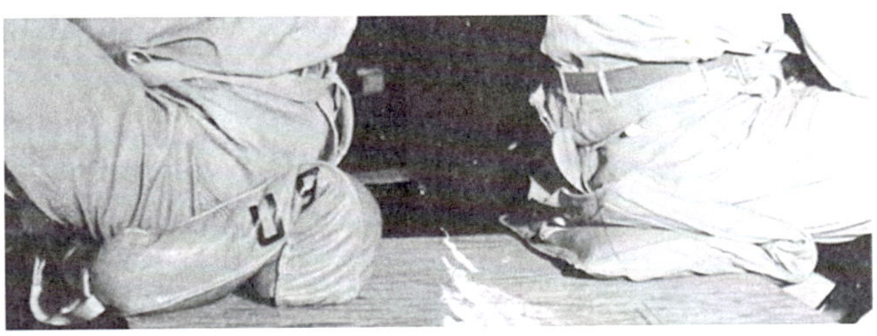

*(Right) Two variants of the back cushions Type A-3 and/or Type A-4, these are secured by two integrated slide fastener metal hooks to the seat pan. The Type A-2 utilized lift-a-dot fasteners. (Jeff Klein Collection)*

*(Left) The Type A-1 seat cushion (bottom) remained unchanged.*

In conjunction with life preserver cushions, both aeronautical services were standardizing life rafts in the 1920s. For individual flotation devices, kapok filled life jackets were the preferred standard.

The earliest origins of a standard model could be found within the USAAC on 10 November 1921, when they standardized a life preserver titled Coat, Flying, Non-Sinkable, Type A-1, under Specification No. 94-3017. The type A-1 was described as a kapok filled coat with full length sleeves and a collar that buttoned up. On 30 September 1931, Specification No. 94-3017 was revised, deleting the word "coat" from the nomenclature and replacing it with the word "vest" since the sleeves had been removed.

Life preservers during the 1920s and 1930s were only used when made mandatory, and even then they were not universally worn, as they were found too bulky and undesirable for wear under parachutes. The USAAC did not have a large requirement for flotation equipment at that time, and only required that the aircraft be equipped with flotation aids of some sort, for when flying over large bodies of water. Such means may have included life preserver cushions and/or a life raft.

The A-1 was procured in limited quantities, with a reported inventory of 1,500 as of August 1932. Five sizes were available under stock numbers 8300-961000 through 8300-967000. The A-1 coat would remain the standard flotation device until superseded by the Type B-3 pneumatic life preserver.

A new specification was drafted and released 15 September 1934, for a new type of kapok life preserver, titled Life Preserver Vest, Kapok, Type A-2, under Specification No. 94-3063. It was described of being made of a yellow, waterproof cotton duck, in the form of a vest, having three kapok stuffed compartments, one on each side and one across the shoulders.

This design eliminated the tendency to float face down (which the A-1 coat had tended to do). The A-2 was service tested from late 1935 into early 1936. Testing concluded that the pneumatic life preserver was superior to any kapok types and the A-2 was not standardized. However, at the time of its original inception it was developed under the premise that there were no guarantees of pneumatic devices, and that based on early testing results of inflatables, a kapok vest should also be available.

The Type A-2 was adopted as limited standard and available in five sizes utilizing stock numbers 8300-968000 through 8300-9720000.

*(Above) Images from the Spalding Aviators Clothing and Equipment catalog illustrating the kapok non-sinkable flying coats. The first two images show the same vest but the image far left includes sleeves, best conforming to the original specification No. 3017 Type A-1. The middle is also described as being an Army type with sleeves removed as per amendment of 1931. The far right image is described as being the standard Navy flying type with a distinctive kapok collar.*

*(Left) Three USAAC aircrew members of a Keystone LB-6 light bomber wearing Type A-1 kapok life vests, Luke Field Hawaii c.-1929. Wearing these vests under the parachute harness created considerable bulk. The Type A-1 was revised 30 September 1931, removing the sleeves and changing the nomenclature to Vest, Flying, Non-sinkable. Removal of the sleeves worn by these individuals was a local modification. (USAF)*

*(Left) Flotation suits were tested by both aeronautical services dating back to WWI, with the US Navy abandoning the concept early on. The USAAC did continue experiments into the early 1920s, with the first experimental model, designated Type C-1, manufactured by the Marshall-Field Co. The C-1 was constructed of gabardine wool with an inner liner of foam padding.*

*(Below left) The Type C-1 non-sinkable suit being field tested, c.-1924.*

*(Right) In 1925, the USAAC began field testing a modified lightweight one-piece suit utilizing the same foam padding, designated Type C-2. The C-2 was standardized 23 October 1928, but only procured in limited quantities. Ultimately, flotation suits were found too heavy and bulky, proving impractical for use. The concept was revisited late in WWII, in the form of anti-exposure suits. (USAF)*

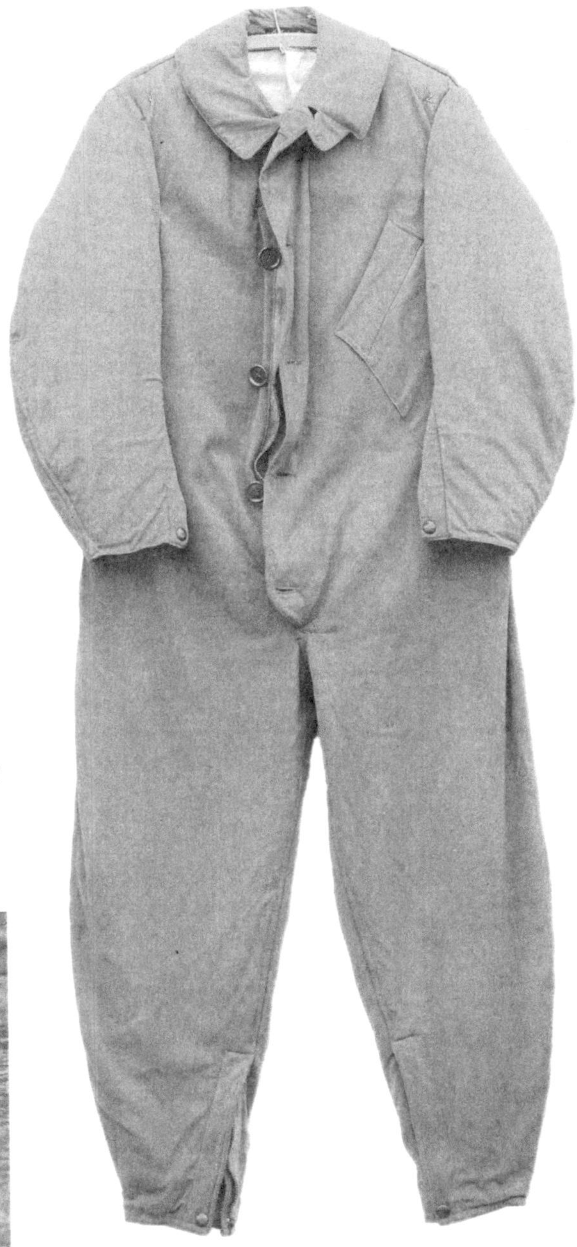

*(Left) Chief Aviation Machinist Mate W.T. Scott poses for a photograph, prior to attempting to surpass the long standing parachute jump record of 24,000' at NAS Anacostia, September 1929. As part of his safety equipment, he wears a kapok life preserver. This is an example of standard equipment worn by Naval aviators and aircrew of the era.*

*(Above) Naval aviation personnel of VF-2 welcome the pilot of a Vought VE-7 biplane aboard the USS Langley, c.1925. Two of the members are wearing kapok life jackets. Wearing of the jacket would have been optional at this time. Note the bulk of the kapok life jackets.*

As the USAAC set course in the development of kapok life preservers, the BuAer was investigating pneumatic types by the early 1920s, both in the form of belts and vests. They operated under the same parameters – life preservers were not mandatory for aircrew. The principle focus was not towards the individual, but rather the provisions for the aircraft.

As operational situations expanded, so did the demand for personal equipment. A variety of pneumatic life preservers and belts were tested and utilized for naval use, often procured locally.

After much trial and error, a life preserver manufactured by the Airubber Sales. Co. was universally found to be the most effective in supporting the user. Their patent, No.1625582, was filed in November 1924, not specifically as a life preserver, but for the method of creating such an article.

These vests were manufactured of a rubberized khaki jean cloth, vulcanized around the edges and manually inflated rather than having automatic pressurized systems. Two valves for inflation were located at the bottom of the collar, below the user's chin. Provided with favorable service test results, the BuAer adopted the Airubber vest under NavAer Specification M-5 on 18 November 1929, titled Jackets, Life, Pneumatic (Aircraft Use) and available under stock number 37-136-300.

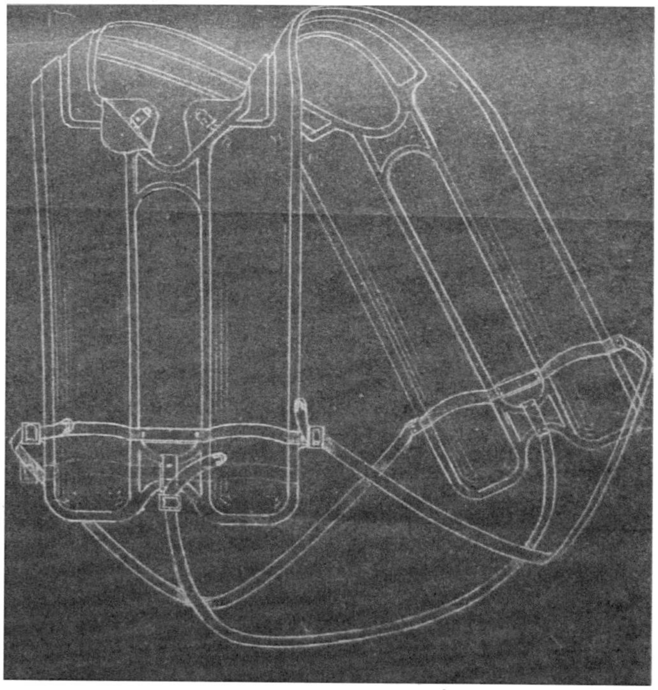

*(Above) The drawing that accompanied Specification M-5.*

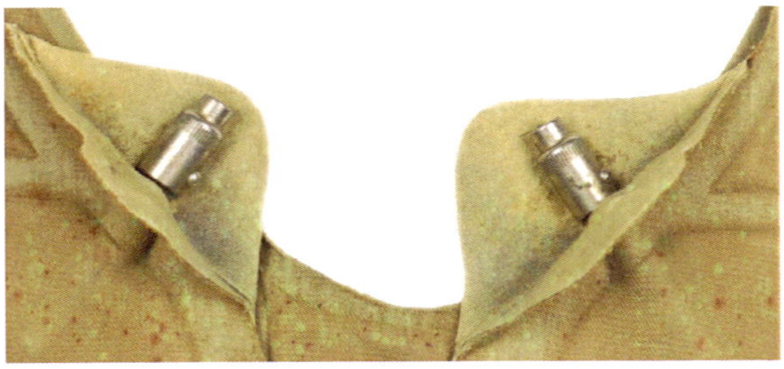

*(Above and left) The Airubber Sales Co. pneumatic life preserver as procured by the BuAer under Specification M-5. Early commercial examples did not have a crotch strap, required by NavAer specifications.*

*(Right) A volunteer, who has just completed service testing a parachute, shown wearing an M-5 life vest. (National Archives)*

*(Above) A group of aviators wearing the Airubber life vest, 1920s. These are commercial models, as BuAer specifications required the fitting of a crotch strap. (National Archives)*

*(Left) M-5 life preservers being worn by LCDR Ralph Wood and RADM William A. Moffit aboard the USS Saratoga, July 1929.*

*(Bottom left) On 27 May 1930, the M-5 life preserver was revised to include nickel plated brass snaffle hooks and D-rings for connecting the waist and crotch straps (the original specification had included slide buckles), designated as the M-5A. Reverend Dobsen Peacock and pilot Lt. George Richards shown wearing the M-5A with snaffle hooks, March 1932.*

*(Below) Drawing accompanying Amendment-A. (National Archives)*

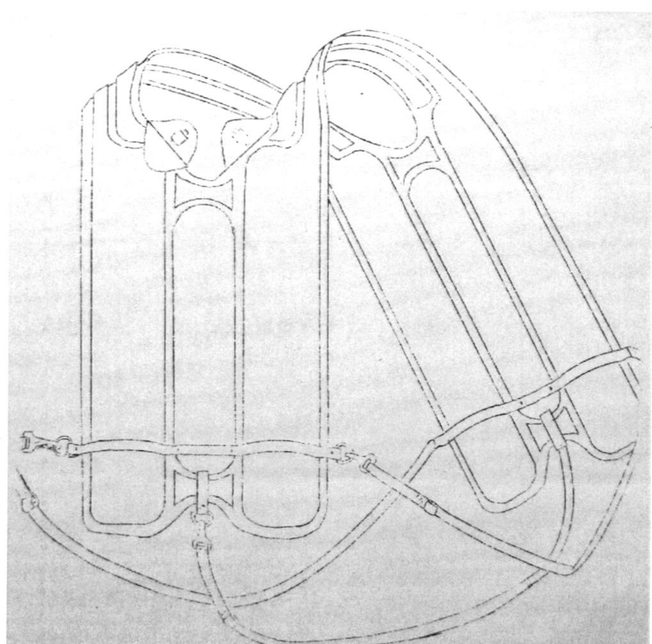

The Airubber vest was in use by units operating in the Panama Canal zone prior to its standardization. Adjoining USAAC units became interested in the vest as a substitute for the exceedingly bulky and hot kapok vests they were using at the time. Long sleeved kapok coats were not optimal in tropical regions, which caused an overwhelming reluctance to wear them. Requests from Panama were submitted to the Materiel Division, enquiring into the possibly of standardizing something similar. The USAAC field tested the Airubber vest from November 1929 to July 1930, concluding that they were superior to the kapok type, and proposing adoption for pilots flying over large bodies of water. They were adopted for use in Panama.

By the beginning of 1930, both services were transitioning to the use of pneumatic life preservers, but there was doubt in their dependability, flotation, and inflation characteristics. A new type of life preserver was presented, from which all immediate future developments would derive. An inventor named Peter Markus developed a series of inflation systems and vests utilizing compressed gas for inflation. Samples were submitted both to the BuAer and the USAAC for evaluation. By late 1930, initial tests proved favorable for the design as it automatically inflated, was comfortable to wear and had improved flotation characteristics, keeping the head and face above water.

*(Below) An existing example of the Peter Markus air inflated life preserver, truly in the form of a vest, as patented under No.1798430 in March 1931. It was presented to the aeronautical services in 1930, as manufactured by his company Markus Devices Inc. He had originally filed the for patent in February 1930. It was composed of rubberized khaki jean fabric with a single cylinder to discharge a CO2 cartridge, inflating a single compartment vest.*

The BuAer was the first to make a move towards revising the design. Principle improvements included creating two air compartments with separate CO2 inflation cartridges, the addition of emergency oral inflation tubes and a back strap.

Manufacture of the new vests was transferred to the Hodgeman Rubber Co. utilizing patents by Marcus. The BuAer adopted and standardized the vest as NavAer Specification M-72 on 26 May 1931.

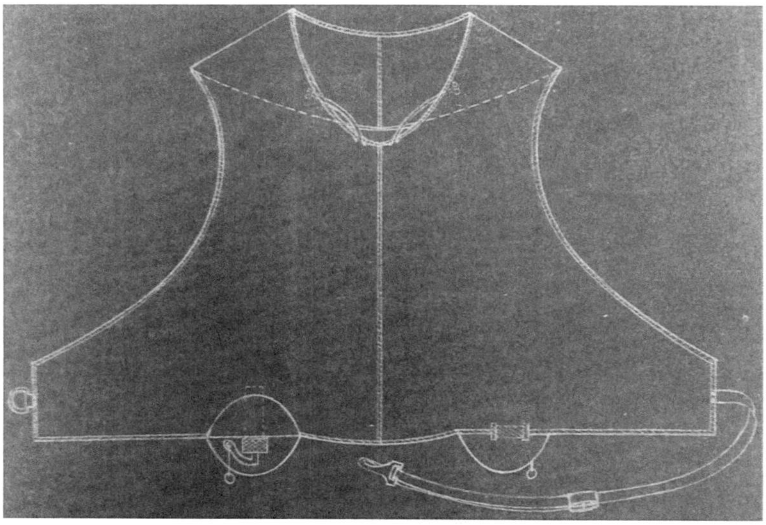

*(Above) Drawing accompanying specification M-72, which superseded the M-5 life preserver. The overall likeness to the original design by Markus is clear. The rubberized jean fabric was retained, but in an olive drab color.*

*(Below) A Naval aviator demonstrates the M-72 pneumatic life preserver, October 1935. (National Archives)*

*(Above) A USAAC officer demonstrates the M-72/B-1 life preserver at Rockwell Field, California, November 1933. The USAAC field tested the M-72 life preserver from November 1931 to April 1932, along with other samples received previously, but found this model superior to all other designs. On 11 April 1932, the Materiel Division authorized the M-72 to be adopted as Substitute Standard, assigning it the designation as the Type B-1 Life Preserver, meaning it was authorized for use only if standard kapok vests were not available. The USAAC was not satisfied with the inflation system and wanted to look into other de-signs before agreeing on a standardized pneumatic type. (National Archives)*

Development continued in both services, with each encountering similar issues with the M-72/B-1.

The primary deficiency was that, in the event of one bladder deflating or failing to inflate, the buoyancy became lopsided. In a worst case scenario, this could force an unconscious user over onto their face. The USAAC essentially dropped the overall design features, going back to the drawing board by designing a completely new type. They had an ulterior motive; a totally new design would be clear of any patent infringements by potential manufactures, since to date the vests had been based on patents held by Marcus.

The Materiel Division constructed a new concept for a pneumatic life preserver.

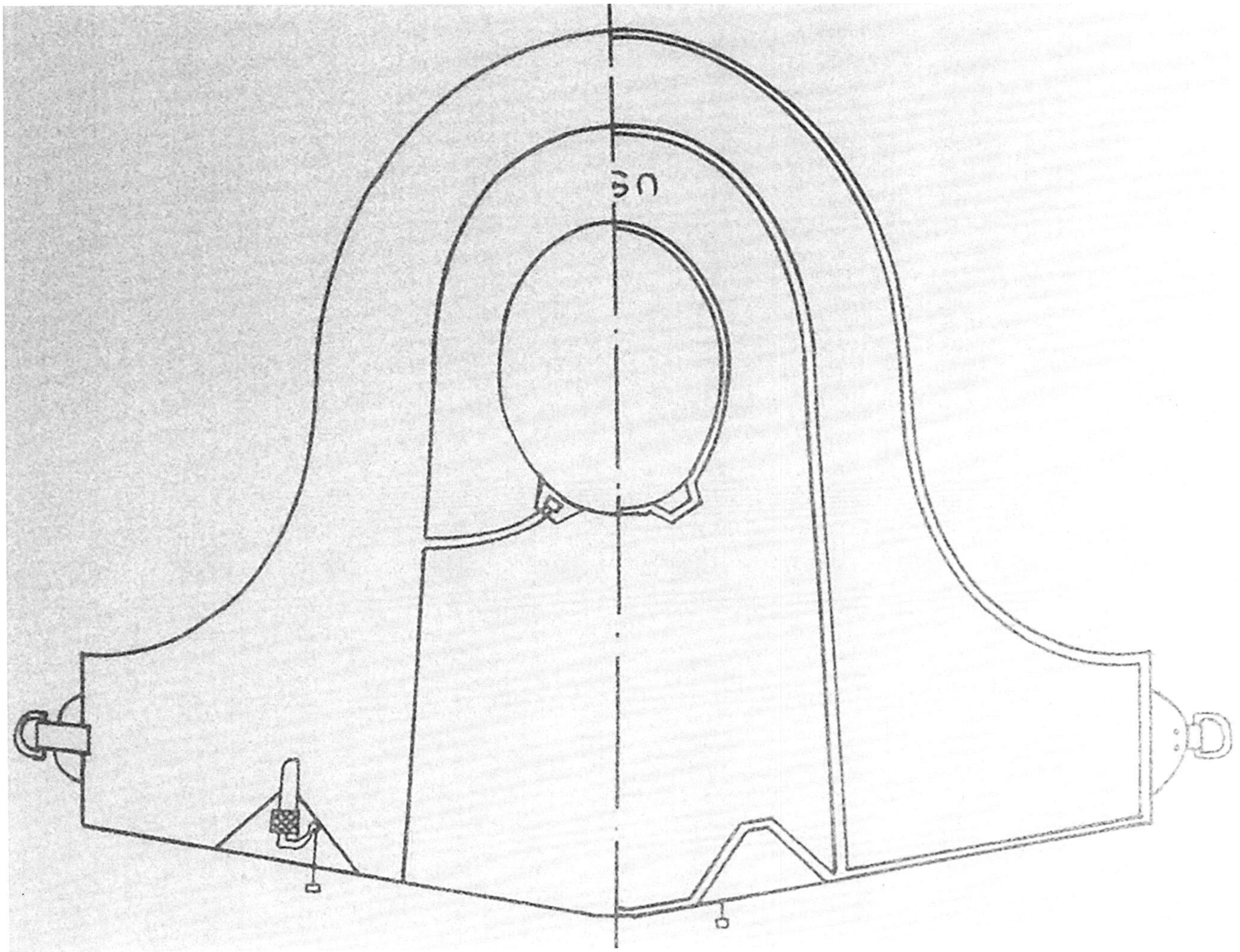

*(Above) The BuAer amended the M-72 to better distribute buoyancy. The two bladders were no longer separated centrally, instead being manufactured with equal halves, essentially an inner and outer, side-by-side bladder. If either one of the bladders were to deflate, the user would still be held on their back and balanced. Drawing of the revised vest as shown in NavAer Specification M-72B, effective 2 February 1933. It retained the O.D. colored fabric but detailed that a 1-1/2" USN stencil be placed on the collar.*

Shortly after issuance of the new revised M-72B life preserver, a service-wide order was distributed, officially authorizing the mandatory issue of self-inflating life jackets as standard life saving equipment for flights over water, detailed in BuAer Technical Note 18-33, dated 25 May 1933.

The BuAer was still open to other suitable designs, expressing interest in developments and procurement taking place with the USAAC.

On 23 March 1936 the M-72, for reasons unknown, was re-designated under Navy Department Specification 23P6 (INT) with the title Preservers, Life, Jacket-Type, Self-Inflating. Two contracts were awarded for unknown quantities, c.1937, to the Hodgeman Rubber Co. under Navy Department contract numbers No. 61156 and No. 63985.

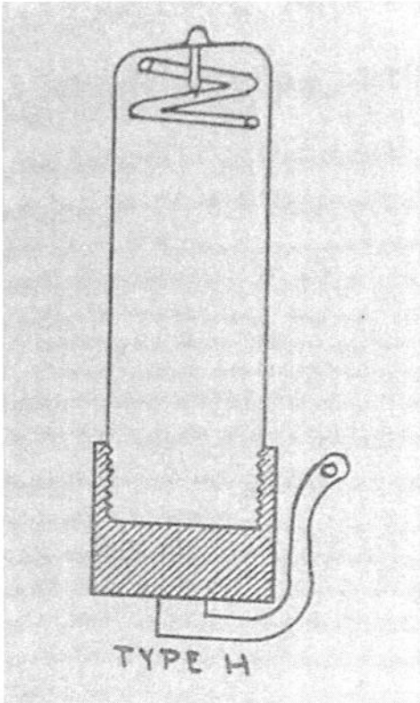


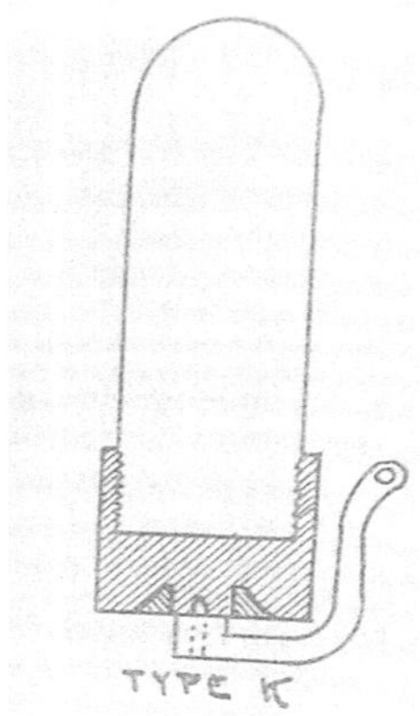


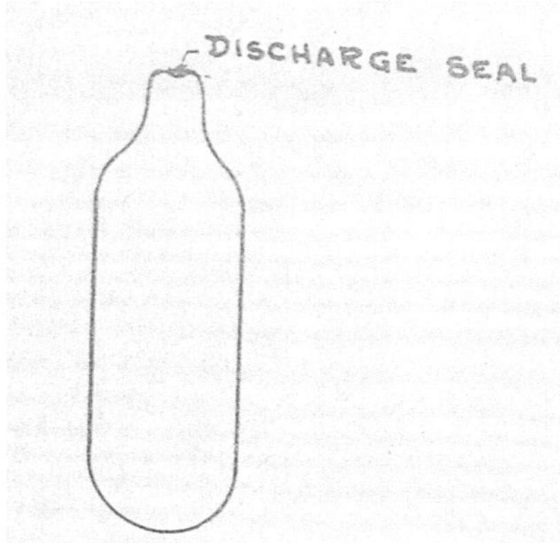


*(Left, top three pictures) Illustrations from Technical Order 8-38. The M-72 and 23P6 were manufactured using two types of cylinder for discharging the $CO_2$. Each used a $CO_2$ cartridge referenced under the commercial trade mark Sparklets. The Type H and Type K differed in the firing pin placement. In the Type K it was in the cap while in the Type H it was at the base of the cylinder. BuAer Technical Order 8-38, distributed 5 February 1938, addressed discharge issues and instructed inspection to ensure the cap was screwed on properly, as there was limited tolerance to properly rupture the seal on the cartridge.*

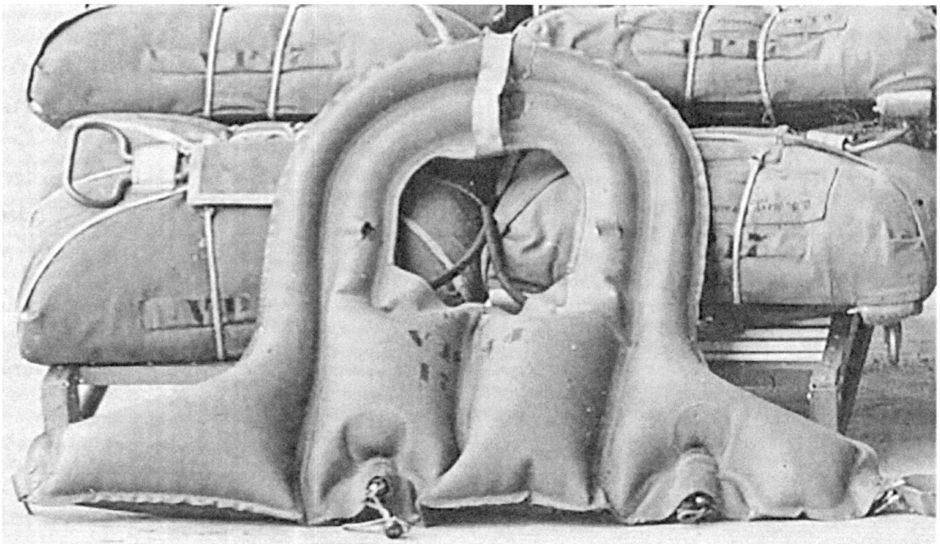

*(Left) M-72B life preserver shown inflated, displayed with an equipment layout provided on a PBY-2, property marked to VP-17, c.1937. Of interest is how the oral inflation tubes would be secured away from the users face by loops on the left and right of the collar opening. This feature would be amended in 23P6 to a single loop between the tubes located on the bottom edge of the collar. (National Archives)*

*(Below) Pilots of Marine Observation Squadron VO-8M or Marine Scouting Squadron VMS-2 "Ace of Spades" wearing a mix of 23P6 and M-72 life preservers aboard the USS Saratoga, c.1936. The four individuals on the right wear the M-72. During this era, designations were changed, hence the possible squadrons noted above. (National Archives)*

*(Right) In specification 23P6, the rubberized jean fabric was required to be dyed an orange yellow.*

*Production of the vests would utilize three styles of CO2 inflators, two of which were illustrated previously. The third inflator provided was a type by Peters & Russel Inc. This cylinder was a late production addition.*

*(Below) Even in the early stages of the life span of the 23P6, it had deficiencies in fitting the user. Technical Note 30-36, dated 2 July 1936, was distributed, instructing that the jacket was to be worn loosely, because it was reported individuals were attempting to fit it snug, like a piece of clothing. This would be the last of the three technical orders or notes detailing requirements for the 23P6.*

*(Right) Each life preserver would be stamped with the contractor name, contract number, Navy acceptance stamp, and the four pertinent patents applied to the fabrication of the vest – all awarded to Peter Markus.*

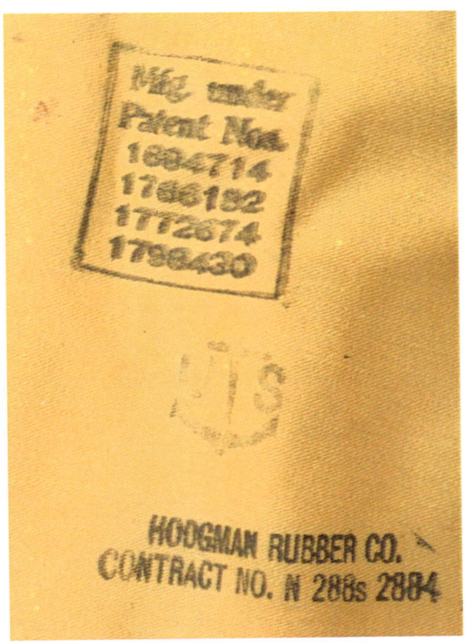


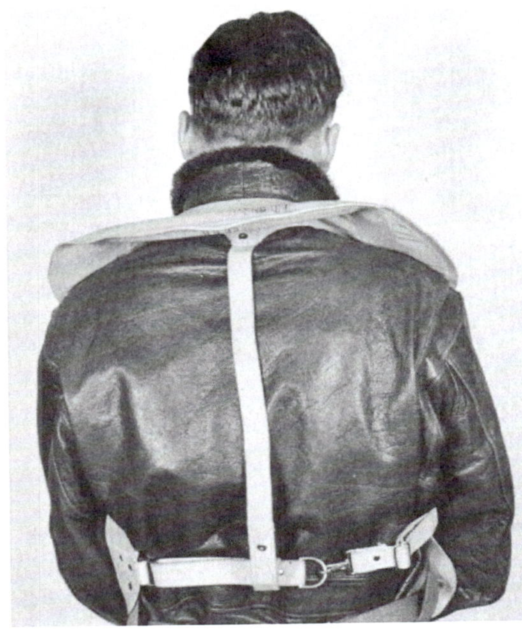

The 23P6 was accepted as an interim, or temporary, specification and would never be officially standardized in its service life, remaining unchanged by specification from 1936 to 1942.

In the late 1930s, the BuAer experimented and field tested other models of vests such as the USAAC Type B-2 and B-3 life preservers, but ultimately decided that the 23P6 would be the standard. It would remain the standard issued model until officially superseded by the AN-V-18 life preserver in 1943 available under stock number 37-J-150.

The vest was known officially as the Navy Aeronautical Specification 23P6 (INT) Preserver, Life, Jacket Type, Self-Inflating.

It should be noted that this life preserver was never referenced as the MK-I in any nomenclature, nor loose jargon in prewar or wartime documentation.

| CONTRACT | AWARD DATE | CONTRACTOR |
|---|---|---|
| N156-19272 | Circa 1940 | New York Rubber Corp. |
| No.69645 | Circa 1940 | New York Rubber Corp. |
| No.88462 | June 1941 | Hodgman Rubber Co. |
| No.89035 | July 1941 | Firestone Tire & Rubber Co. |
| No.90599 | August 1941 | U.S. Rubber Co. |
| 288s-113 | October 1941 | U.S. Rubber Co. |
| No.95770 | December 1941 | Firestone Tire & Rubber Co. |
| No.99662 | February 1942 | Hodgman Rubber Co. |
| No.99863 | February 1942 | Durkee Atwood Co. |
| 288s-2883 | March 1942 | New York Rubber Corp. |
| 288s-2884 | March 1942 | Hodgman Rubber Co. |

*(Above) Additional known contracts for the 23P6 life preserver. Quantities unknown.*

*(Above) Pilot Lt. jg Donald Hill of VGF-29 wears the standard 23P6 aboard the USS Suwannee, January 1943. (National Archives)*

*(Above right) 23P6 worn by pilot O.C. Morgan of VF-1 on USS Yorktown, September 1943. This picture illustrates a common complaint of the vest – its bulk would crumple under the parachute harness, creating a hindrance. Morgan would be killed in action shortly after this photo was taken over Marcus Island. (National Archives)*

*(Right) Pilot Ens. Greeg of VC-13 wears a 23P6 life preserver aboard the USS Core, August 1943. He is also wearing a life preserver belt. (National Archives)*

*(Far right) Pilot Lt. William F. Eadie of VS1-D-14 wears a 23P6 with squadron and personal property markings. Eadie piloted the plane that spotted Eddie Rickenbacker, and other survivors, and flew its occupants to the nearest hospital, December 1942. (National Archives)*

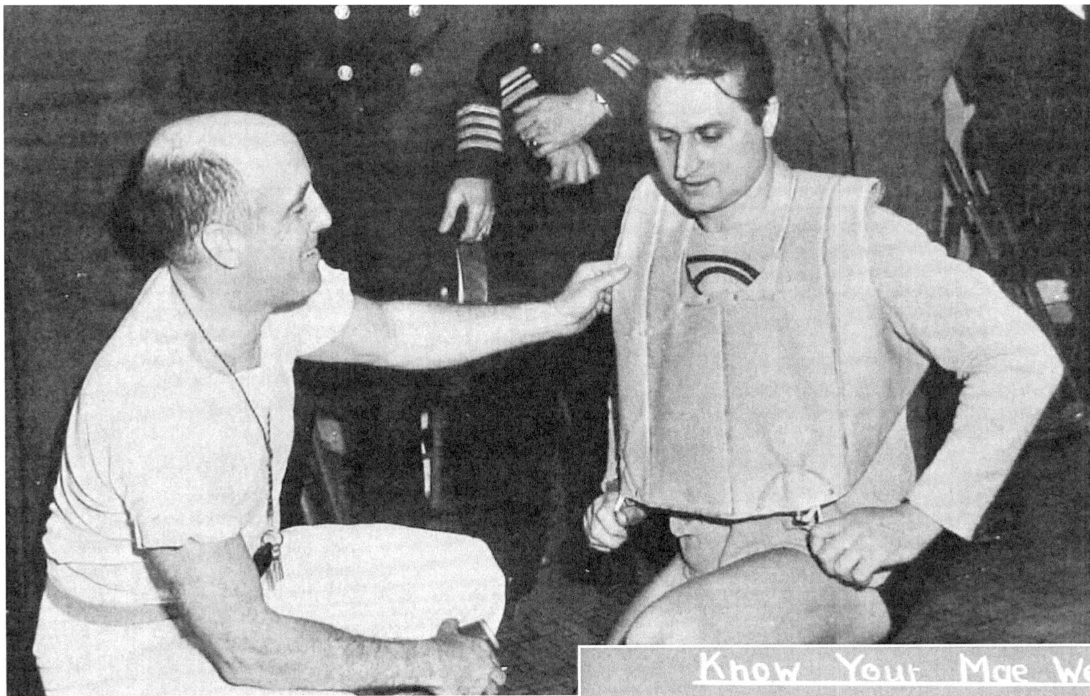

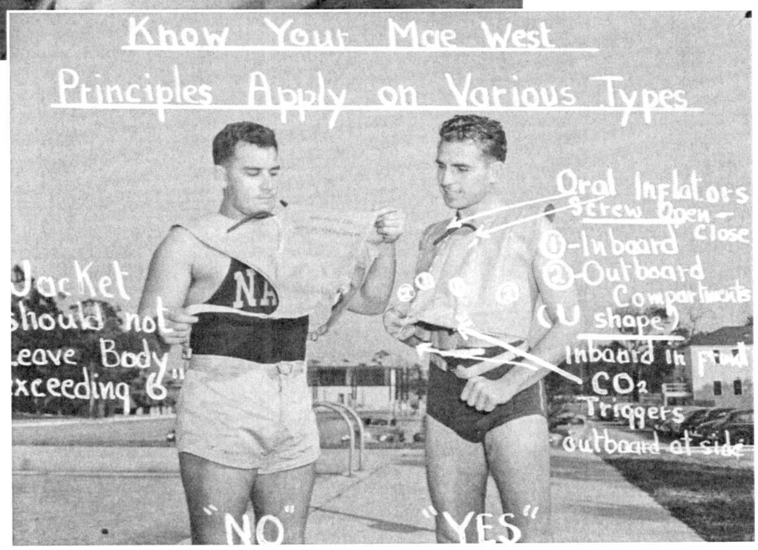


(Above) Ens. Peter Fick is demonstrating the 23P6 life preserver to NROTC members of the University of North Carolina, February 1943.

(Right and below) Proper wear of the 23P6 as detailed in T.O. 30-36.(National Archives)

*On 23 March 1943, an all service BuAer circular was distributed, with the directive to modify the 23P6s in service with the addition of a crotch strap. Production of life preservers had all been converted to the manufacture of the USAAF specification, Type B-4, several months prior, the modification therefore would not be applied by the contractor. Attached to the bulletin was a simple one-page drawing, detailing the attachment of the crotch strap.*

*(Above right) Two methods of attaching the crotch strap were outlined in the drawing. The first was by cementing a rubberized patch, as seen on life preservers worn by aircrew of VD-5. (National Archives)*

*(Below) VB pilot Lt jg Chester Spray also wears a crotch strap attached to a cemented rubberized patch, August 1944. (National Archives)*

*(Right) The second method of attachment was by means of a rivet, as seen on the 23P6 worn by Lt jg Donald Eli of VF-5 aboard the USS Yorktown, October 1943. (National Archives)*

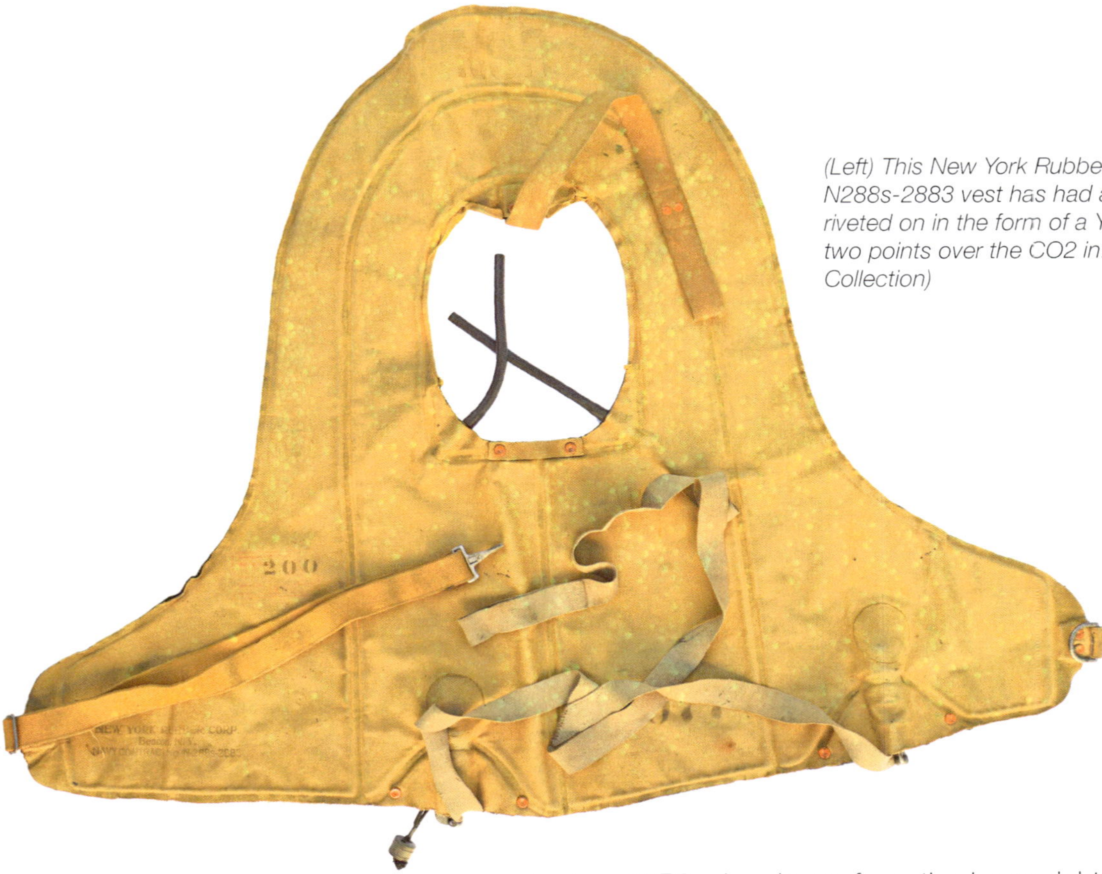


*(Left) This New York Rubber Corp. 23P6 contract N288s-2883 vest has had a crotch strap system riveted on in the form of a Y strap, attached at two points over the CO2 inflators. (Robin Bartel Collection)*

The unsatisfactory nature of the 23P6 became apparent early in naval operations in the PTO, as complaints were received stating it was too bulky and did not provide a good fit. It was also too loose under the arms and back of the head, which caused discomfort and flapping of the jacket when flying with the cockpit open. It overlapped so much on the shoulders that it interfered with the freedom of movement, and, because of its bulkiness, the vest tended to catch on projections in the cockpit.

To address this issue, the BuAer evaluated the Type B-4 life preserver under development at the time by the USAAF, with procurement contemplated. By the fall of 1942, the issue of the 23P6 had become over-abundant, and the decision was taken to procure the life vest as detailed in USAAF specifications in order to overcome the deficiencies outlined above.

Specification 23P6 (INT) was amended on 21 November 1942 to conform to USAAF Specification No. 94-3135 for the Type B-4 life preserver, filed as 23P6A (INT).

The ASO negotiated the changeover on the existing contract 288s-2884 to the Hodgman Rubber Co. in March 1943, for the manufacture of a two compartment, inverted U-shaped vest, utilizing existing material that was to be used in the manufacture of the 23P6, as detailed in NAF Dwg. No.69690 and detailed in specification 23P6A. An inverted "U" shape compartment formed both the front and back profile of the vest. If a line were to be drawn from the lower right hand corner up and over the collar to the opposite corner it is in the shape of a "U". Two separate compartments, back to back, were adjoined around the inside of the collar and a seam down the center of the vest. Each compartmentalized bladder was inflated by separate CO2 cylinders.

*(Right) The vest manufactured under the change order shared characteristics of both the 23P6 and the Type B-4, which can be clearly seen.*

*The 23P6A utilized the old-style Type-H CO2 cylinder (illustrated on page 206). Most vests examined have the stamp STYLE 5648. It is not clear what this is referencing; whether it is a contractor part, drawing number or serial number of the change order. The contract for Hodgman was not extended, but manufacture of this vest – until the agreed quantity under the original terms – was completed, making for a limited run of this style. All previous contracts for the 23P6 were completed by the time negotiations were initiated, making Hodgman the only known manufacture of the STYLE 5648 vest. This was the last vest manufactured by Hodgman for the US Navy, and once their 1942 contracts for the USAAF Type B-4 were completed, they declined all further inquiries to bid, stating that their facilities were consumed with other government contracts.*

*(Above) Three officers of VC-36 wear the distinctive Style 5648 23P6A vest on the flight deck of the USS Mission Bay, December 1943. (National Archives)*

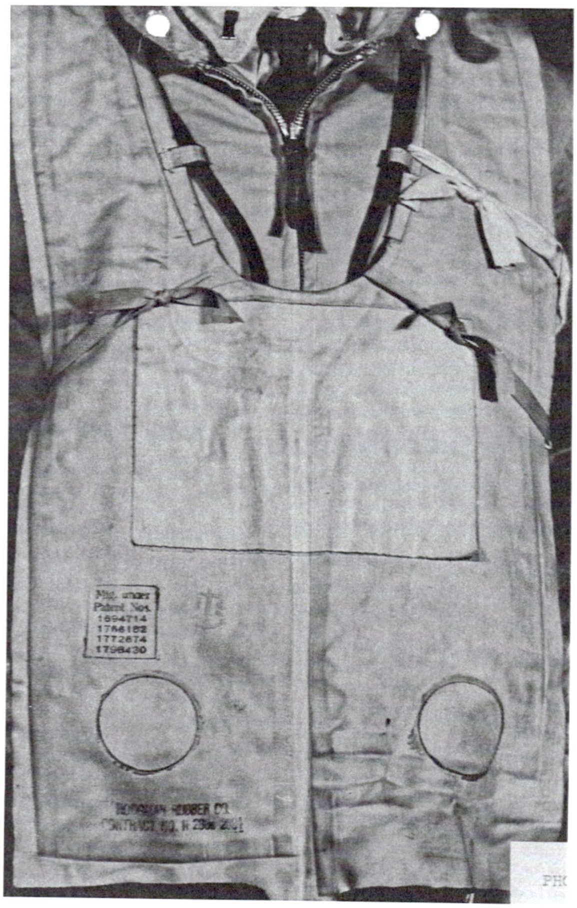

*(Left) Photograph, taken in 1943, illustrating a 23P6A vest that does not have the stamp STYLE 5648. (National Archives)*

*(Right) VF-14 pilot Lt. Wiley T. Williams lends another look at a 23P6A life preserver on board the USS Wasp, August 1944. (National Archives)*

To overcome the deficiencies of the Type B-1, The USAAC Materiel Division developed a new concept for a pneumatic life preserver, detailed under Dwg. No.32G6084 and titled Vest, Life Preserver, Type B-2, Specification No. 94-3037, dated 5 May 1932,

Samples of the new vest were manufactured by the Dayton Rubber MFG. Co. and available for field testing in early 1932. The B-2 was described as being manufactured of rubberized fabric colored chrome yellow, with three inverted "U" shaped compartments, with one waist and two leg straps. A single inflation cylinder was provided with three tubing outlets to inflate each of three compartments. This was soon amended to two inverted "U" shaped air compartments, 25 August 1932, detailed under Amendment-A.

One style consisted of one large cylinder that would inflate both compartments via a single "Y" shaped manifold. The other style included two smaller cylinders – one for each compartment. This vest was considered a major improvement over the B-1 in that it did a better job of holding the users head above the water, and maintained buoyancy balance if one bladder were to deflate.

The Type B-2 was approved for Limited Standard status on 16 May 1933 in the two cylinder configuration, available under stock number 8300-977000. In this case, it was more specifically orientated as a limited standard, for use only in pursuit type aircraft, due to the space restrictions.

After approval, limited procurements were made, the vest manufactured by Air Cruisers Inc., with the CO2 valves and cylinders procured through the Walter Kidde & Co. The B-2 would be covered in the first draft of Technical Order 13-1-3, covering the inspection and maintenance of a pneumatic life preserver, dated 10 March 1934.

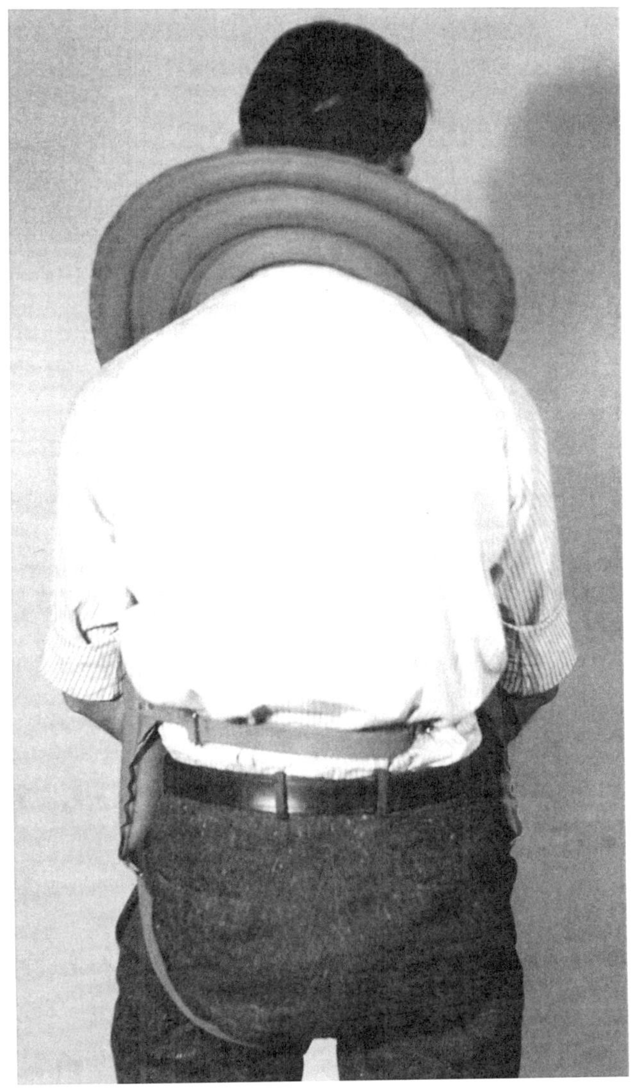

*(Above) A volunteer demonstrates the original design of the Type B-2 life vest. The inflation manifold is not present, but would had been located in the location of the rubber patch at the lower right corner of the user. The manifold was to consist of a brass cylinder cap, fitted with three rubber covered copper tubing outlets – one for each compartment. The flap on the inside of the collar was to contain three separate oral inflation tubes. (USAF)*

*(Above) Four B-10 aircrew members of the 31st Bombardment Squadron are wearing Type B-2 life preservers at Hamilton Field, California c.1935. The inflation compartments of the B-2 shared many of the same characteristics with the M-72B. The version shown utilized a single CO2 cartridge with a manifold to inflate both compartments. Two oral inflation tubes were mounted on one side of the collar. The two A-rings for attachment of the leg straps can be seen. (USAF)*

Development continued by the Materiel Division, utilizing the B-2A as the frame work. Its buoyancy characteristics were its best feature, but it lacked comfort. The Airubber Sales Co. had no further influence on life preserver development after the designs by Peter Marcus were introduced.

An individual by the name of Frank G. Manson, who was a civilian employee of the Materiel Division, Engineering Section, had influence in flotation development as early as 1931. Manson was the architect of the B-2 life preserver, he also ultimately held a dozen or more patents all pertaining to military applications on safety equipment. Manson took the basic concept of a bladder type construction being used for life rafts and applied it the B-2 life preserver, creating a soft pliable vest.

Preliminary specifications were drafted 1 April 1935, as USAAF Specification No. Y-3065 designated Vest, Life Preserver, Type B-3. Air Cruisers Inc. and Walter Kidde & Co. coordinated to update design features. A design was completed by January 1936, followed by a limited procurement for field testing. Following successful tests and positive comments, overcoming the deficiencies of earlier designs, and proving durable for service use, approval for standardization was authorized as USAAF Specification No. 94-3065, 7 March 1936.

There was still some doubt about the use of pneumatic life preservers, so for assurance the vest was procured in minimal quantities for service testing

through 1937. Through Air Corps Policy #159, dated 7 April 1937, personal life preservers became a mandatory standard.

Policy #159 stated that all airplanes, except pursuit and basic combat, be equipped with flotation devices (seat cushions) and pneumatic rafts. All pilots and aircrew in all airplanes were required to wear a kapok or pneumatic life preserver for over water flying. Following this policy, the pneumatic vest became the preferred type establishing the standard until well into the modern era.

*(Facing page, bottom) This unknown individual is demonstrating the new Type B-3 life preserver as originally procured at Wright Field under specification Y-3065. This would have been the vest service tested from 1935 into 1938. Manson filed for patent in June 1935 for this exact vest. Several features of this vest include a single pull cord to actuate the two CO2 cylinders, the orientation of the oral inflation tubes relative to the snap, and D ring of the waist strap. A quantity of twenty vest were procured from Air Cruisers Inc. under Purchase Order 35-4048. (National Archives)*

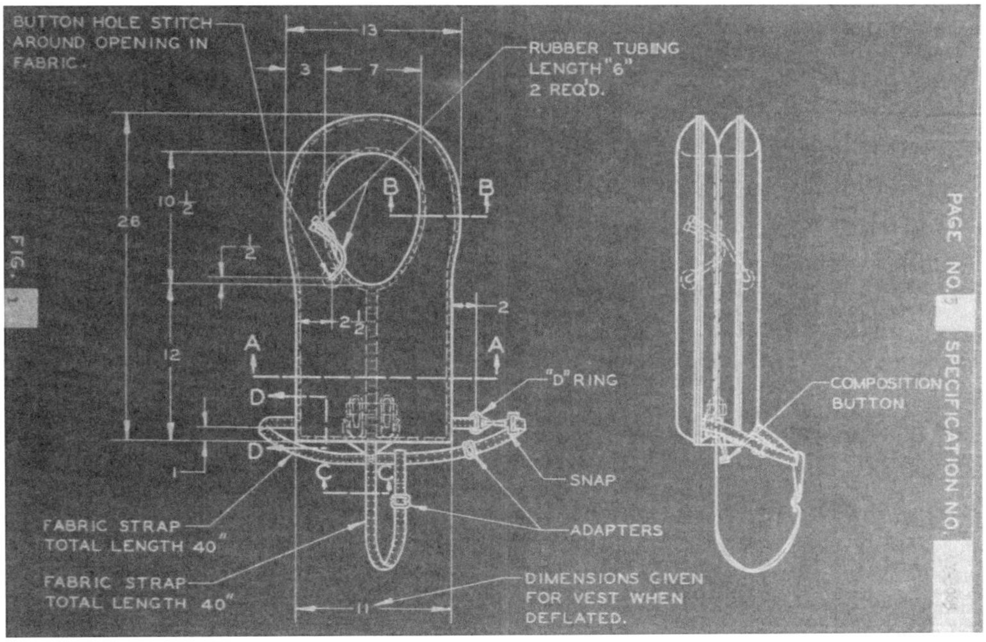


*(Right) Drawing that accompanied specification Y-3065. (National Archives)*

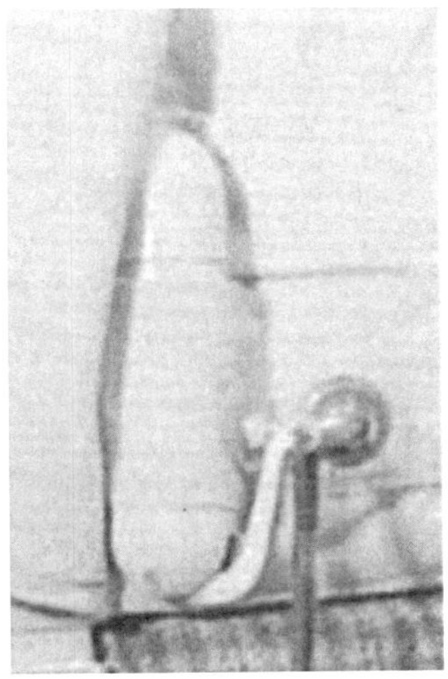

*(Left) Upon standardization, the vest was reconfigured for the more prominent right-hand user, with the oral inflation tubes moved adjacent to the waist strap connection on the right. To prevent wear of the parachute chest snap, a 6" x 8" leather reinforcement patch was secured to the front, and the lanyards for activating the CO2 cylinders were separated under Amendment-A, 14 July 1937.*

*(Above) The original designed Walter Kidde & Co. inflation cylinder or holder. Two procurements of the B-3A were executed under purchase orders 26-5631-P for a quantity of 1000, and 37-1450-P for a quantity of 500. (National Archives)*

(Left) Under Amendment-B of Specification No. 94-3065, made effective 19 September 1938, a new type CO2 actuating cylinder and elbow manifold was incorporated into the vest and remained the standard until 1942. A USAAF sergeant demonstrates the new inflation system. (National Archives)

(Below) The Type B-3 life preserver specification 94-3065B would be the standard for the USAAC until superseded by the Type B-4 in 1942. Available under stock number 8300-978000, it consisted of two superimposed compartments extending around the neck and over the head, held in place by a single waist and crotch strap. It was constructed of a cotton twill fabric outer shell containing latex rubber bladders, with separate automatic inflation and emergency mouth inflation tubes.

(Bottom left) The B-3 featured new CO2 cylinders designed by the Walter Kidde & Co. utilizing Schrader elbow valves.

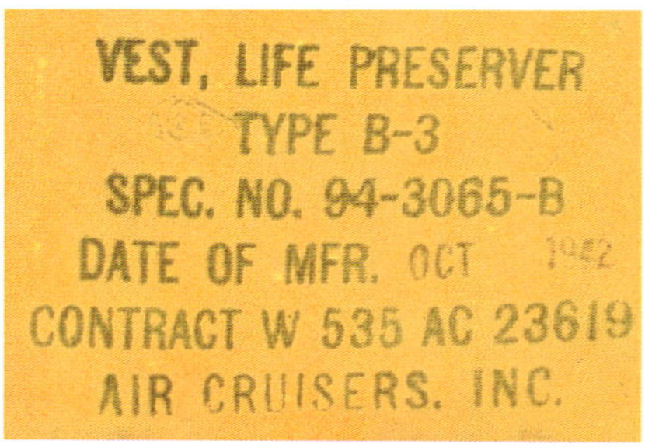

(Left and below) Air Cruisers Inc. was the principle contractor for the Type B-3 until the Goodyear Tire & Rubber Co. was approved for manufacture in July 1940. Technical Order 13-1-3 on inspection and maintenance was revised, deleting the Type B-2 and replacing it with the Type B-3, 20 October 1936. Shown is an example of a Goodyear Type B-3 vest with the distinctive red gum rubber oral inflation tubes.

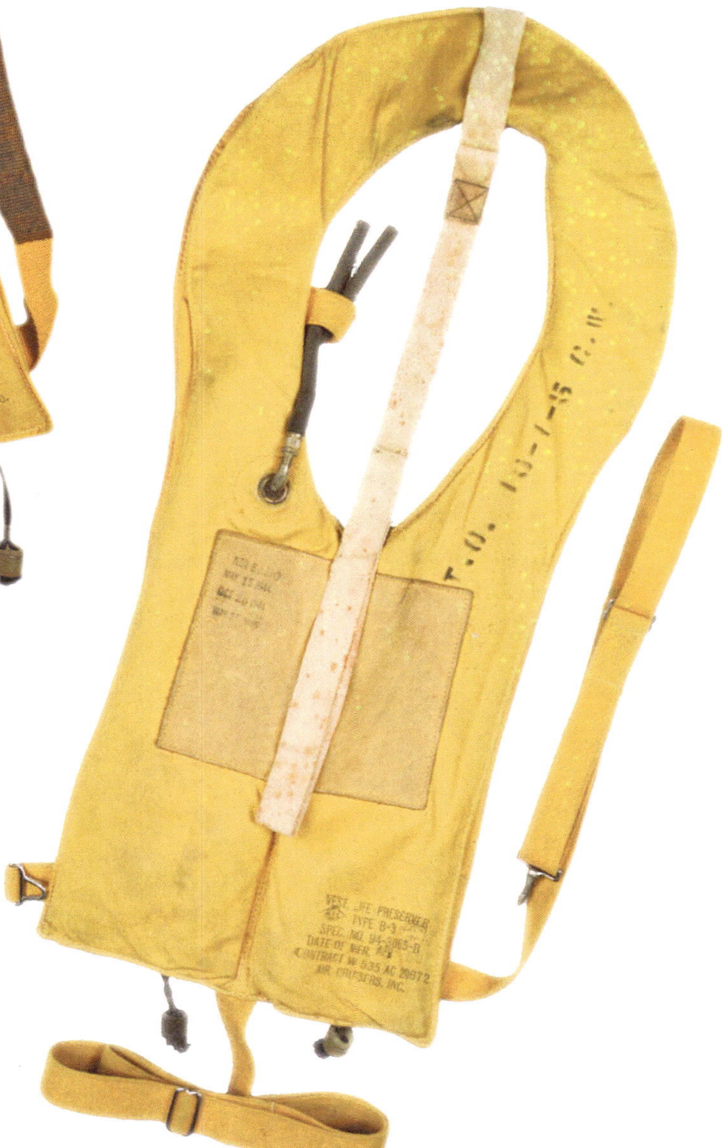

(Right and below) This Air Cruisers Type B-3 shows all the typical signs of being a veteran. Life vests were required to receive a full inspection on a semi-annual basis, as can be seen from the dates stamped on the front leather patch. It has been updated as per technical order for the addition of the back strap and sea marker. Type B-3 vests procured prior to fiscal year 1939 were to be disposed of three calendar years after the date of manufacture; five years for vests manufactured after fiscal year 1939.

*(Above) 5th AF A-20A Douglas Havoc crew members Capt. Roger Dunbar, S/Sgt. G.F. Brauswell and Pfc. George Gann wearing B-3 life preservers in New Guinea, March 1943. The Type B-3 would be the prominent life preserver used from 1942 through 1943 until sufficient quantities of the Type B-4 entered service. It would remain in supply until it was declared unserviceable. It was only replaced due to the manufacturing requirements of the consumption of rubber and latex. (National Archives)*

*(Below left) Color photograph of a B-17 waist gunner in the ETO, T/Sgt. Mexico Barraza illustrates the purpose of the leather patch on the front. (National Archives)*

*(Below right) A member of the 376th BG receives last minute details from the intelligence officer while wearing a B-3 life preserver in the operations tent. North Africa, 1943. (National Archives)*

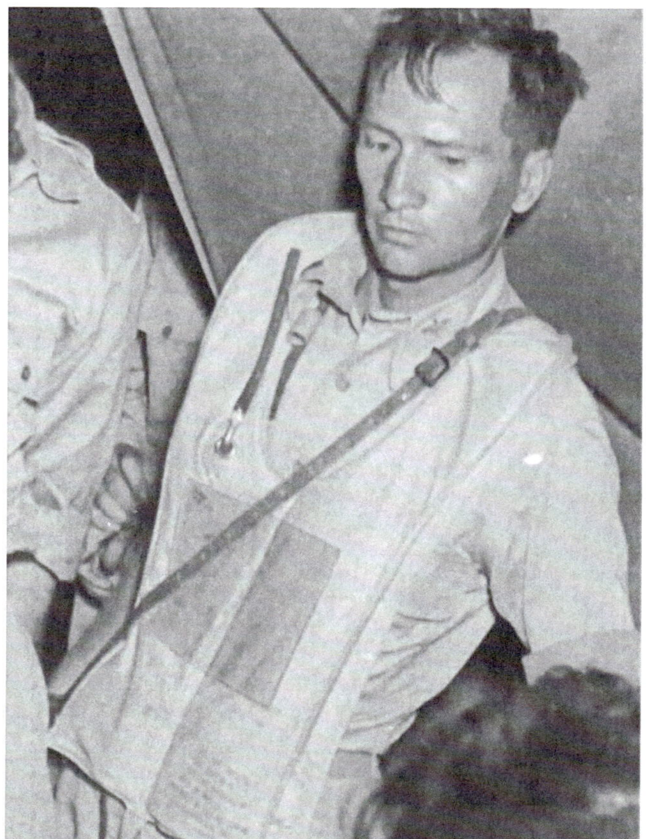

*(Above) c.1936, the BuAer contracted for a vest utilizing the basic drawing of the Type B-3 established by the USAAC for service testing purposes. Contracted through the Hodgeman Rubber Co. the vest consisted of two inverted U-shaped compartments superimposed, but with rubber coated internal and external surfaces and colored chrome yellow. This was the same application as used on early production life rafts and gave the fabric a sheen. The design addressed the shortcomings of the M-72. This color photo shows the chrome yellow finish, complying with specification 23P6 and USAAC requirements. Note also the most unusual application of rivets on the inside of the collar. (National Archives)*

*(Top center) The same style of vest worn by a pilot of VF-6 in May, 1942. (National Archives)*

*(Top right) In February 1943, this style of vest appears on the flight deck of the USS Enterprise being worn by a gunner of VB-10 John Jennings ARM/3c (National Archives)*

*(Right) For evaluation and service tests, the BuAer made several procurements of the B-3 life preserver. The first occurred in 1937 for what was described as a "Lot" under Navy Dept. contract No.54627, fabricated to USAAC specifications. This purchase would had been for the Type B-3A. Lt. Stewart of VGS-30 has just completed his first carrier landing on the USS Charger in February 1942, he wears a B-3A life vest. (Paul Faulkner Collection)*

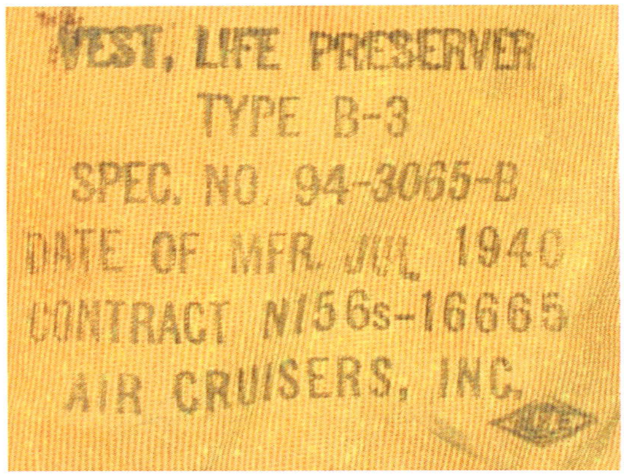


*(Left and right) The other known Navy contract for the B-3 was issued through the NAF under contract N156s-16665 c.1940. This contracted vest was fabricated identically to USAAC specification 3065-B, the only difference being the use of a D ring, versus the A ring on USAAC vests. The quantity for this contract is not known, but in comparing known quantities for USAAC contracts of that era the number would be significantly lower. (Paul Faulkner Collection)*

*(Right) Lt jg. William Totherow aboard the USS Saratoga is wearing a USAAF contract Goodyear B-3 life preserver, November 1943. The results of the service tests for the B-3 are conflicting and it appears to have been a bureaucratic decision to retain the use of the 23P6. In tallying the various comments, the B-3 was the preferred vest for comfort, being less bulky and hindering than the 23P6. Combat service requirements would change the perspective in 1942. Units operating overseas began requisitioning the B-3 from USAAF stock, principally from Hawaiian air depots. The BuAer would not make any direct procurements from contractors during the war for basically the same reason as the USAAF. Due to the shortage of critical war materials, the USAAF was forced to the manufacture of the B-4, and by the time the BuAer transitioned to the approval of that style, all contracts were diverted to the fabrication of B-4 specifications. All use of the B-3 by the Navy would be by direct requisitions from USAAF channels, except for existing stock of the procurements made from 1937 to 1940. (National Archives)*

*(Left) Pilot Lt. H.H. Labyak of the newly commissioned VT-4, aboard the USS Ranger, wears a Type B-3 life preserver in January 1942. (National Archives)*

*(Above) Pilots from VF-5 in the ready room aboard the USS Yorktown, October 1943. A good quantity of B-3s are hanging on the wall as well as being worn. Type B-3 life preservers were commonly used by many naval aviation units and highly sought after when available.*

*(Below) Pilot M.C. Hoffman wears a USAAF contract Air Cruisers Inc. vest. Note how compact his B-3 vest is under the parachute harness as compared to the 23P6. (National Archives)*

*(Below) Known USAAF contracts and quantities for the Type B-3 life preserver. No further contracts were awarded after the standardization of the Type B-4, however those open orders were allowed to process to completion. B-3s have been found with manufacture dates as late as spring 1943.*

| CONTRACT | AWARD DATE | CONTRACTOR | QUANTI-TY |
|---|---|---|---|
| 535-AC-10584 | December 1937 | Air Cruisers Inc. | 2,400 |
| P.O. 39-3022 | January 1939 | Air Cruisers Inc. | 800 |
| 535-AC-13947 | February 1940 | Air Cruisers Inc. | 4,000 |
| 535-AC-16149 | October 1940 | Air Cruisers Inc. | 4,000 |
| 535-AC-16150 | November 1940 | Goodyear T & R Co. | 5,500 |
| 535-AC-18155 | March 1941 | Goodyear T & R Co. | 9,000 |
| 535-AC-18157 | March 1941 | Air Cruisers Inc. | 15,000 |
| 535-AC-20671 | August 1941 | Goodyear T & R Co. | 20,000 |
| 535-AC-20672 | August 1941 | Air Cruisers Inc. | 25,000 |
| 535-AC-23619 | December 1941 | Air Cruisers Inc. | 36,000 |
| 535-AC-23620 | February 1942 | Goodyear T & R Co. | 31,000 |

By 1940, each aeronautical service had established its own standard for pneumatic life preservers that were mandatory for all over water flights.

The BuAer utilized the 23P6 and the USAAC utilized the Type B-3. The principle factor affecting the future of these life preservers was the materials used in their construction.

The bladder in the B-3 consisted of liquid latex and the 23P6 required 75% crude or new plantation rubber. These materials would soon become critical, and were rationed upon the United States entry into WWII. Natural rubber was imported from Southeast Asia, but the supply was cut off by the rapid expansion of the Japanese war machine.

Latex was considered three times more critical than rubber in 1942. Its use became restricted and prohibited in the manufacture of the Type B-3 after June 1942, switching to a formula of 50% crude and 30% reclaimed rubber, meeting the requirement of a minimum 80% rubber hydrocarbon content. This formula followed the guidelines established in USAAF Specification No. 94-3135 for the new Type B-4 life preserver, standardized at that time.

The Type B-3 was a critical item for combat activities, but its production could not be sustained due to the volume of natural materials it consumed. In January 1942, the Materiel Command set forth an objective to reduce the use of rubber in the fabrication of life preservers, and this was found in Navy specification for the 23P6. This specification outlined the use of a rubber coated fabric. Reaching out to existing manufacturers and potential contractors for life preservers, samples were fabricated adopting the Navy method of a rubberized fabric with vulcanized edges, but conformed to the basic drawing of the Type B-3. Samples were then submitted and given preliminary tests which proved favorable.

The conclusion was to waive actual service testing and adopt the design as standard, temporarily, until a suitable replacement of latex or new plantation rubber could be found to resume production of the Type B-3 bladder type vest. Specifications were then drafted and approved, 28 April 1942 as Vest, Life Preserver, Type B-4, USAAF Specification No. 94-3135.

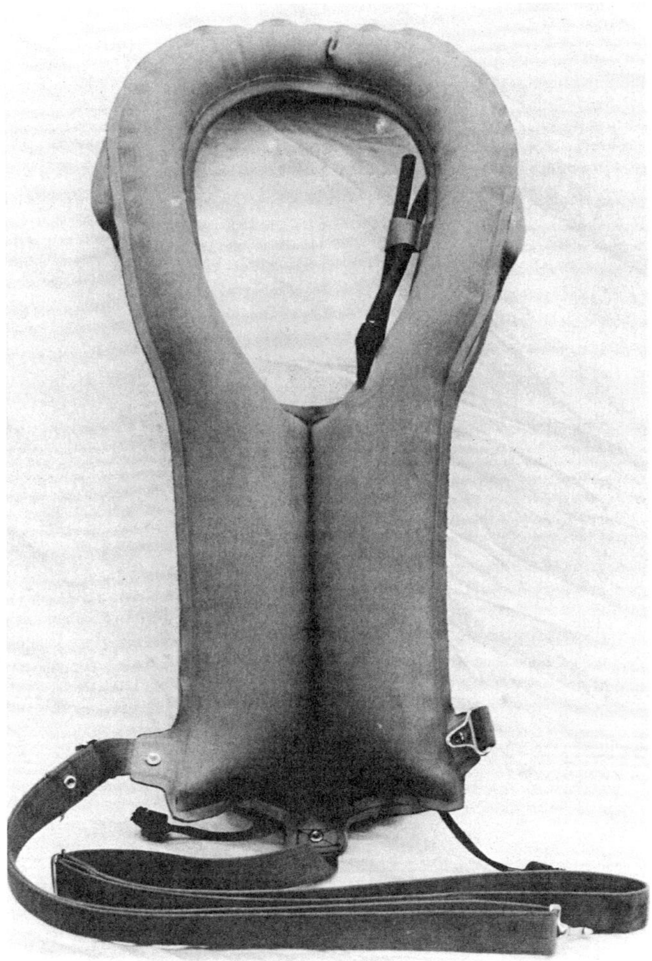

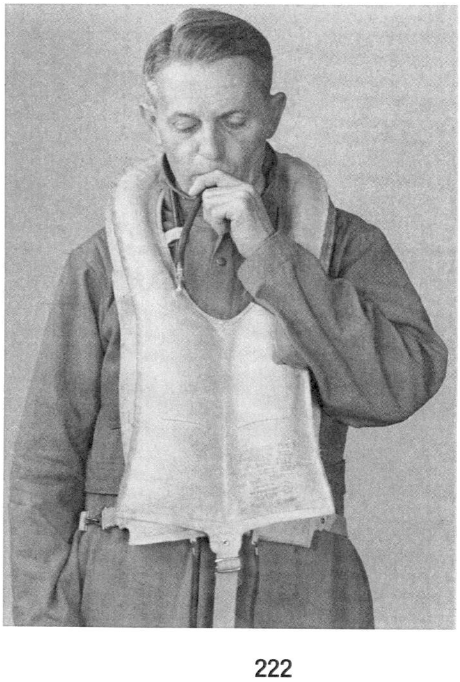

*(Above) One of the first experimental life preservers presented by the Goodyear Tire & Rubber Co. manufactured to the new specifications of a rubber coated fabric, April 1942.*

*(Left) The new standardized Type B-4. A volunteer demonstrates the proper wearing of the vest, deflated and inflated. He is also illustrating the procedure for manual inflation. (USAF)*

*(Left and below) The Type B-4 was officially standardized 6 May 1942, and assigned stock number 8300-978200. In June 1942, eleven contracts were awarded to ten different companies for the mass production of the vest. The vest comprised two superimposed compartments of rubber coated fabric, retaining all other features of the Type B-3. The vest illustrated is an example of the first production run of the B-4 that includes the Schrader elbow valve and Walter Kidde & Co. inflator.*

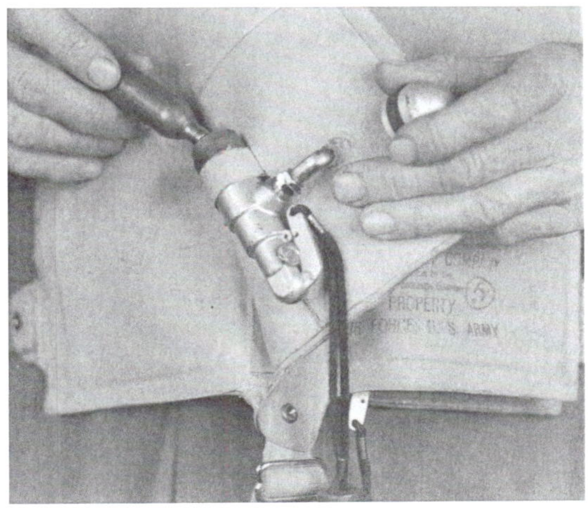

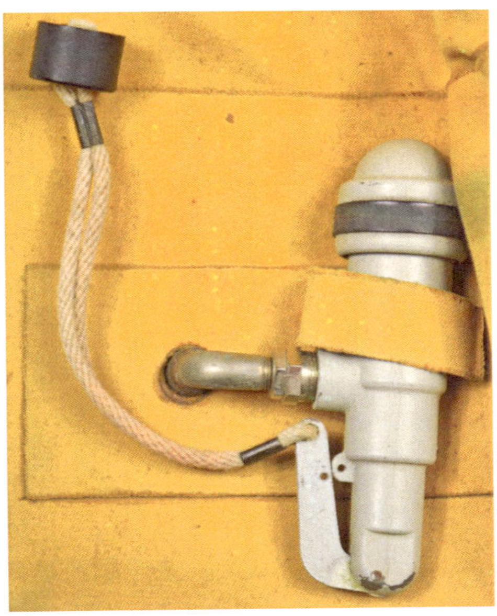

*(Left) Detail of the Schrader elbow valve and Walter Kidde & Co. inflator.*

*(Right) The first significant change to the visual appearance of the Type B-4 occurred in Amendment-4, dated 1 September 1942, authorizing the change to a new inflator and valve system by Peters & Russel Inc. Existing stock of the old inflators were utilized until exhausted. Shown is an example of the Peters & Russel Inc. inflator design that would be the staple in all pneumatic life preserver manufacture from 1943 onward.*

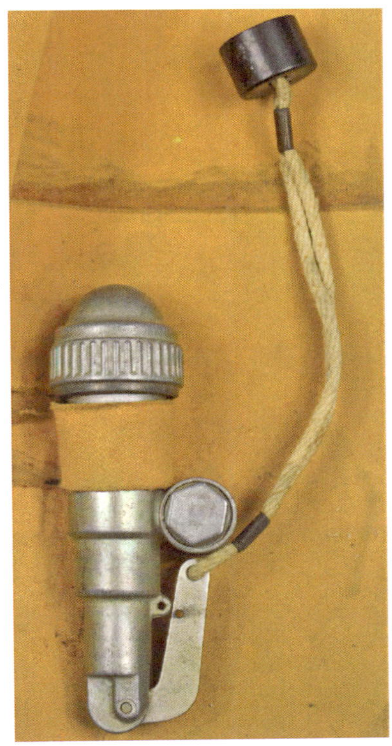

Initial production was slow, as it took time for contractors to tool their facilities and await supplies from sub-contractors. Production would eventually reach an average of 400-500 vests per day (22,000 per month). Manufacture was also prolonged with significant defects in vests delivered due to the rush to meet delivery demands.

By 1943, significant depletion of natural rubber pushed development into synthetic materials. Neoprene, Buna-S, and Buna-N were some of the synthetic rubber materials tested by the Materiel Command and the BuAer through 1943. Sample vests coated with these materials were subjected to a series of controlled exposure tests to determine its durability and characteristics compared to new planation rubber. Neoprene ultimately proved superior to natural rubber and other synthetics.

The first application for the use of Neoprene was the fabrication of the oral inflation tubes, replacing the gum rubber types utilized on the Type B-3 and early production of the Type B-4. By September 1943 some contractors were transitioning to the complete use of Neoprene, authorized as an acceptable substitute from specification. Specifications would continue to outline the use of the combination of new plantation and reclaimed rubber.

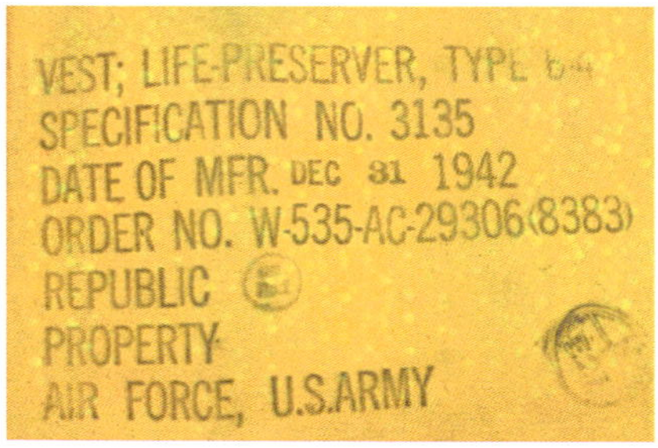


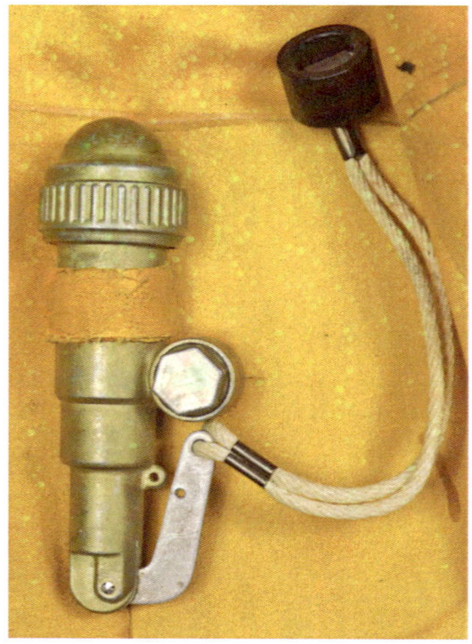

*(Left, above and right) Another early production Type B-4. This vest includes the Peters & Russel Inc. Inflator, and this contractor utilizes a different style D ring. All contracts awarded in June 1942 would extend into 1943 for manufacture.*

*(Right) P-38 pilots Capt. Miller and Lt. Col. Johnson of the 77th FS wearing a Type B-4 and Type B-3 life preserver, England c.1944. The Type B-4 would slowly integrate into service use, eventually overcoming the availability of the B-3 in quantity.(National Archives)*

The BuAer officially adopted the specifications for the Type B-4 as outlined in specification 23P6A in November 1942, however, it was more orientated toward amending existing production of the old type, as per NAF Dwg. No. 69690 conforming to USAAF specifications.

At that time the ASO did engage in negotiations for contracts for new production vests under USAAF Specification No. 94-3135.

*(Left and far left) The ASO awarded their first two contracts in January 1943, one to the New York Rubber Corp. and the other to the U.S. Rubber Co. Shown is an example from the U.S. Rubber Co. The only differing features between the services were the printed contract data and property marks. Of interest is the marking SPECIFICATION NO. 3135-A, which was the active amendment at the time of the award of contracts in January 1943. No. 3135-A was made effective 7 October 1942, and was merely a consolidation of information from the original specification and its subsequent four amendments. These new vests were referenced from the ASO under stock number R37-J-150-75.*

*(Right) Lt. jg Arthur Whiteway of VF-16 wears an early production US Rubber Co. life preserver. Photo taken aboard the USS Lexington, December 1943. (National Archives)*

*(Right and above) In February 1943, the BuAer's Aviation Engineering Division distributed their specifications to contractors, which was a direct copy of the USAAF 94-3135, but requiring the property mark USN be placed on both sides of the collar. The example pictured here, manufactured by the New York Rubber Corp., complies with BuAer specification. Not all contractors adhered strictly to the requirement (as can be seen on the vest by the U.S. Rubber Co., previous page).*

*(Left) Ens. Robert Halsin of AG-20 wears a New York Rubber Corp. B-4 vest, aboard the USS Lexington, March 1945. (National Archives)*

*(Bottom left) LCDR. Richard Gray at NAS Jacksonville Florida, September 1943. He wears a BuAer contract Type B-4 life preserver manufactured by the Hood Rubber Co. 5,000 vests were reported to have shipped from Hood to Aviation Supply Annex Oakland in May 1943. (National Archives)*

*(Bottom right) This photo of VB-8 pilots Lt. J. Ward, Lt. M.D. Carmody, and Lt. M. Kilpatrick illustrates the transition of the integration of USAAF type life preservers. Each is wearing a representative of the standard types 23P6, Type B-3 and Type B-4, USS Intrepid October 1943. In 1943, demand exceeded availability and this photograph gives some insight to the nonconformity and limited supply. Through 1944, the 23P6 would be phased out of service, along with the Type B-3, with the Type B-4 becoming the typical standard by 1945. (National Archives)*

The Types B-4 and the B-3 rapidly became the preferred life vest for the majority of aviation personnel serving in the combat zone. The ASO maintained a steady procurement of the Type B-4 through 1943 to meet growing service demands.

In November 1943, Commander Fleet Air, South Pacific, directed that the new Army type life preserver, Type B-4, be supplied to all units in theater and to all units destined for the South Pacific. It was further recommended that aviation units draw these vests from Army supply depots wherever and whenever possible, if not available through standard ASO supply channels.

*Examples of USAAF procured life preservers that had been requisitioned for Naval Air use. The vests pictured both have USAAF contracts and property marks.*

*(Above) A gunner of VB-9 Henry Eley ARM2/c aboard the USS Lexington, February 1945. (National Archives)*

*(Left) An unknown VBF-17 night fighter pilot preparing for strikes on Japan aboard the USS Hornet, February 1945.*

In short, the Type B-4 was an expedient measure to meet wartime requirements, and is one of the rare occurrences where a superseding item is a purposeful regression in design, dating back a decade to the Type B-2 developed in the early 1930s.

With the vest rushed into production and with further pressure placed by BuAer demands contractors were overwhelmed and plagued with production delays. As the vests entered service they were tried under combat conditions, creating opposing points of views by both services in what changes should be made. They were in turn urged to reach an agreement for better economy and efficient production for contractors.

Specification No. 94-3135 was redesignated under joint Army-Navy Specification AN-V-18 standardized on 10 December 1943 under Dwg. No. AN6519, Part No. AN6519-1.

The only significant change was the use of Sateen cotton replacing the woven cotton type outlined in specification 94-3135. Multiple weaves were utilized and approved for contractors of the woven cotton fabric making for a variety of appearances on the early production B-4s.

(Left and above) The BuAer were the first to award contracts under joint Army-Navy specifications, illustrated is the first of two awarded in February 1944. The Seamless Rubber Co. was awarded the first, followed by an award to the U.S. Rubber Co. Being an early award, after December 1943 it does not completely conform to specification. It utilizes the Sateen fabric as outlined by specification and it includes the stock number but still retains the older practice of USAAF nomenclature and service-specific property marking. Technically all contracts awarded after December 1943 were to conform to specification AN-V-18, but not all contractors initially complied, leaving some variation within early 1944 manufactured vests. Detailed in specification AN-V-18 vests were to be legibly marked with AN-6519-1, service stock number, and property marked US Government. The back strap on this vest was an applied addition later in service.

(Right and below) The BuAer vest illustrated here conforms to specification AN-V-18 after August 1944. Vests were to include the Army-Navy part number in the nomenclature, the stock number, a back strap, and be property marked U.S. GOVERNMENT. Procurement of these vests was executed under the part number, hence it being reflected in its nomenclature. The specification AN-V-18 was merely a guideline for contractors in its manufacture.

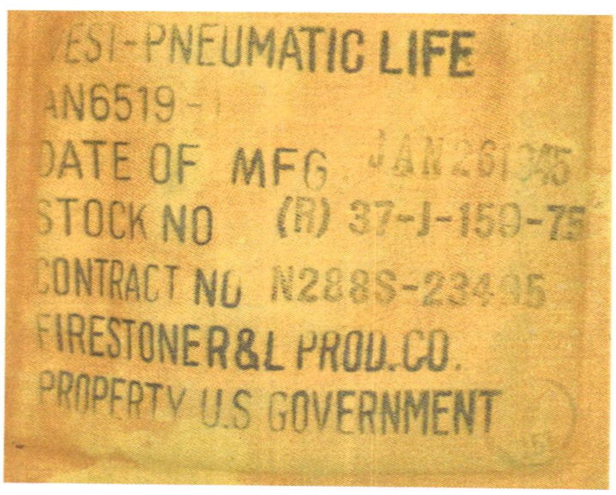

(Left) The distinctive large snap hook, as found on late production AN-V-18 life preservers, can be seen on the vest of pilot Lt. jg. Armand LaMarre of VC-85 aboard the USS Lunga Point, March 1945. (National Archives)

(Above) Pilot Lt. jg. David Horton of VBF-83 wears an AN-6519-1 life preserver aboard the USS Essex, March 1945.

(Right and below) Due to the substantial number of vests procured by the USAAF in June 1942, in conjunction with B-3s in stock, no further procurements were made for exactly two years, meeting all projected requirements for all combat theaters. Procurement resumed for the USAAF in June 1944,u awarding three contracts. The first contract awarded under AN-V-18 specifications was to the Firestone Rubber & Latex Co. as illustrated. This vest completely conforms to specification. Having a February 1945 manufacture date, it has been updated with the large snap hook and factory applied back strap.

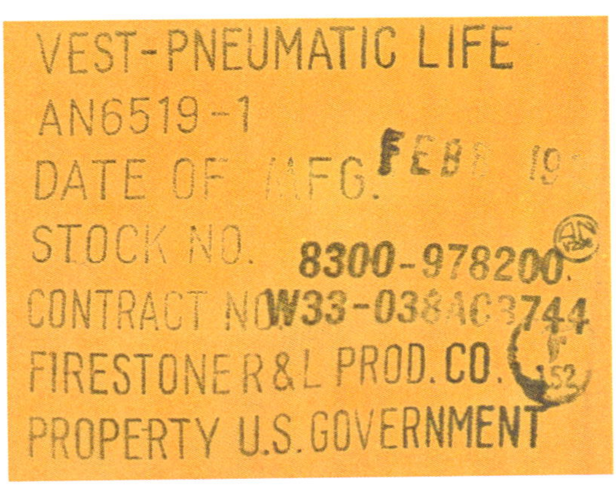

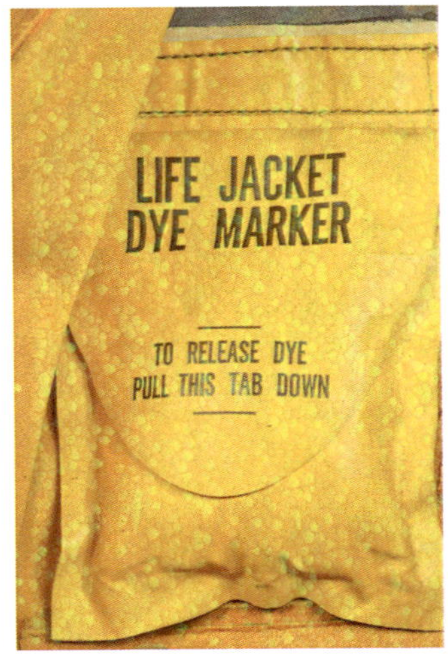


*Above and right) The vest illustrated is the last contract awarded by the USAAF for the AN-6519-1 life preserver. As per USAAF requirements, the sea marker was to be applied by the contractor. This example complies with all details except the inclusion of the stock number in the printed data on the front of the vest.*

*(Left and right) An experimental vest, field tested in the spring of 1944, addressing harness configurations. This was an attempt to improve the comfort of the user in the water and to prevent slippage of the inflated vest over the head. This design concept was integrated into the Type B-5 life preserver.*

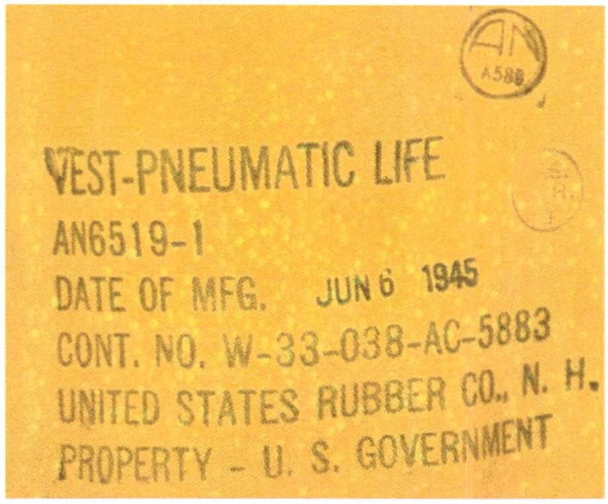

*(Left, above and right) Vest manufactured in error by the U.S. Rubber Co. and coinciding with the adoption of the Type B-5 life preserver. Tentative revisions were drafted to modify existing contracts of the Type B-4 to conform to the Type B-5. The U.S Rubber Co., which was involved with the development of the Type B-5, believed the modifications were all but certain and began the process to convert B-4s on their new contract for the AN6519-1. The update mimicked details of the Type B-5 to include two pockets and oral inflation tubes between the bladders, and place the automatic inflators in a horizontal position. Like the Type B-5, the left pocket was to contain a sea marker and whistle, with the shark chaser in the right pocket. Authority was not officially granted for the conversion, and after months of deliberation it was found too impractical, as the modifications would delay deliveries. Specifications were drafted titled as the Vest, Life Preserver, Type B-4A dated 1 May 1945 under Dwg. No. 45G18840.*

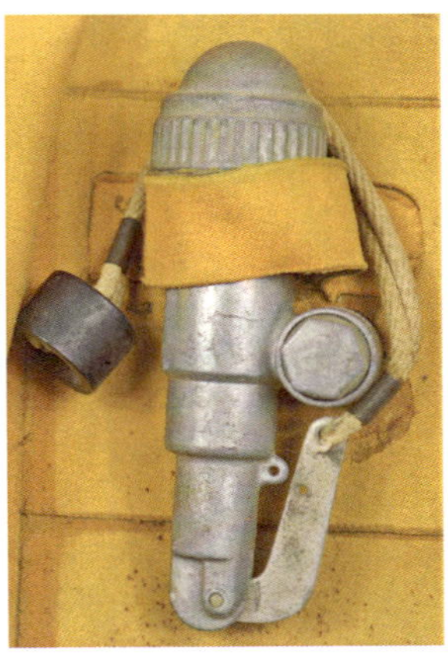

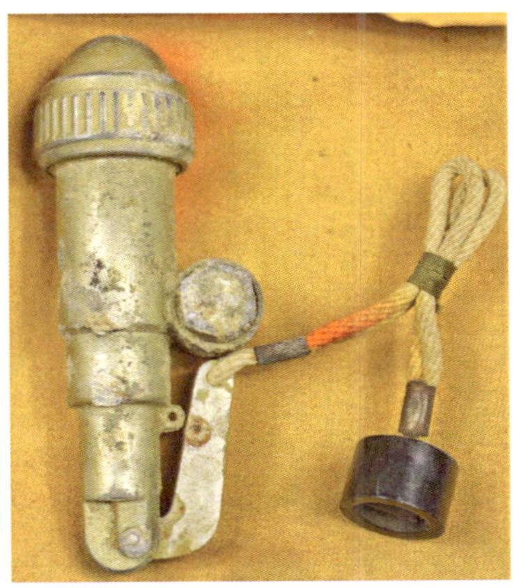

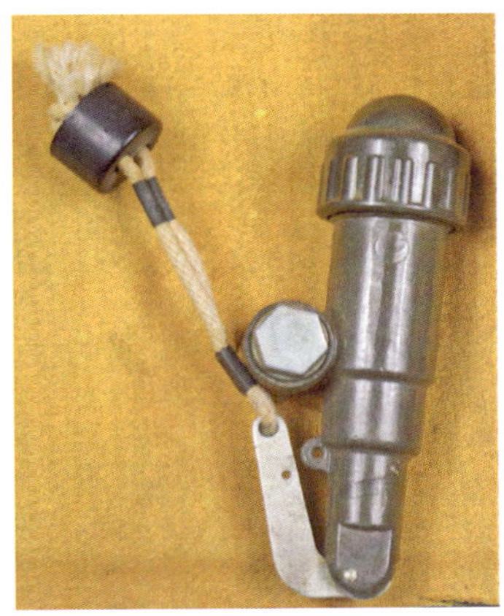

*(Right) The first-generation inflators manufactured by Walter Kidde & Co. were made of aluminum, but the new Peters & Russel Inc. were made of zinc, die cast with an integral manifold. Either style was provided with a 12" braided cord, double looped, leaving a finished length of 6". Pre- and inter-war cords were provided with fiber composition pull buttons which were found to be sensitive to moisture, becoming soft and breakable. In March 1942, Walter Kidde & Co. introduced a molded acetate button, which soon after became the standard. By 1944, the Peters & Russel type inflators were being manufactured by Walter Kidde & Co., Pro-phy-lactic Brush Co., and E.A. Laboratories. The pull cord, suspended well below the vest, often got snagged causing premature inflation. A typical field expedient was to wrap the cord around the inflator, as pictured, which interfered with proper functioning in the event of an emergency. The USAAF amended their technical order 13-1-3 on the inspection and maintenance of life preservers dated 17 September 1944, outlining that the cord was to be folded in half and sewn as pictured. The BuAer distributed Technical Order 117-44, dated 14 September 1944, directing that the lanyard was to be shortened, leaving only a one-inch exposure below the vest. Specification AN-V-18 was amended under Amendment-1, dated 16 August 1944, for contractors to shorten the lanyards to three inches.*

*(Far right and right) When the vest left the factory, the actuating lever would be safety-tied to the inflator with a breakable wire, as prescribed by specification. These were removed during inspection periods and not replaced. The USAAF did not officially direct that the safety wire be reapplied. To further prevent premature inflation, the BuAer distributed Technical Order 44-44, dated 1 April 1944, for the application of a breakable thread as illustrated right.*

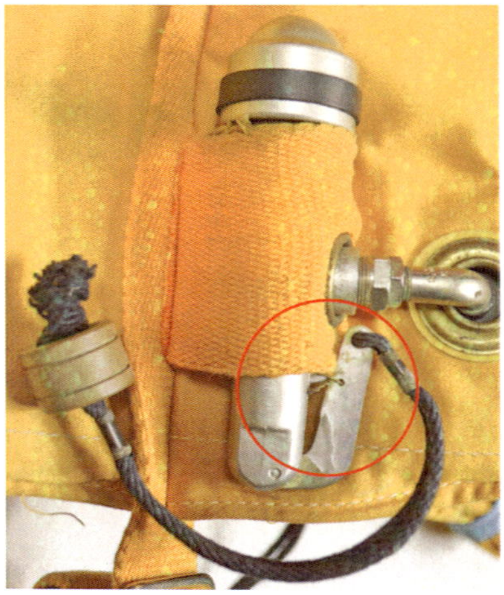

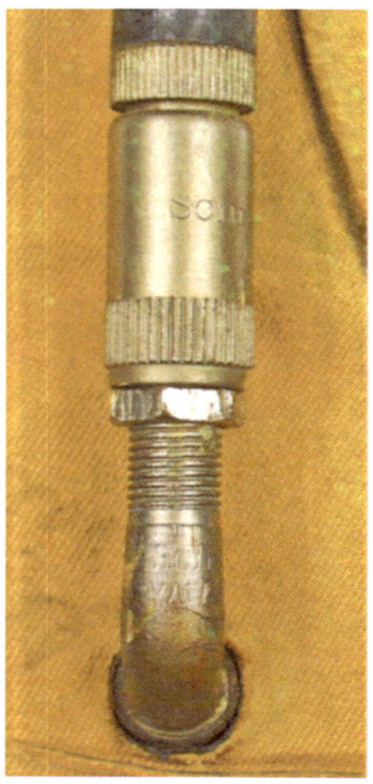

*(Left) Three variations of the oral inflation valves, left to right: Schrader & Sons, Dill MFG. Co., and unknown. Included on the Schrader and Dill valves were locking nuts, these were later additions to prevent the valves from loosening and leaking air. USAAF Technical Order 13-1-3 of September 1944 directed that locking nuts were to be added. They were outlined in Amendment-1 to specification AN-V-18, August 1944, and the BuAer outlined the addition in Technical Order 25-45 on the inspection and maintenance of life vests, dated 2 March 1945.*

*(Far right) Each service had their initial preference for the style of D rings, the BuAer favored an actual D shape, while the USAAF utilized an A shape ring. The manufacture of the Type B-4 for both the USAAF and BuAer contractors used both types, with eventual formalization of the D shaped ring under specification AN-V-18.*

*(Right) The progression of snap hooks. The small hook was included on pre- and early-war manufactured vests, then the medium sized hook under AN-V-18, and finally a large style hook under Amendment-1 of AN-V-18 in August 1944. The small hooks were found difficult to clip and un-hook wearing gloves or with cold hands, in or out of the water. The large hook helped alleviate this problem.*

Both services reported the tendency of the vest to slip over the head of the wearer, especially when inflated. This factor affected weak or unconscious users in the water, often leading to fatalities.

To help alleviate this issue, and to provide more comfort in supporting the head, each service distributed directives to modify vests in service.

The USAAF outlined the attachment of a back strap in Technical Order 13-1-17 dated 25 May 1944, and the BuAer through Technical Note 55-44, dated 26 June 1944. Contractors were additionally notified to apply the back strap prior to deliveries, at the time of distribution of the service directives.

The back strap was officially addressed in Amendment-1 of specification AN-V-18 in August, 1944.

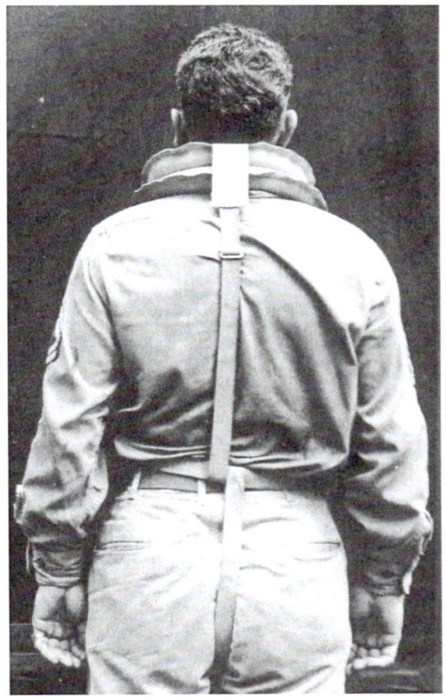

*(Above left) The benefit of the back strap. The vest placed unnecessary pressure on the back of the neck, pushing the head forward, causing discomfort.*

*(Center and right) With the back strap in place, the head is able to retain a more natural position, providing better support in the water. Images from HQ 2nd Central Medical Establishment (National Archives)*

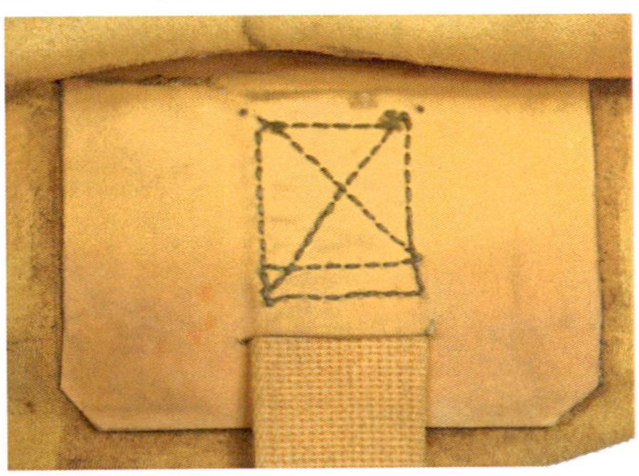

*(Above left) Service applied modifications as per Technical Order 13-1-17 and Technical Note 55-44. An approximate 3" x 2" rubberized fabric patch was to be applied securing the non-elastic cotton tape or strap.*

*(Above middle) Variant utilizing a white cotton strap.*

*(Above right) A factory applied back strap.*

*(Left) B-29 crew member in the center of the picture (holding his Type B-4 life preserver by the collar), with the back strap applied as per technical order, but it has been attached to the external surface of the vest, Saipan November 1944. (National Archives)*

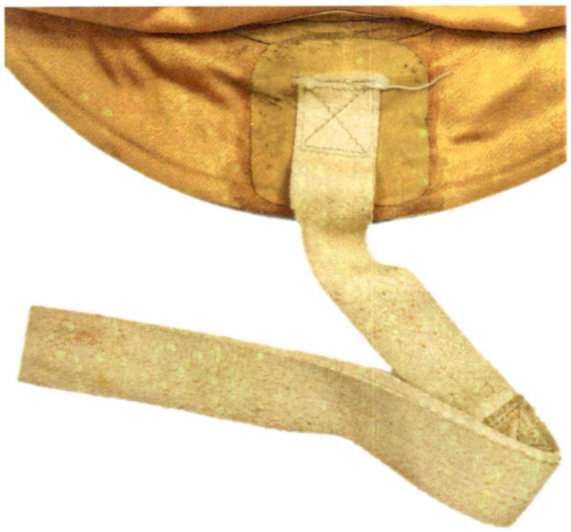

*(Left) BuAer Technical Note 55-44 prescribes the exact same application. Shown is a BuAer contract vest with a service applied back strap modification.*

*(Right) Ultimately many subtle variations exist in these applications. Vest worn by an unknown VF-11 pilot in the ready room of the USS Hornet, February 1945. (National Archives)*

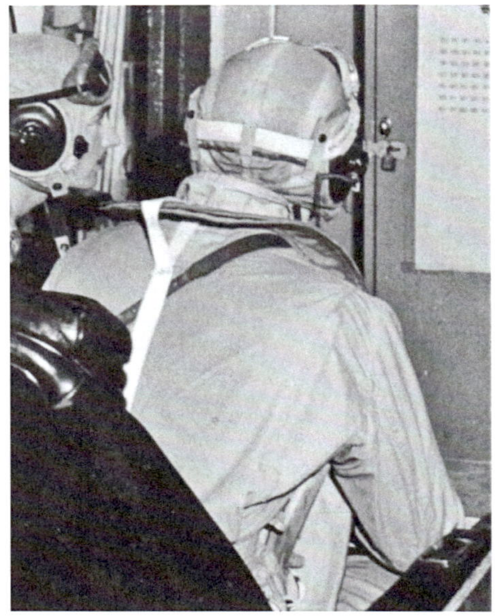

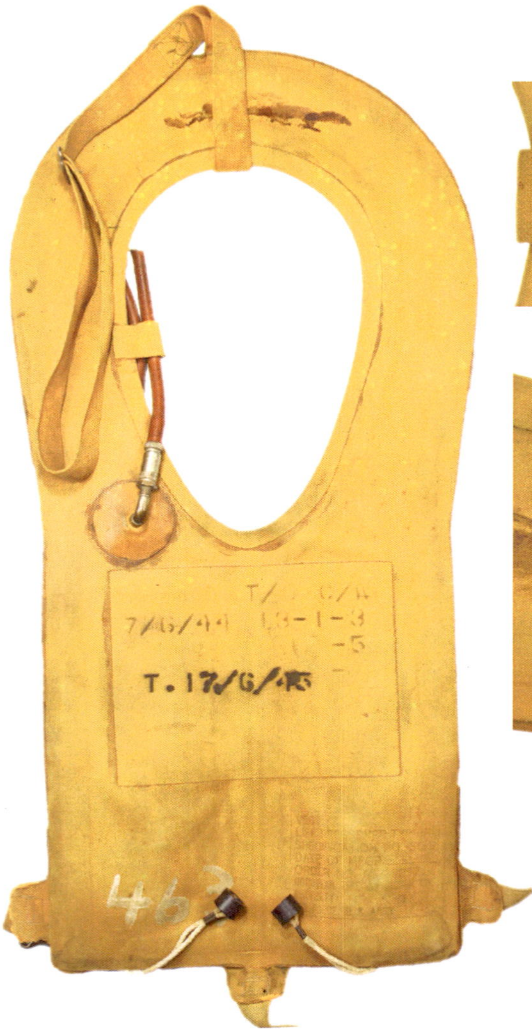

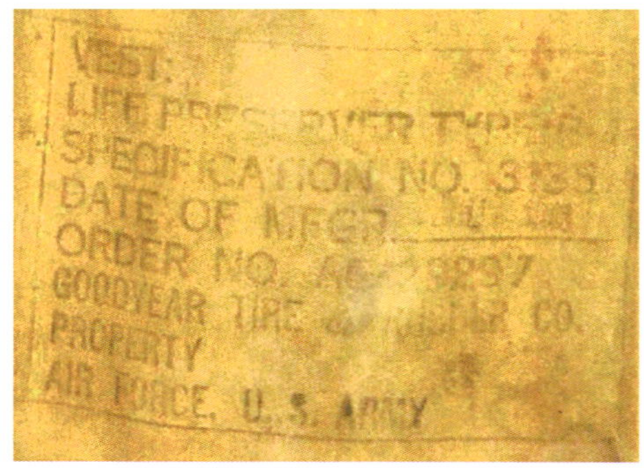


*(Left and above) An example of the first USAAF contract of the Type B-4 life preserver awarded to the Goodyear Tire & Rubber Co. It has an alternate method of applying the back strap, which is simply looped around the collar and tacked to the edge of the vest to prevent slippage.*

*(Right) Brig. General Emmett O'Donnell goes over last minute details with pilot, Maj. Robert Morgan, before air strikes over Japan. Morgan was the former pilot of the famous B-17 Memphis Bell. Each is wearing a B-4 with the looped back strap modification. Saipan, 1944. (National Archives)*

*(Left) Two B-17 crew members of the 390th BG wear B-4s with painted serial numbers on their vests. These numbers were applied by Personal Equipment Officers for issuing and accountability purposes (see color image on previous page). England, April 1945. (National Archives)*

*(Below left) Life preservers for both services were required to undergo daily inspections prior to and after flight operations checking for external damage, inflation valves, and CO2 cylinders. Life vests in storage were required to undergo semiannual inspections. Personnel assigned the responsibility would inflate the vests, checking for leaks, and make any modifications as necessary. The sergeant is checking the oral inflation valves and below on the table after the inspection is complete he will attach the sea marker as per technical order. Upon completion of the semiannual inspection stamps will be applied reflecting that it has been serviced.*

*(Below) B-24 radio operator T/Sgt. Nelson Keefer wears a Type B-4, illustrating additional variations of inspection stamp applications. England, 1944. (National Archives)*

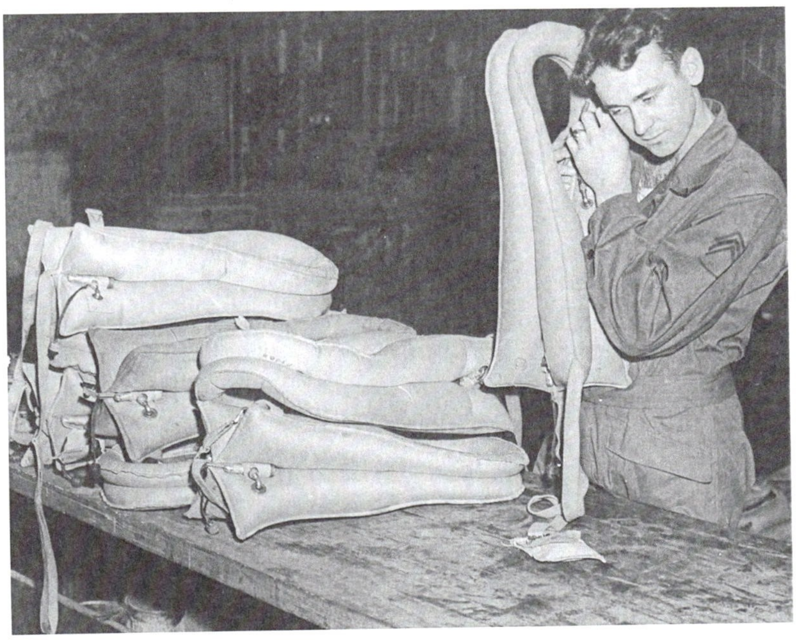

*(Above) Examples of official stamps on USAAF vests that underwent their required inspection cycles. They are stamped with the date of inspection and which technical orders have been complied with.*

*(Left) This USAAF vest reflects that it underwent general inspection as per Technical Order 13-1-3, has had the sea marker installed as per Technical Order 13-1-5, and has had the back strap applied as per Technical Order 13-1-17. The C.W. signifies that the technical order has been Complied With.*

The USAAF maintained a steady distribution of Technical Order 13-1-3 since 1934, with subsequent revisions through 1944 on the inspection and maintenance of life preservers.

The BuAer, on the other hand, essentially left its use to the discretion of the user. However, the technical order and notes from the 1930s were found sufficient in their basic use and maintenance addressing the fitting and CO2 systems. These were eventually found outdated by 1944, with the widespread use of the Army type vest, and superseded by Technical Order 134-44 addressing maintenance and repair in addition the care to check inflation systems.

Unlike their USAAF counterpart, they did not prescribe inspection stamps, which is why typical BuAer contract vests are void of any such markings.

*(Left) Inspection tags were the preferred method of the BuAer. Illustrated to the left are two examples, tied to the oral inflation valve, each having a hand-written inspection date. After each inspection, a new tag would replace the old one, if still present.*

*(Right) As part of the inspection, the checking of oral inflation valves was emphasized. A common leakage issue amongst vests was improperly tightened valves on the stem, hence the addition of locking nuts. Caution tags were a common application to vests, as illustrated to the right.*

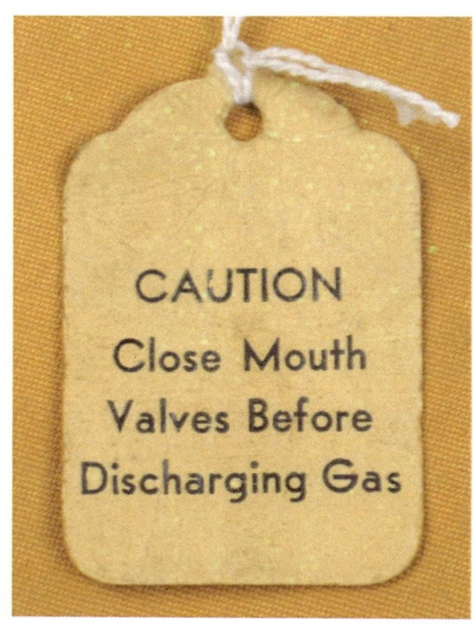


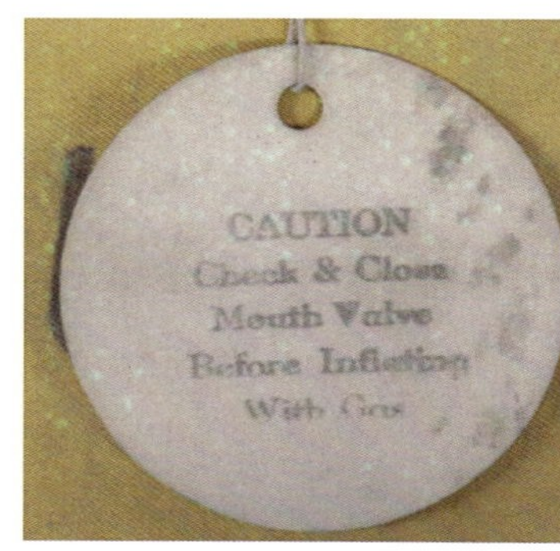


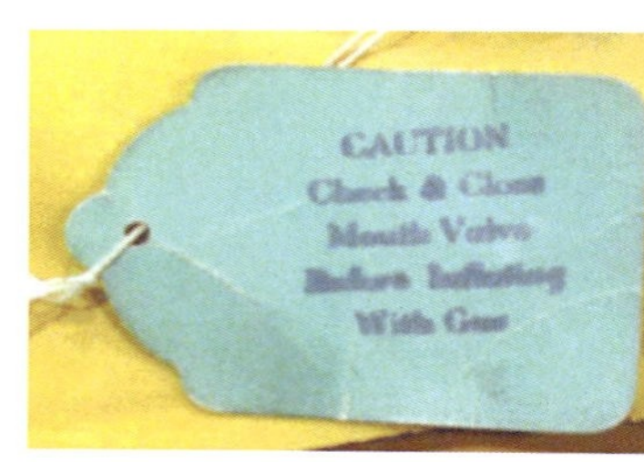

*(Left) Ens. T.G. McClelland of VF-5 aboard USS Yorktown, October 1943. Note inspection tag attached. (National Archives)*

*(Right) Two BuAer PBM-5 crew members wear Type B-4 life preservers with an example of the tags, illustrated above, hanging from their oral inflation valves. Tanapag, Saipan, April 1945.*

(Left) It was a common practice to write or stencil the user's name on BuAer life vests. With the USAAF, ground unit's life preservers were stored in rooms, in lockers, or in racks - essentially having dressing rooms prior to operations. Life vests, like parachutes, were fitted to the user and stored in the ready rooms aboard carriers for the BuAer. Above the left shoulder of VC-20 pilot Lt. R.H. Moore is a parachute and life vest in the ready room of the USS Marcus Island, January 1945. With a row of equipment on the wall, personal property marks assured each crew member grabbed the right equipment. Moore has his name hand written on his life vest.

(Below) Additional examples of life vests both hand written and stenciled with user's name. (National Archives)

(Right) Life preservers were technically squadron property, and often marked as such. VT-14 pilots Ens. Jack Cochran and Ens. E.F. Haffman wear Type B-4 life vests with personal property and squadron property marks, USS Wasp, August 1944.

(Below) Additional examples of squadron property markings applied to vests of Night Torpedo Squadron 90 (VTN-90) and Airship Patrol Squadron 32 (ZP-32).

*(Above) The practice of adding personal property markings to life preservers was at the discretion of the squadron Personal Equipment Officer.*

*(Left) Type B-4 being worn by B-17 crew member Lt. R.V. Crane of the 401st BG, England March 1944.*

In June 1943, the New York Rubber Corp. submitted samples of a life preserver to the USAAF that incorporated camouflage properties. The vest was described as the standard yellow on one side and sea blue on the other.

It incorporated a swivel mechanism at the junction where the waist and crotch strap intersect. The idea was that the user, while still wearing the vest, had the option to either have the yellow or blue side out by simply slipping the vest over the head, pivoting 180 degrees, and placing it back over without unfastening the straps. There was a thought at the time that the yellow, though effective for visual recognition purposes, could also attract the attention of the enemy as rumors of Japanese strafing survivors began to spread.

The USAAF field tested the vest and concluded that there was little value in the camouflage, since only a small area was visible from the air. The BuAer was also aware of the suggestion, and it too concluded that the vest offered little value, and that changing specifications would ultimately affect urgently needed deliveries from contractors. However, this did not affect camouflaging applications in theater.

*(Below) The camouflaging of life preservers appeared to be proprietary to the BuAer, as no evidence of this application has surfaced within the USAAF. Pilot Ens. Humphrey wears a camouflaged 23P6 life vest aboard the USS Core, August 1943. (National Archives)*

*(Right) This is the earliest example observed of a camouflaged vest, 23P6, in use by a pilot of VGS-29 going over reports in the ready room of the USS Santee preparing for support operations over North Africa, November 1942. (National Archives)*

*(Above) Cmdr. E.M. Snowden reads a manifesto to all flight personnel aboard the USS Lexington, May 1944. He is wearing a camouflaged B-3 life preserver. (National Archives)*

*(Top center) VT-2 Avenger crew member R.A. Twarowski, AMM2/c, is a wearing a camouflaged Style 5648 life preserver aboard the flight deck of the USS Hornet, June 1944. (National Archives)*

*(Top right) Combat veteran and now instructor Lt. Reichel is wearing a camouflaged Type B-4 life vest while training new pilots at NAS Melbourne, January 1945. It is likely this vest served him in the combat zone. (National Archives)*

*(Above right) The camouflaging of vests appeared to be more of a personal preference, with no consistency amongst squadrons as to whether it should be worn with the yellow or blue side out. This image of members from VF-2 illustrates the contrast of a camouflaged vest versus the standard yellow. Lt. jg. C.L. Carlson is wearing the camouflaged vest, USS Hornet July 1944. (National Archives)*

*(Right) A pilot of VF-16 wears a camouflaged life preserver in the ready room of the USS Lexington, November 1943. The color differential can be noted at the crotch strap and between the bladders. (National Archives)*

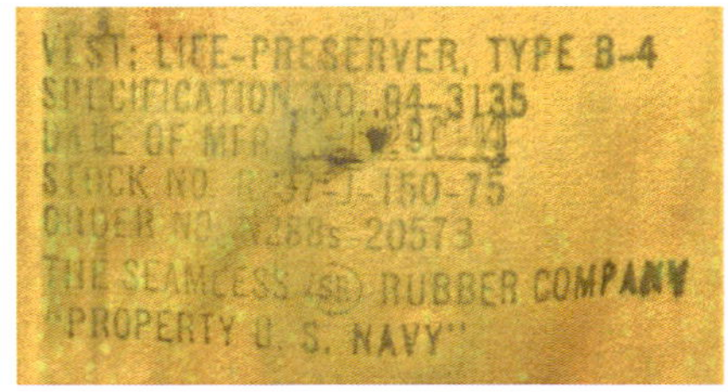


*(Left and above) The typical color applied was known as non-specular blue, either hand-painted or sprayed to one side of the vest. Non-specular blue was a standard color on naval aircraft and would be readily available from the repair facilities. The example pictured here had the unique addition of seven emergency panel signals as per Technical Note 91-44.*

*(Left and above) This camouflaged example is property marked to a USN/USMC Lt. Craddock. This vest was a requisition from USAAF depots as evidenced by the contract data.*

(Right) Commercial packaging for the Sparklets CO2 cartridge and its distinctive red band. These were commonly used for carbonating soda, and were also utilized in the medical field. Introduced in the 1920s, they presented the perfect apparatus for discharging compressed gas for inflating life preservers. Sparklets was the principle supplier until Walter Kidde & Co. began distribution. These red banded cartridges are often found in government life preservers but it is not clear if they were placed there in their civilian life, after surplus, or if they were government procured.

(Left) The USAAF established specification 94-40238, outlining requirements for CO2 cylinders for life preservers, with the development of the Type B-2. Pictured are two variations procured by the USAAF. Each box contains six bulbs. They are identical, except one box is labeled "For Type B-3" and the other "For Type B-4", referenced under stock number 8300-163000. These examples were procured through Walter Kidde & Co.

(Right) To regulate the manufacture of bulbs used in BuAer vests, the NAF distributed Dwg. No. 1135, Part No.1135-1, to contractors. They had the same requirements as the USAAF but stipulated that the cylinders be stamped USN as an identifier that the manufacture had complied with the requirements. Referenced under ASO stock number R83-C-93500. Pictured are some variations of the boxes containing 6 bulbs.

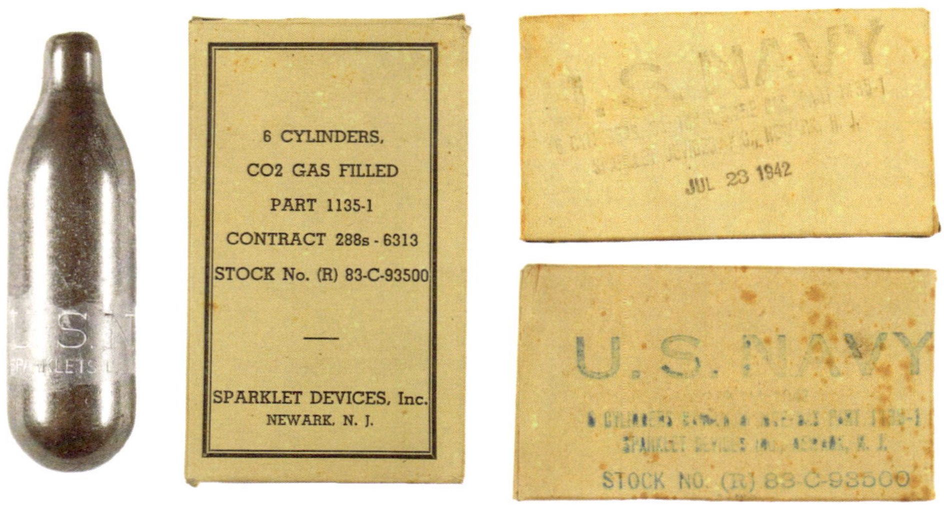


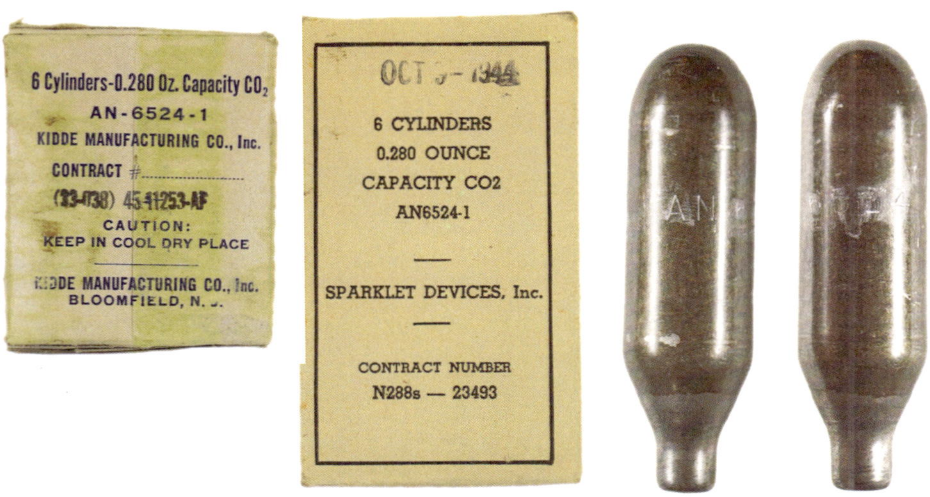


(Left) On 24 June 1943, both the USAAF specification and NAF drawing were re-designated under joint Army-Navy Specification AN-C-105, Dwg. No AN-6524, Part No. AN-6524-1. This was a consolidation for efficient manufacture. The only differing aspect outlined was that the cylinders were to be stamped with the drawing number AN-6524. A procurement example from each service is shown, USAAF on the left and BuAer on the right.

*(Right) In 1944, service reports indicated that the CO2 cylinders contained insufficient gas to inflate vests. It was discovered that when in storage for long durations and exposed to extreme climate changes the sealing caps in the cylinders under specification AN-C-105 leaked around the washer. To eliminate this condition, the BuAer adopted a method developed by the Knapp-Monarch Co. called electric welding. This method prevented any loss of gas after exposure. The new cylinders were standardized under NavAer Specification M-660, 2 December 1944. It was also adopted by the Navy Department for use in life belts. As per specification, each cylinder was to be stamped USN EW (Electric Weld). The system took time to revise and for the company to tool up for mass production. Quantity production did not commence until the spring of 1945. The new cylinders were available from the ASO under stock number R51-C-1952 and from Navy Department stores under stock number 51-C-123. As outlined by Technical Order 81-45, dated 12 September 1945, AN-C-105 cylinders were to be removed from service and replaced with the M-660.*

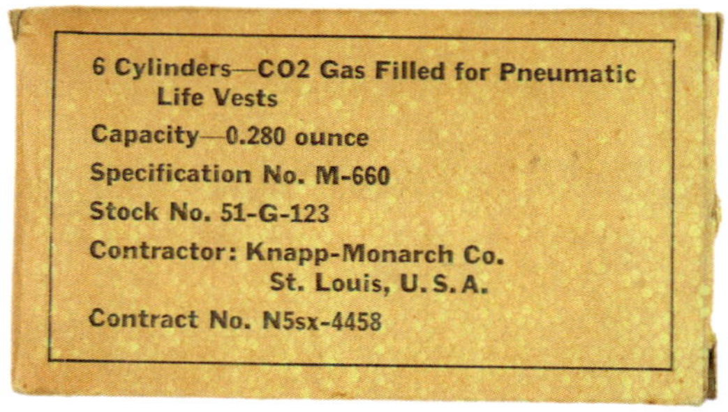

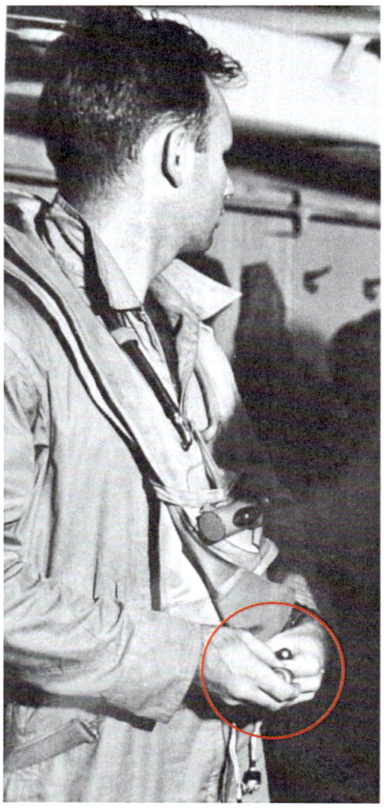

*(Left) All pilots were instructed to check the caps and cylinders in their life vests prior to flight operations. Pilot Lt. Richard Gale of AG-87 checks his automatic inflation system as part of his pre-flight preparations, before air strikes over Japan, in the ready room of the USS Ticonderoga, July 1945. (National Archives)*

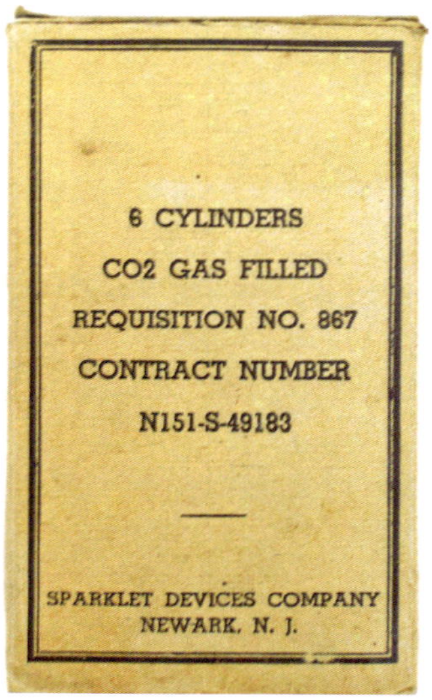


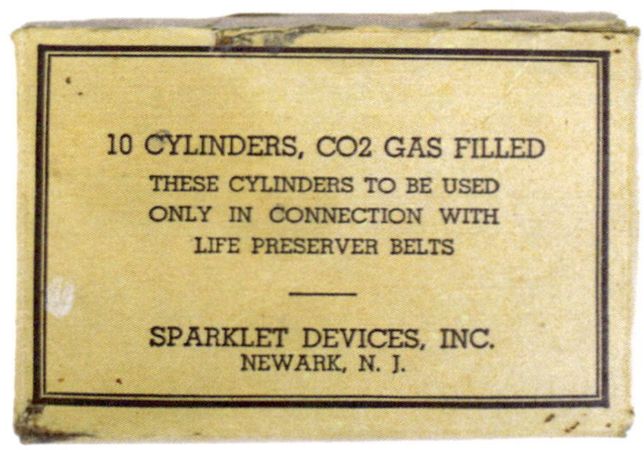


*(Above) The Navy Department standardized the cylinders under their specification 51C24. The cylinders were the same as designated under specifications 94-40238, AN-C-105 and NAF Drawing 1135. Their intended purpose was for use in the pneumatic life preserver belts but, due to availability, they were also used in BuAer life vests.*

*(Right) Cylinders can be found painted and unpainted, typically green. This color was used on all sizes of containers that held compressed gas.*

*(Left and below) Vest manufactured to USAAF Specification No. 94-3135, but procured by the Army Transportation Corps. These vests would include the contract prefix 33-092-TC and were procured to supply commercial airlines and commercial sea going vessels under government contract. 140,000 vests were procured under 7 contracts, with the bulk awarded in September 1943. Often they included the ATC ships wheel acceptance stamp, but can be found with the typical USAAF inspection stamps and technical order updates. The reason for their entering service with USAAF was the vests were often diverted to meet delivery schedules for the USAAF, taking precedence over supply from contractors. The example pictured was procured under AN-6519-1, other examples would have the old nomenclature stampings. The ATC did not require property markings.*

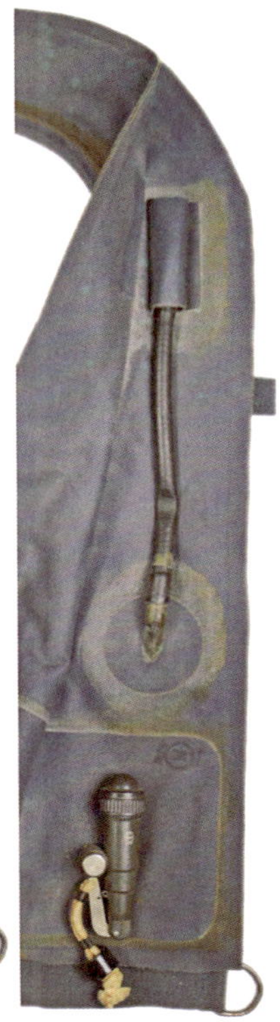

*(Above and left) The Navy Department additionally procured life preservers under USAAF Specification No. 94-3135 with modifications. Procurements began in the spring of 1943 and continued through the duration of the war, available under stock number 23-P-169, titled Self-inflating, Vest Type, Life Preserver. This vest was intended to supplement the standard Kapok type aboard ship, having the advantage of being compact. It was intended for use in confined or heated spaces below decks. It also required less storage space and could be carried readily from place to place. The principle differences were the omission of the crotch strap and addition of a chest strap. Two manufacture variations are pictured, both made from neoprene.*

*No photographic evidence has yet surfaced showing their use for aviation.*

The principle changes and variations of life preservers have been highlighted, but there were many subtle variations between manufacturers.

When samples were submitted for approval, they only had to meet very basic criteria. Specifications and drawings were distributed, which were highly detailed, but as long as the item met the parameters of the design it would often be approved as a deviation from specification. Each contractor would have its own signature, such as overall dimensions, the shape of the horse collar, and other assembly features.

*Two variations of the cotton twill used in the manufacture of the Type B-3 life preserver.*

*(Far left) Pre-war material which is coarse, or ribbed, for a better description.*

*(Left) For wartime manufacture a tighter weave was specified.*

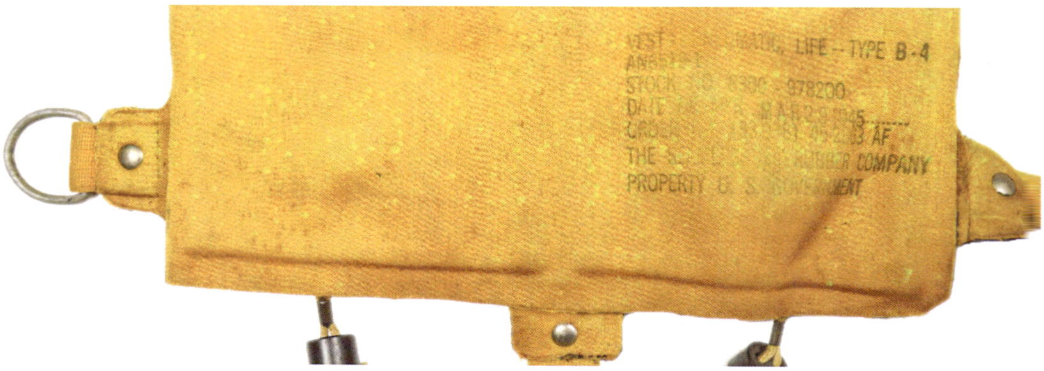

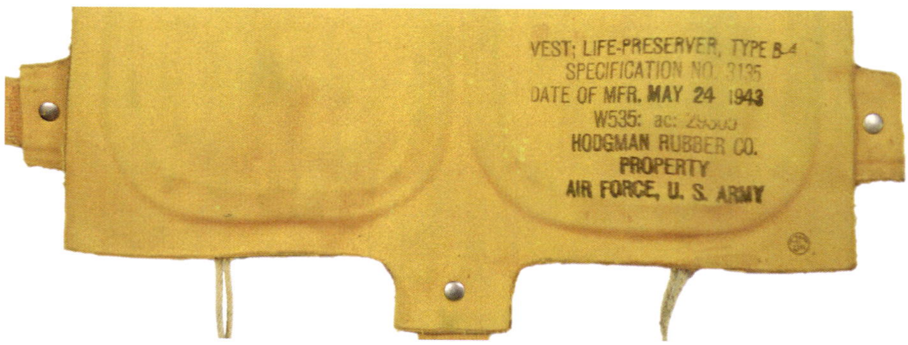


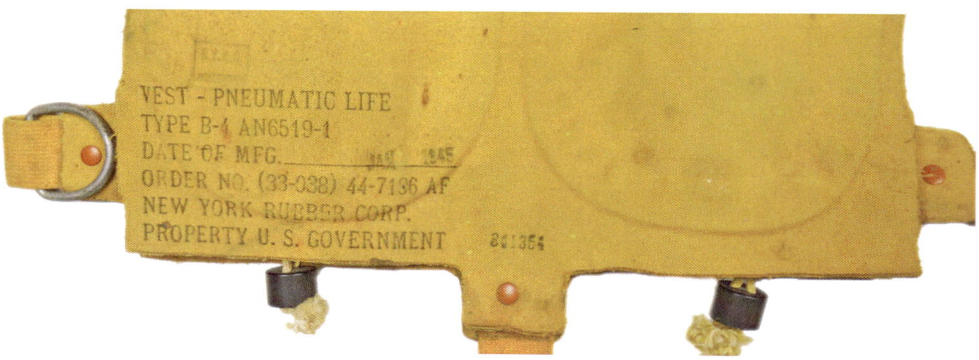


*(Above and right) Some additional variable construction features of the Type B-4. As per drawing and specification, rivets were to be used to reinforce the connections of the waist and crotch straps. The New York Rubber Corp. preferred a painted rivet whereas the bulk of other contractors used a nickel-plated rivet. The Goodyear Co. eliminated the use of rivets and had the straps sewn (above left) which was an approved deviation from specification. One of the major defects in vests received was that when tested, they would blow out under the prescribed PSI. The typical construction is illustrated (top right) by simply vulcanizing the edges along the contour of the cut. The Hodgeman Rubber Co. strengthened the bladders by creating a rounded corner radius, which better distributed pressure. The New York Rubber Corp. incorporated this concept into their vests in late production runs, additionally placing the nomenclature on the right hand side of the vest.*

*(Below) List of known USAAF contracts for the Type B-4/AN-V-18 life preserver.*

| CONTRACT | AWARD DATE | CONTRACTOR | QUANTITY |
|---|---|---|---|
| 535-AC-29297 | June 1942 | Goodyear Tire & Rubber Co. | Under 5,000 |
| 535-AC-29298 | June 1942 | Firestone Rubber & Latex Prods. Co. | Under 5,000 |
| 535-AC-29299 | June 1942 | U.S. Rubber Co. | 98,000 |
| 535-AC-29300 | June 1942 | Air Cruisers Inc. | 100,000 |
| 535-AC-29301 | June 1942 | New York Rubber Corp. | 50,000 |
| 535-AC-29302 | June 1942 | Seamless Rubber Co. | Under 5,000 |
| 535-AC-29303* | June 1942 | B.F. Goodrich Co. | Under 5,000 |
| 535-AC-29304 | June 1942 | B.F. Goodrich (Hood Rubber Div.) | 78,000 |
| 535-AC-29305 | June 1942 | Hodgeman Rubber Co. | 50,000 |
| 535-AC-29306 | June 1942 | Lee Tire & Rubber Corp. (Republic Div.) | 50,000 |
| 535-AC-29361 | June 1942 | Air Cruisers Inc. | Under 5,000 |
| 33083-AC-3744 | June 1944 | Firestone Rubber & Latex Co. | 45,000 |
| P.O. 44-7195 | June 1944 | Seamless Rubber Co. | 2,000 |
| P.O. 44-7196 | June 1944 | New York Rubber Corp. | 13,000 |
| P.O. 45-2863 | October 1944 | Seamless Rubber Co. | 20,000 |
| 33038-AC-5883 | December 1944 | U.S. Rubber Co. | 85,000 |
| 33038-AC-5884 | December 1944 | New York Rubber Corp. | 87,000 |

** The B.F. Goodrich Co. additionally made vest vests under sub-contract for Air Cruisers Inc.*

*(Below) Relevant USAAF Technical Orders pertaining to life preservers and the dates of the known subsequent amendments.*

| | | |
|---|---|---|
| T.O. 13-1-3 | Life Preserver Vests Types B-3 and B-4 (Maintenance and Inspection) | 10 Mar. 1934, 20 Oct. 1936, 10 Sept. 1940, 15 Aug 1942, 8 Dec. 1943, 27 Sept. 1944 |
| T.O. 13-1-5 | Addition of Sea Marker to Life Preserver Vests | 14 Aug. 1943, 23 Nov. 1943 |
| T.O. 13-1-17 | Modification of B-3 and B-4 Life Preserver Vests (Back Strap) | 25 May 1944 |

*(Right) List of known Aviation Supply Office (BuAer) contracts for the Type B-4/AN-V-18 life preserver.*

** Contract was cancelled, no vests under this contract are known to have been manufactured.*

| CONTRACT | AWARD DATE | CONTRACTOR | QUANTITY |
|---|---|---|---|
| 288s-10125 | January 1943 | New York Rubber Corp. | 7,000 |
| 288s-10126 | January 1943 | U.S. Rubber Co. | 7,000 |
| 288s-12349 | March 1943 | B.F. Goodrich (Hood Rubber Div.) | 8,000 |
| 288s-12415 | March 1943 | Goodyear (Barr Rubber Prods. Co.) | 8,000 |
| 288s-13876 | May 1943 | Seamless Rubber Co. | 8,000 |
| 288s-13877 | May 1943 | U.S. Rubber Co. | 7,500 |
| 288s-13878 | May 1943 | B.F. Goodrich (Hood Rubber Div.) | 8,000 |
| 288s-14859 | June 1943 | New York Rubber Corp. | 8,000 |
| 288s-20573 | February 1944 | Seamless Rubber Co. | 30,000 |
| 288s-20574 | February 1944 | U.S. Rubber Co. | 35,000 |
| 288s-23494 | June 1944 | U.S. Rubber co. | 22,000 |
| 288s-23495 | June 1944 | Firestone Rubber & Latex Prods. Co. | 45,000 |
| 288s-23500 | June 1944 | Seamless Rubber Co. | 25,000 |
| 288s-27633 | December 1944 | New York Rubber Corp. | 40,000 |
| 288s-32082 | June 1945 | Firestone Rubber & Latex Prods. Co. | 25,000 |
| 288s-32083 | June 1945 | New York Rubber Corp. | 30,000 |
| 288s-34286* | August 1945 | Firestone Rubber & Latex Prods Co. | 145,000 |

*(Right) Known BuAer Technical Orders pertaining to the life preserver.*

| | | |
|---|---|---|
| 8-38 | Life Jackets | 5 February 1938 |
| 44-44 | Life Vests-Provisions for Safety Loops for CO2 Actuating Lever | 1 April 1944 |
| 117-44 | Life Vests-CO2 Actuating Lever Lanyard Modification of | 14 September 1944 |
| 134-44 | Life Vests– Inspection and Maintenance of | 17 November 1944 |
| 25-45 | Life Vests-Inspection and Maintenance of (Supersedes 134-44) | 2 March 1945 |
| 30-45 | Life Vests Specification AN-V-18 Use of | 16 March 1945 |
| 81-45 | CO2 Life Vest Cylinders –Replacement of | 12 September 1945 |

*(Right) Known BuAer Technical Notes pertaining to the life preserver.*

| | | |
|---|---|---|
| 18-33 | Inflatable Life Jackets-Information Concerning | 25 May 1933 |
| 30-36 | Life Jackets Self Inflating Type | 2 July 1936 |
| 77-42 | Fluorescein Dye Markers for Life Jackets | 27 October 1942 |
| 44-44 | Fluorescein Dye Markers for Life Jackets | 25 May 1944 |
| 55-44 | Pneumatic Life Vests-Installation of Back Starp | 26 June 1944 |
| 15-45 | Life Vests-Distress Smoke Hand Signals-Attachment of Pockets for | 23 February 1945 |
| 87-45 | Proper Use of Shark Chaser | 17 September 1945 |

From extensive use in the combat theaters, the same deficiencies of the Type B-4 were consistently reported from both services.

The Type B-4 was an expedient measure to meet wartime demands and it was understood further development was warranted. By the close of 1943, enough data and experimentation had been collected to begin the process on a new generation of life jackets.

Some of the principle complaints reported were the discomfort around the neck from inside the collar and, most importantly, its buoyancy characteristics. The vest would not support heavily-laden wearers and had a tendency to list, causing the user's face to submerge, in some cases resulting in drowning.

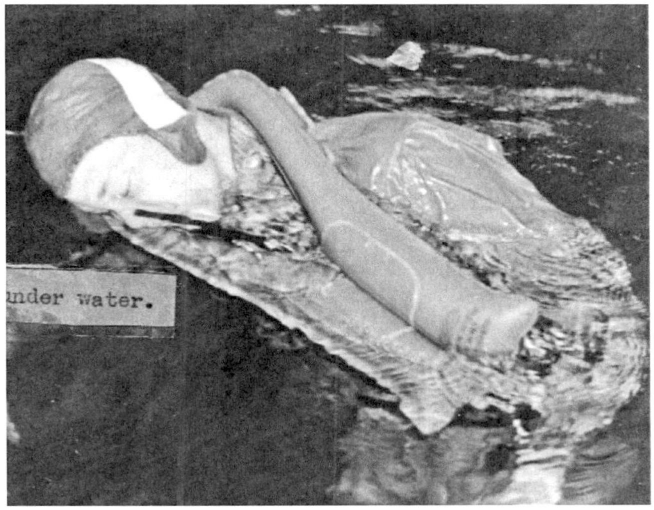


*(Right) Demonstrating the listing properties of the Type B-4 through pool tests being conducted by the BuAer. The British and USAAF also performed the same simulations demonstrating the same results. This was an important factor in future engineering assuring that a life vest remains upright. (National Archives)*

*(Left) Units operating in the ETO often utilized the RAF 1941 Pattern life preserver for its comfort and better flotation properties. Three C-47 crew members of the 439th Troop Carrier Group wear RAF life preservers on an Airfield in England, July 1944. (National Archives)*

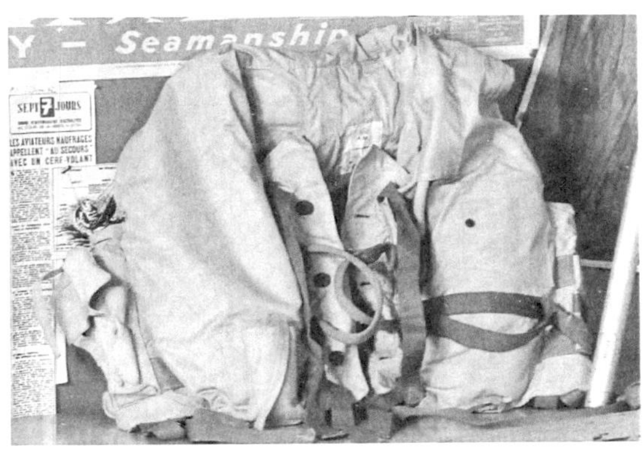

*(Right) The RAF 1941 Pattern life preserver included inflation bladders and kapok pads. It also included accessory pockets for a recognition light, skull cap, sea marker and whistle.*

The U.S. Strategic Air Forces Europe initiated the process of developing a new life preserver to meet service requirements in February of 1944.

Prompted by the request, the Air Technical Service Command began the project, in March, by conducting comparative experiments on the AN-V-18, British and German life vests. The purpose was to compare flotation properties, bulkiness, comfort, and method of suspension for an in-depth study of engineering.

Upon conclusion of the tests, the British RAF 1941 Pattern vest performed the best but was unfavorable in bulk.

Engineering guidelines were established and submitted to prospective contractors for the fabrication of samples, stipulating that a design should be like that of the Type B-3 life preserver.

Beginning July 1944, contractors submitted their samples for testing and evaluation with continued refining of the designs. A model submitted by Air Cruisers Inc. and improved upon by the U.S Rubber Co. was found to be the most favorable. Tests samples were submitted in November 1944. Following these tests, the vest was recommended for standardization.

Drawings were approved 1 December 1944 under Dwg. No. 45J9080 and standardized as the Preserver, Pneumatic, Life Vest, Type B-5 under USAAF Specification No. 94-3240, dated 12 December 1944, available under stock No. 8300-667760.

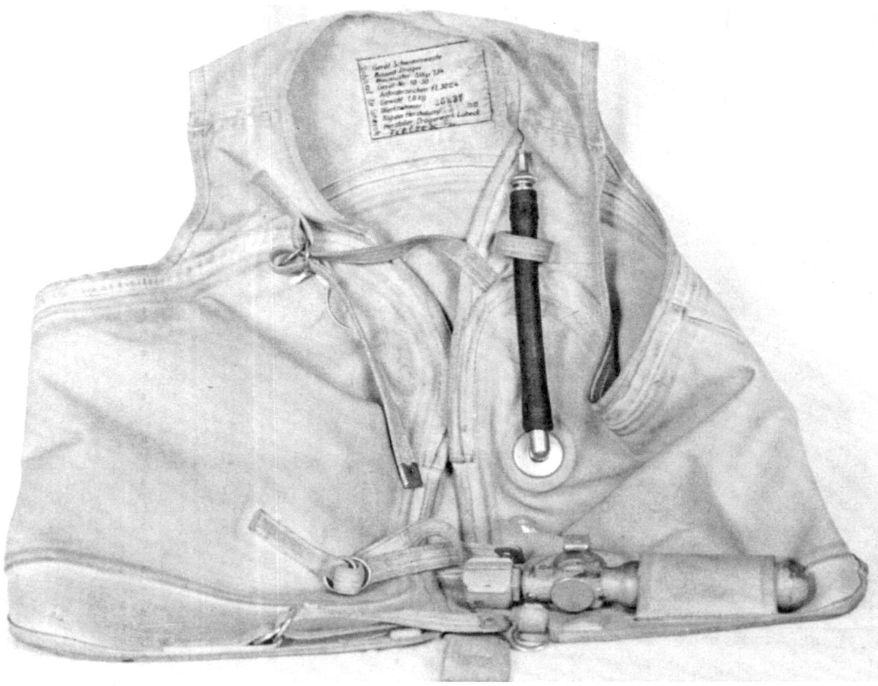

*(Left) The German Model 10-30 life preserver was evaluated during these comparative tests, only narrowly being out-performed by the RAF vest with its best feature being compactness.*

*(Left) Volunteers wearing experimental examples of the Type B-5 life preserver prior to and upon adoption. Note the oral inflation tubes on the external surface, utilizing the standard type elbow valves. These would be significantly amended prior to production. (USAF)*

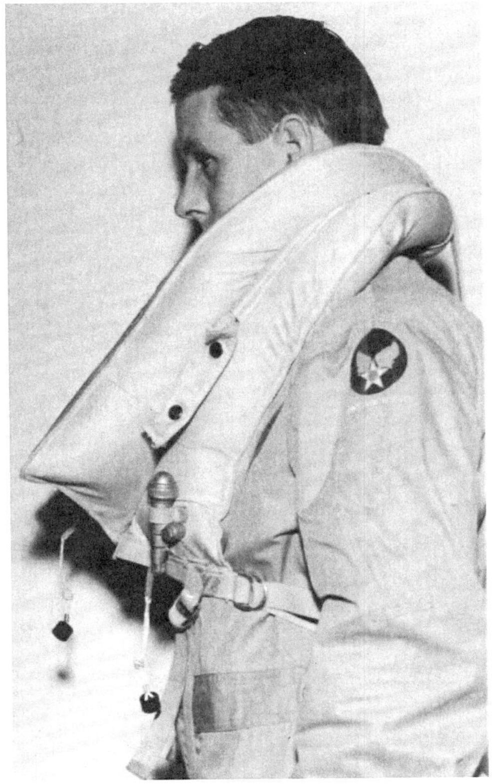

*The Type B-5 consisted of an outer fabric casing containing two separated, superimposed neoprene bladders. The harness system incorporated a back strap, two leg straps, and a waist strap. The back strap was designed to lift the wearer out of the water. Double D rings provided quick and simple adjustment, with an additional D ring on the waist strap for the attachment of the pararaft lanyard. Two pockets were provided between the bladders - the left containing a sea marker and whistle supplied by the contractor, and the right pocket contained a shark chaser installed by the service prior to issue. Later recommended additions included a life preserver light and signaling mirror.*

*(Below) By the close of WWII, the B-5 was to contain a shark chaser, sea marker, whistle, life preserver light and signaling mirror. This image from the early 1950s has all those articles in updated form, but the same concept applied in WWII. (USAF)*

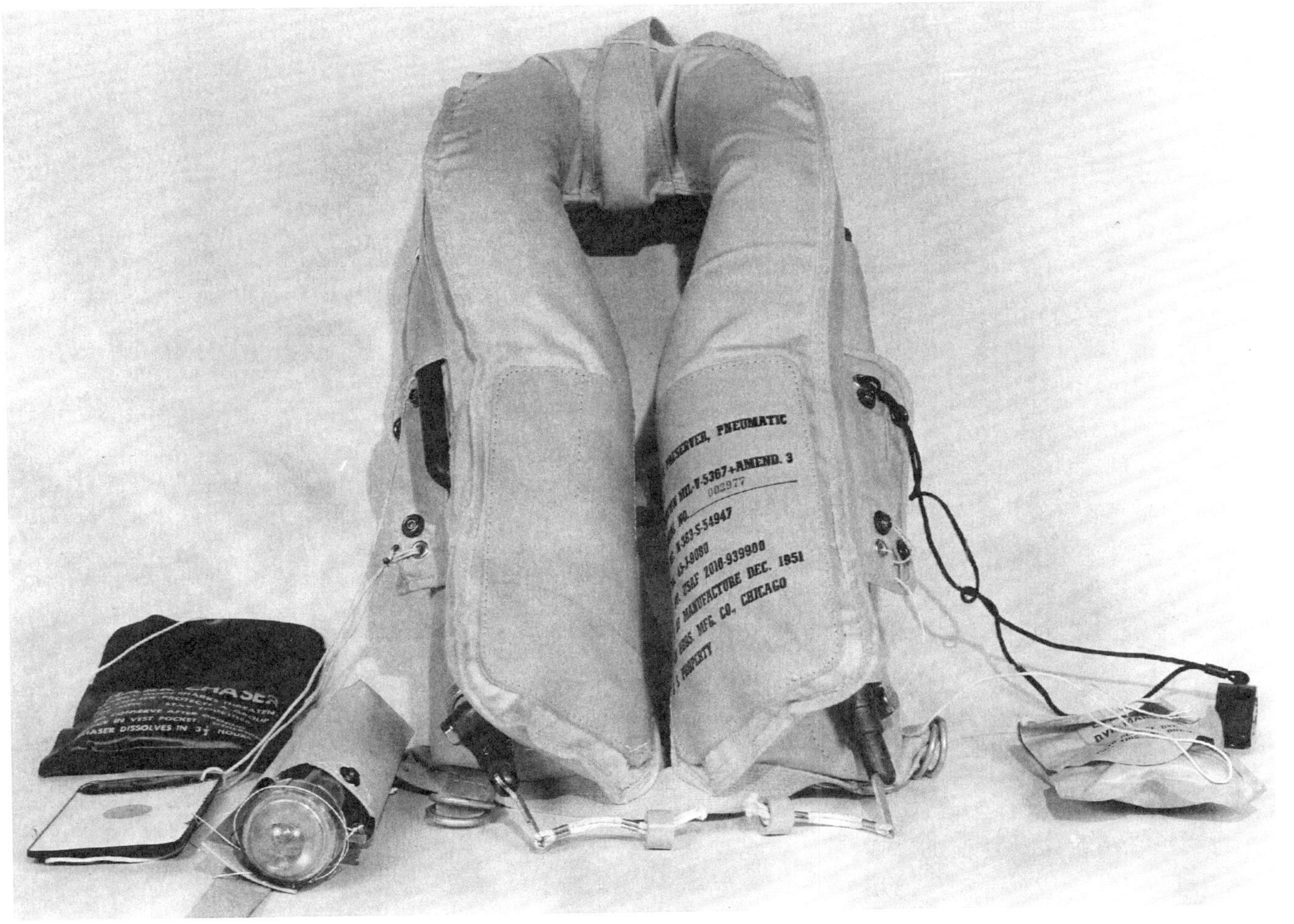

*(Right) As part of its design features, the vest was self-righting, firmly supporting the user's face above the water. The harness system provided additional support, increasing comfort while suspended in water.*

*(Left, right and below left) An example of the Type B-5 life preserver. The back strap is looped and reinforced for rescue craft recovery.*

*(Below) Detail of the pockets, CO2 inflators and oral inflation tubes between the bladders. For maintenance and repair purposes, the neoprene bladders were designed to be removable by means of a slide fastener across the top.*

*(Left) Though standardized in December 1944, the award of contracts was delayed to await supplies from sub-contractors. The first contracts were awarded in March 1945. P-51 pilot of the 318th FG Lt. Eugene Lancaster wears a B-5 life preserver at an airfield on Ie Shima, Ryukyu Retto, July 1945. (National Archives)*

*(Below) The Type B-5 was compact, adding little additional bulk under the parachute harness as illustrated by Maj. Robert Moore of the 7th Fighter Command on Iwo Jima. (National Archives)*

*(Below) B-29 air crew of the 20th AF issued with new gear including the C-1 emergency vest and Type B-5 life preserver, Guam, c.1945. (National Archives)*

*(Right) List of known WWII contracts for the Type B-5 life preserver. Quantities of each contract are unknown but the accumulative quantity is 90,000. Due to the substantial quantities of the AN-V-18 and Type B-5 in stock, no further procurements were executed until 1949 under new specifications MIL-V-5367.*

| CONTRACT | AWARD DATE | CONTRACTOR |
|---|---|---|
| 33038-AC-8347 | March 1945 | U.S. Rubber Co. |
| 33038-AC-8348 | March 1945 | General Tire & Rubber Co. |
| 33038-AC-8349 | March 1945 | Firestone Rubber & Latex Prods. Co. |
| 33038-AC-11063 | May 1945 | Firestone Rubber & Latex Prods. Co. |
| 33038-AC-11065 | May 1945 | General Tire & Rubber Co. |
| 33038-AC-11066 | May 1945 | Durkee Atwood Co. |
| P.O. 45-7390 | Unknown | Seamless Rubber Co. |
| P.O. 45-7630 | Unknown | Seamless Rubber Co. |

The BuAer began work simultaneously with the USAAF, conducting their own investigation and reviewing the findings of tests conducted by other services. Initially, the experimental Type B-5 models were submitted for joint Army-Navy coordination and eventual standardization, but the BuAer opted to pursue other avenues.

They viewed the prospective Type B-5 models as complicated and not very durable. They worked in coordination with the Goodyear Tire & Rubber Co. building a vest that would be lighter in weight, more comfortable for the wearer for extended periods of time – whether inflated or deflated, provide better support keeping the head out of the water, reduced metal parts, and offer provisions for lifting the user out of the water. The fabric cased bladder construction of the Type B-5 was found to be susceptible to damage. To help prevent punctures and abrasions, the BuAer adopted a method of a neoprene coated nylon. The process would include some of the construction process of the AN-V-18.

Engineering aspects progressed through 1944, with several designs ready for field testing in February 1945. It was decided to standardize the vest in a service conference held in March 1945, followed by instructions to the Goodyear Co. to finalize engineering aspects for production.

The vest was officially standardized under NavAer Specification M-701, titled Vest, Pneumatic, Life, Mark-2, and dated 10 May 1945, available under stock number R37-V-710. A Contract for 25,000 vests was awarded to the Goodyear Tire & Rubber Co. under contract NOA-6864, which was in turn sub-contracted to the Wooster Rubber Co. for manufacture. This was considered a small production run, so that any manufacturing difficulties could be corrected on future contracts.

The Mark-2 life preserver would not become available for service use until after VJ-Day, in December 1945.

A subsequent revision was initiated on 2 January 1946, making minor changes as outlined in revised specification M-701A. It would then be re-designated under a new aeronautical specification format 28 January 1947 as the 23P17 (AER).

The USAAF field tested the Mark-2 in May 1945, at Eglin Field, with the representatives, concluding the Mark-2 to be far superior to the Type B-5, and recommending it for joint Army-Navy standardization. It was not approved, stating that it did not integrate with the standard parachute.

The debate continued through 1946 and 1947, with the conclusion from both services that the Mark-2 was superior in all respects. Joint Army-Navy specifications were drafted under AN-V-30 dated 21 May 1947. They were submitted but the USAAF did not accept them. With this vest adopted as the Mark-2, it had led to the erroneous designation of the 23P6 as the Mark-1 (MK-I). In NAVPERS manual 10352 titled Aircraft Survival Equipment, it details the Mark-1 to be that of the AN-V-18. Mark-1 was not an official designation.

*(Opposite page) The Mark-2 life vest as procured under the initial contract in June 1945. The vest consisted of neoprene coated nylon fabric forming three bladders, two automatically inflated and one orally inflated. The oral inflation tube was copied from the German 10-30 life preserver. On the surface of the vest, tabs were provided for the attachment of two sea markers and a life preserver light. Two pockets were provided for either the MK-1 MOD-0 hand held smoke or the MK-13 MOD-0 distress signals. Integrated into the nylon harness was a lifting strap that folded over the right shoulder and snapped onto the front of the vest for storage. Technical Order 90-45 was distributed 26 September 1945, titled as Mark-2 Life Vest– Description, Inspection and Maintenance of.*

VEST LIGHT ATTACHMENT
HOIST LINE
DYE MARKER TAB
ORAL INFLATION CONNECTOR LOCK
EQUIPMENT
EQUIPMENT
CO2 HOLDER COVER

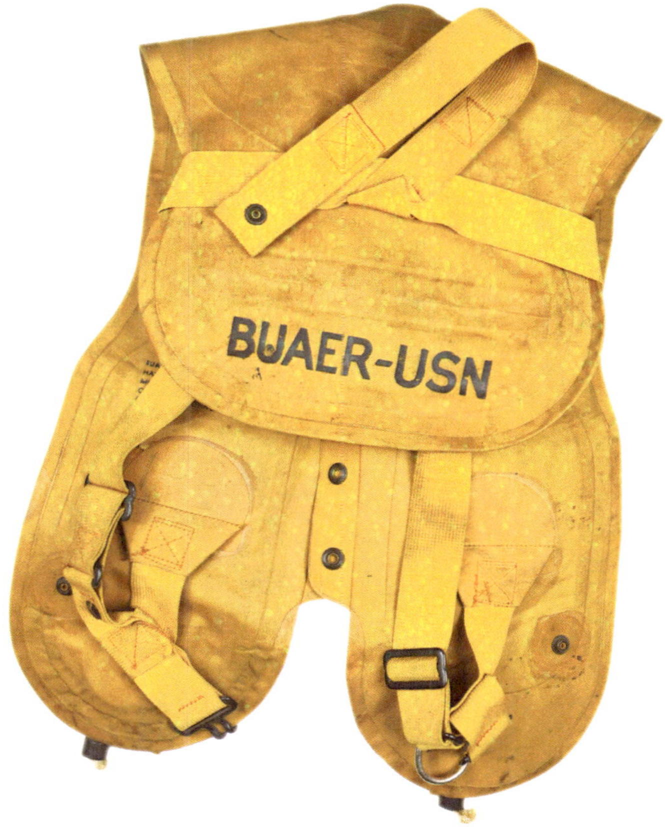
BUAER-USN

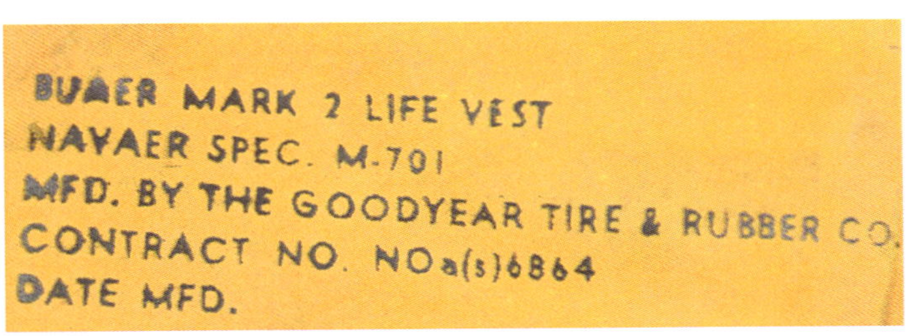
BUAER MARK 2 LIFE VEST
NAVAER SPEC. M-701
MFD. BY THE GOODYEAR TIRE & RUBBER CO.
CONTRACT NO. NOa(s)6864
DATE MFD.

*(Top left) Features of the Mark-2 life vest*

*(Top right) The head is firmly supported above the water. Both the Type B-5 and Mark-2 were designed to have self-righting buoyancy.*

*(Bottom left) The internal harness secured. An early design flaw was the weakness of the lift-a-dot fasteners, which tended to come unfastened. This was remedied in M-701A. Further revisions would be made to the Mark-2 during its service life. (National Archives)*

*(Bottom right) The proper wearing of accessories on the vest.*

# Afterword

In statistical studies conducted at the close of the war, the life preserver and life raft saved many more lives than did the parachute.

Emergencies requiring survival and rescue became routine occurrences in World War II, especially in combat. Today, Air Sea Rescue is considered an integral facet of all military operations. Complete survival instruction is an integral part of the training of aviation personnel, as well as key to the preparation of the aviator's equipment.

# Acknowledgments

This work could not have reached its potential without contributions and support from Robin Bartel and Paul Faulkner. Robin Bartel has been a friend and a peer for more than 25 years, and has contributed many artifacts for reference as well as solid support for the project, consistently urging me on. Paul Faulkner is a long-time friend who shares the same interests, and his knowledge of Naval aviation is unequalled. I would also like to thank Kriss Usherwood, Charlie Flick, Frank Trzaska, Bob Baldwin, Mathieu Bianchi, Remi Bouche, Jeff Klein, Jerry Kolakowski, Ben Major, Steve Milnthorpe, George Petersen, Gregory Pons, Geoff Pringle and Mick Prodger for their contributions and graciously shared information. Significant assistance was provided by Brett Stolle of the National Museum of the United States Air Force, Bob Thomas of the National Naval Aviation Museum and Maranda Gilmore of the Air Force Historical Research Agency.

# Author's Note

I have maintained an interest in WWII for as far back as I can recall, playing army in the woods in my home state of Oregon. It was on my 13th birthday, back in 1987, that I established myself as a collector. My gift was an M1 Carbine. Rummaging through my trunk of field gear, I separated the WWII items from everything else, and those, together with the rifle, formed the basis of my collection.

My interest only grew, accumulating more and more items until today, 30 years later, all aspects of United States WWII memorabilia remain a passion of mine. Some 20 years ago I obtained an empty US Navy M-592 emergency back pad kit – my first piece of WWII survival equipment – and so began my transition solely into that area. This initiated a quest to find out and obtain what went in the kit. I quickly found myself frustrated at the lack of information available, gathering more questions than answers. As the investigation moved forward I became more and more interested in the overall subject, accumulating artifacts but still operating with a very limited reference library. Influenced by peers in my local collecting community, I eagerly joined in conversations about their adventures across the country, networking and traveling to various archival holdings. It soon became apparent that I would need make those pilgrimages in order to find the answers to questions that I, and many others, shared. After a relatively short time I decided that if I was going to make that effort, I should put my findings in a reference book for anyone who may be interested.

My initial research efforts were at local university libraries then inter-library loans, until the environment allowed me to visit the National Archives. I have since been back to the NARA multiple times, accumulating documents and images. Not every answer to every question was found, though it was certainly through no lack of effort. Those answers simply may not exist.

It wasn't mandatory for government agencies or facilities to retain all documents, once they were determined obsolete or irrelevant, often ending in their destruction. What survived post-war downsizing and finished up in repositories was at the grace and courtesy of individuals at these government installations, for which I am eternally grateful! However, many records are incomplete, and it can be viewed as a stroke of luck that they still even exist. Information from various sources was sometimes contradictory but through continued accumulation of data, a compilation of redundancy pointed towards the facts. I have attempted to detail each genealogy as clearly as possible.

The overall experience has been very educational for me, and hopefully those who read these studies will find them helpful.

Dustin Clingenpeel
May, 2019

# Index

# Also Available

## Survival & Rescue Equipment of World War II:
### Army Air Forces and U.S. Navy

**VOLUME 2**

Distress Signals
Personal Aids Kits
Protective Clothing
Survival Training

The companion book in this series by noted collector and historian Dustin Clingenpeel. Volume 2 offers a comprehensive study of survival and rescue equipment used at sea by pilots and aircrews of the U.S. Army Air Forces and U.S. Navy and Marine Corps prior to and during World War II. Sea markers, smoke and pyrotechnic signals, mirrors, flashlights, whistles, sails and transmitters, as well as personal aids kits, protective clothing and survival training are all examined in meticulous detail. Featuring over 1000 photographs and detailed descriptions including government contract information and developmental history. 248 pages.

ISBN 978-1-943492-61-9 Available from booksellers or directly from **elmgrovepublishing.com**

www.ingramcontent.com/pod-product-compliance
Lightning Source LLC
LaVergne TN
LVRC101547220926
848486LV00041B/390

* 9 7 8 1 9 4 3 4 9 2 6 0 2 *